Praise for *Man of Iron*

'Absorbing ... [Glover] has done a fine job in his first book of establishing his subject's credentials both as a man and as a maker. His clear, concise style gives the plain-speaking Telford ... a fitting literary treatment' Alexander Larman, *Observer*

'Glover has ... been digging in the archives with Telfordian energy to produce a readable book about an overlooked man' Damian Whitworth, *The Times*

'Glover has done detailed research and any other author will struggle to improve on his life of Telford ... *Man of Iron* is a competent, interesting book about an engineer whose star ... deserves to shine a little brighter' Simon Heffer, *Daily Telegraph*

'[An] engaging, evocative and very readable tale' Kaye Wiggins, *Financial Times*

'A lively, well-researched book that will help to put Telford back where he belongs – in the Valhalla of British national heroes, an evergreen role model for ambitious, skilled boys' *Country Life*

'A strikingly clear portrait of the man who helped shape Britain ... A beautifully written biography, reading almost as a work of classic literature' *Engineering & Technology*

A NOTE ON THE AUTHOR

Julian Glover works in journalism and politics. Previously a columnist for the *Guardian*, he was appointed chief speech-writer to David Cameron before being made special adviser on transport policy. He is married to *The Times* columnist and former MP Matthew Parris and lives in Derbyshire.

MAN OF IRON

*Thomas Telford
and the Building of Britain*

Julian Glover

B L O O M S B U R Y

LONDON · OXFORD · NEW YORK · NEW DELHI · SYDNEY

Bloomsbury Paperbacks
An imprint of Bloomsbury Publishing Plc

50 Bedford Square   1385 Broadway
London   New York
WC1B 3DP   NY 10018
UK   USA

www.bloomsbury.com

BLOOMSBURY and the Diana logo are trademarks of Bloomsbury Publishing Plc

First published in Great Britain 2017
This paperback edition first published in 2018

© Julian Glover, 2017
Map artwork © Emily Faccini

Julian Glover has asserted his right under the Copyright, Designs and Patents
Act, 1988, to be identified as the Author of this work.

British Library Cataloguing-in-Publication Data
A catalogue record for this book is available from the British Library.

Library of Congress Cataloguing-in-Publication data has been applied for.

ISBN: HB: 978-1-4088-3746-7
PB: 978-1-4088-3748-1
ePub: 978-1-4088-3747-4

2 4 6 8 10 9 7 5 3 1

Typeset by Newgen Knowledge Works (P) Ltd., Chennai, India
Printed and bound in Great Britain by CPI Group (UK) Ltd, Croydon CRO 4YY

To find out more about our authors and books visit www.bloomsbury.com. Here you
will find extracts, author interviews, details of forthcoming events and the
option to sign up for our newsletters.

To Matthew

CONTENTS

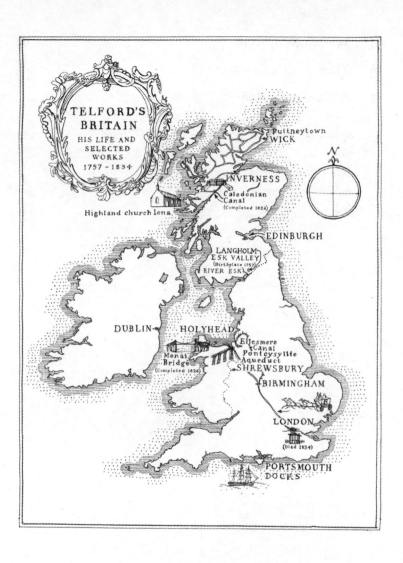

TELFORD'S
BRITAIN

HIS LIFE AND
SELECTED
WORKS
1757 – 1834

Pultneytown
WICK

INVERNESS
Caledonian
Canal
(Completed 1822)

Highland church Iona

EDINBURGH

LANGHOLM
ESK VALLEY
(Birthplace 1757)
RIVER ESK

DUBLIN   HOLYHEAD
Ellesmere
Canal
Pontcysyllte
Menai   Aqueduct
Bridge   SHREWSBURY
(Completed 1826)
BIRMINGHAM

LONDON
(Died 1834)

PORTSMOUTH
DOCKS

Menai in deep winter, well past midnight. To the east, the cold heights of Snowdonia; to the west, Anglesey and the Irish Sea. The wind was up, the tide was racing on the rocks below and rain was on its way. It was ordinarily no place for people without thick cloaks, tall boots and a pressing need to be out of their beds: but this night was different. There was a crowd, shouts, jostling, the stamping of hooves and the velvet smell of sweating horses. Bright lights reached down to the dark water that had separated Anglesey from the mainland for perhaps 4,000 years until this moment but would never do so again.

The modern world was arriving in North Wales that night in 1826. Everything was changing. And Thomas Telford was there. Thomas Telford was the man who made it happen. It was his new bridge people had come to see. His bridge they wanted to cross for the first time. And what they saw stretching into the darkness was indeed extraordinary: a structure as bold as anything built in Britain since the Romans, like a great blade cutting between water and sky. Technology was being tested to the limit that night. So was courage. No one could know for sure if the iron chains which held the bridge high above the Menai Strait would be safe; if Telford's design would be proved; if a mathematical dream would come true.

And now, from the road to the east, came the sound of hurry: iron-rimmed wheels and horses' hooves galloping on rough gravel, racing to keep time. The Royal London and

Holyhead coach was heading for the port: its scarlet sides and the royal crest proclaiming the right to carry the precious mail-bag for Dublin. As it pulled up outside the Bangor Ferry Inn its exhausted horses were unharnessed and a fresh team attached. A man stepped forward with an air of command. He climbed quickly onto the top of the coach and sat down next to David Davies, the coachman on duty that night.

The man's name was William Provis and he had overseen the building of the Menai Bridge under Telford's instruction. Now, he broke his momentous news. This was to be the first coach to cross the new structure. Briefly, there was confusion: even an argument. Davies and his guard, William Read, resisted. Orders were orders, they said. They insisted they needed written instructions to change their route from the old Menai ferry. But Provis held firm. The bridge, he told them, was to open that night. That was final. He was taking command of the coach.

And now the excitement increased. A note was sent down to the ferryman on the Bangor shore to inform him that his boat was not needed that night. It would never be again. At about 1 a.m., with Provis by his side, the coachman flicked his reins, the horses strained and the coach set off on the short journey from the inn to the bridge. It was, one witness remembered, quite a sight, 'the high mounted steeds mantling their proud crescent necks, as if conscious of the triumphant achievement'.[1]

As the coach reached the bridge many more passengers tried to scramble aboard, wanting to be among the first to cross. The weather was wild. 'Amid glare of lamps (a heavy gale of wind blowing by this time),' Provis recalled later, the passengers were joined by 'as many more as could be crammed in or find a place to hang by'. Then, with the overloaded coach creaking, the iron gates that guarded the crossing were swung open for the first time. 'A crack of the whip put the horses in motion and we were quickly conveyed to the opposite end, amidst the cheers of the men around us and the shrill whistling of the gale.'

And so, without more fanfare than this, on a horrible night at 1.35 a.m. on 30 January 1826, the Menai Bridge opened. It was the finest moment of Thomas Telford's career: at sixty-eight, he had built the most daring structure of his life, the first great suspension bridge of the modern age, its central deck almost 580 feet long, slung 100 feet above the tidal channel from sixteen thick chains. The bridge was 'this stupendous effort of human genius', one popular guide of the time declared.[2] 'The spectator cannot but be wrapt in mute astonishment, and adore the attributes of the great Creator, who endues the mind of man with gradual capacity for the construction and achievement of such labours.' A local doctor, in the crowd as the bridge opened, bubbled with excitement. 'The National and splendid specimen of British Architecture, will be a lasting monument to the discernment of the present Government for having calculated into Requisition, the transcendent talents of Mr Telford, who has thus by a positive proof of superior scientific knowledge and taste, signalised himself in this line – The First Architect of His Age.'[3] 'The Menai Bridge,' he concluded, 'will rank, in future ages, as the Eighth Wonder of the World.'

Telford's response to all this was telling. A different man would have sought public glory. A different man would surely have insisted on crossing first that night rather than letting his friends and juniors make the trip before him. A different man would have made a pompous speech or wanted to name the bridge after himself, or perhaps tried to call it after King George IV in the hope it would bring him rewards, just as a different man would have sought honours or riches throughout his life.

Telford wasn't like that, ever. If the design of his bridge exposes one side of his character – relentless, daring, ingenious – then the manner of its opening exposes another – private, perhaps lonely, always travelling. And so the morning that followed that first windswept crossing was dominated by the bridge and not by its designer. The political grandees who had backed the project made sure they were seen, of course. The first private carriage to cross, 'drawn by four beautiful greys', belonged to one of the parliamentary commissioners and the second 'was Mr Telford's'.[4] Soon after, the celebrated Oxonian stagecoach followed, heading for London and reputed to be the fastest in Britain. Then came a scrum of 'numerous Gentlemen's carriages, landaus, gigs, cars, poney-sociables &c &c upwards of one hundred and thirty in number; and horsemen innumerable'.

The Royal Standard flew from the great stone towers, cannons fired and competing bands played on either shore. 'Joy, admiration and astonishment seemed depicted in every countenance on beholding the proportion, symmetry and grandeur apparent to the most common observer, in every part of this unrivalled structure.'

A party began. 'At one time the Bridge was so crowded that it was difficult to move along,' Provis wrote in a letter the next day.[5] 'Most of the carriages of the neighbouring gentry, stage coaches, post chaises, gigs and horses passed repeatedly over and kept up a continuous procession for several hours – the demand for tickets was so great that they could not be found fast enough and many in the madness of their joy threw their tickets away that they might know the pleasure of paying again. Not a single accident or unpleasant occasion took place and everyone appeared satisfied with the safety of the Bridge and delighted that they could go home and say "I crossed the first day it opened." '

The day turned into night, the celebrations went on and bottle after bottle was opened. At a celebratory dinner, Provis recalled, one man burst in 'with a cut cheek on one side'. Normally 'considered a patron of sobriety', tonight he was drunk and babbling. He said he had 'called on a very unpleasant business'.

His 'feelings he said were hurt' – he had been insulted 'by [men] who knew nothing at all about work but to shift a barrow full of rubbish'.

The bridge's engineers were letting their hair down. Provis's response was to roll his eyes. 'Such a specimen . . . gave me a tolerable idea of the situation of the party he had quitted.' 'A good dinner having been provided at . . . a party assembled then to make merry and drink success to the Bridge.' William Hazledine, the famous ironmaster who had designed and forged the chains that held the bridge aloft, 'kept up the life and spirit of the evening with his broad wit and funny stories . . . There being a general feeling to receive of all sorts of good things, they were soon as happy and joyous as could be imagined.'

This is a cheerful letter. 'As far as I learn all went off well but it is very difficult today to know what was going on yesterday,' Provis concluded. But read further and a disquieting mystery emerges. Why did it have to be sent at all? This was not a boastful note to an absent relative. It was a report written to Telford, the man who had been among the first to cross that morning. The letter is still in its wax-sealed envelope, addressed in Provis's heavy pen to 'Thomas Telford, Talbot Hotel, Shrewsbury' – half a day's journey away.

The reason Provis needed to tell Telford what had happened on the day his greatest bridge opened is that by nightfall Telford wasn't there. He had slipped away, soon after crossing in his carriage. He had moved on. He was always moving on. His life was all about movement. Never settled, always restless, always trying to create something more.

And the roots of that restlessness lie in the place he came from and the circumstances of his birth. He was not trying to run away; but he was never at ease. This agitation shaped his country. In his seventy-seven years he worked on 184 big projects, among them ninety-three large bridges and aqueducts, seventeen canals and thirty-seven docks and harbours. He cut one great waterway from sea to sea across the top of Scotland and another on the same scale to the centre of Sweden. He constructed more 1,200 miles of roads and 1,076 bridges to open up the Highlands of Scotland and worked on a dozen more road schemes in England and Wales, including the expressway to Ireland across the Menai Bridge, the best road built anywhere before the coming of the motor car. He was the architect of three churches in Shropshire and thirty-two in Scotland, as well as of houses, a prison and a courthouse. He worked on water works for four cities, improved the navigation of four major rivers and was the engineer of four major drainage schemes. He surveyed the route of three early railways. He shaped the lives of the Victorian civil engineers who followed him and he led the institution which still guides the profession. Almost everything he built is still in use.

His great endeavours, like the phases of his life, cannot be laid out chronologically, end-to-end, to be covered in neat succession by his biographer, for Telford was seldom doing less than a dozen big things at once. Time-lines shoot all over the place, so our narrative must do the same, often leaping ahead to see how one thing ended, then doubling back to describe another's beginning. Sometimes the reader may find this confusing. So did the writer. But I have preferred to pick up a thread and follow it, returning later for the next thread, rather than page through a calendar. Telford's life is a weave, not a list. His achievements cannot be ticked off consecutively.

But behind all these triumphs lay a character whose life draws us into a question beyond design. How did a simple shepherd boy rise so high?

PREFACE

In October 1968 readers of the *Shropshire Journal* were upset. Many backed plans for a new city to be built on industrial wasteland north of the River Severn. But they hated what it was to be called. 'The Minister's choice of Telford as the name for the new town does not seem to fit in at all,' one wrote.[1] 'I must protest at the choice of name for the new area covering Wellington, Oakengates and Dawley,' complained another. Telford: it was an alien title to Shropshire people, picked in London by the local government minister in the hope that the energy of the eighteenth-century engineer would rub off on the site. From now on residents were to find themselves living in what the *Shropshire Star* called 'Swinging Telford', a place marketed as a modern place 'for people on the move'.[2] Both papers printed short biographies of Telford, the man, so that residents of Telford, the town, could understand the life of the Scot whose name now formed an unexpected part of their address.

Today it is a place of sheds, roundabouts, trees and municipal daffodils in spring, arguably better known than the man after whom it is called. For there is no doubt but that we have neglected the new town's namesake. Many people have half-heard of him, but few know much more than that. Proposing him for greatness, one gets a nod followed by a question: what was it that he actually did? He has been left caught in a sort of historical twilight, revered mostly by engineers and industrial archaeologists.

Why? One reason is that his life is hard to sum up. He did so much over such a long time that it is a struggle to sharpen the focus. Paradoxically, he might be better known now if he had died younger or done fewer things and left less of a tangled story over many decades and places.

Even the man himself found his work almost impossible to explain. Reflecting, late in life, on the things he had done, he listed 'architecture, bridge building, road making, inland navigation, drainage, the construction of docks and the improvement of harbours' as his skills – in short almost everything that a newly unified country caught up in a long and technologically demanding European war required.[3] He worked with tremendous force – and almost always success – for over sixty years, creating through the power of his own energy the rules and roles of the modern profession of civil engineer. He was hailed in his own time as the man who designed and oversaw the construction of the finest roads and bridges and canals built in Britain since the Romans.

But this only touches on his sparse artistry, shaping stone and iron into structures of a kind never seen before; for he was never just a utilitarian and the power of his work lies as much in its purity and conscious focus on appearance as it does in the mundane tasks to which it was put. His bridges and roads were not just useful structures but part of the imagery of the late Georgian age, much-pictured, celebrated in writing, shaping the nation. William Wordsworth wrote a sonnet about one of his iron bridges; Lewis Carroll wrote doggerel about another – his White Knight in *Through the Looking Glass* singing:

I had just
Completed my design
To keep the Menai bridge from rust
By boiling it in wine.

Then it all changed. In the years after his death, Telford slipped from our consciousness. The reason is not hard to find. Technology shifted; newer creations came along; the purpose behind the things he built fell away; and the Victorian age found fresh heroes. Even as he died his era was passing. He built roads and canals but the craze which followed his death was for railways. By the middle of the nineteenth century his life's work was a curious but useless dead end to the engineers who followed him. His fine road to Holyhead grew grass once the Irish Mail express started crossing Stephenson's rival Britannia rail bridge instead. His Pontcysyllte Aqueduct, a stunning, thin, high ribbon of iron crossing the River Dee in North Wales, has been a largely purposeless curiosity for most of its existence; the Caledonian Canal, on which he was a leading engineer, running across northern Scotland, has always been of more use to pleasure boats than freighters.

This technological shift was in part a successful consequence of Telford's legacy. It was Telford who pushed the use of iron in construction to new limits; it was Telford who championed the Institution of Civil Engineers in London, which still honours his name and which trained the engineers who succeeded him. Yet it was Isambard Kingdom Brunel who beat Diana, Princess of Wales, to second place, behind Churchill, in a BBC poll of greatest Britons. Not for Telford the larger-than-life iconic photograph with cigar and stovepipe hat or the championing by Jeremy Clarkson. It is telling that, by comparison with Brunel, not so many people have attempted to write about Telford's life; put off, perhaps, by the relentless pace and quantity of his work which risks any book lapsing into a list of achievements until one canal or tunnel blurs into another and the whole life just becomes one great muddy site visit.

Even Telford's elaborate attempt at autobiography, finished by friends after his death, hardly sold: the book was first remaindered and then copies were bought back by the institution he founded to avoid the shame of their loss of value. They were given away as prizes to engineering students for years after his death. There is no picture of him on permanent display in the National Portrait Gallery in London.

This is not to say Telford ever vanished. Samuel Smiles, the Victorian moralist who championed self-improvement through the study of character, elevated him to a place on the shelves of many nineteenth-century homes, as if his life was nothing more than a piece of sustained application. A century later, the engineer Sir Alexander Gibb (the great-grandson of one of his collaborators) wrote about him again and so – more concisely and cheerfully – did the pioneering industrial enthusiast, conservationist and historian L. T. C. Rolt, the admirable postwar rescuer of steam railways and canals.

Telford has been cherished by modern historians, too, notably from the Midlands where he forged his reputation, among them Barrie Trinder who helped found the Ironbridge Gorge Museum, which has done so much to secure the importance of eighteenth-century industrial history. Telford is a presence throughout Trinder's recent history of the industrial revolution, built from a lifetime's learning and love for his subject.

Works such as these have helped bring about a recognition of the importance of Telford's work and the things he built: and some, like Pontcysyllte Aqueduct, have been restored and protected. There are brown tourist signs along the A5 road to Holyhead, marking it as a 'Historic Route 1815–30' – although they do not explain why it is in fact historic or that Telford engineered it or that so many of the original features survive if you slow down and look for them amid the traffic, tarmac and white lines.

To some this new respect has gone too far. In the 1990s the canal historian Charles Hadfield wrote about Telford in polemic terms as a man determined to obliterate the work of others from the record: 'though he became a great engineer, it seems that he could not bear to feel that others might be his equals, even his superiors'. It is, perhaps, the industrial archaeologist's equivalent of reattributing the authorship of Shakespeare's plays.

Other recent academic work has also pulled back from the fallacy that he achieved everything alone. It has pointed out the importance of his collaboration with engineers and contractors; that he did not lead on all his projects and that others deserved more credit than perhaps Telford was willing to give them. The shift in attribution is clear when you visit Pontcysyllte. Of the half-dozen monuments and plaques scattered around the site recording its importance, the early ones name only Telford while the most recent, erected when the structure gained World Heritage status a few years back, is careful also to credit William Jessop, the senior engineer who supervised the construction of the Ellesmere Canal, of which it is part.

But this sharing of tributes and fame should have a limit. Telford was a genuinely great man; an archetype of Carlyle's 'hero in history'. He matters to our age. He knew that what we now call 'infrastructure' shapes lives and nations. His life was made solid by the structures he built and which drove him on: the – to him – conscious physical embodiment of Great Britain. He built things not for private gain but for progressive purpose, with the clear intent of creating a stronger and more united kingdom so that people from even the remotest valleys, such as the one in which he was born, could share in the British adventure of industry and empire.

Because of this his was a political life of a kind we do not always recognise for its utility today, at least in Britain. His politics lay not in ideas but in the creation of solid things. He was explicit in this ambition to shape Britain: it was everything to him and he succeeded by every measure he set himself. He had an absolute pride, as a Scot, in the Union which created Great Britain and he would have found growing support for Scottish independence a strange surprise and a repudiation of the physical links he built between the four nations of the United Kingdom.

These links were the circuitry which powered his country as it moved from industrial revolution to empire and there are lessons to learn from them today. About project management: Telford was a master. About the connection, rather than contradiction, between massive pieces of civil engineering, beauty and land-scape: he never built an ugly thing or a boring one. And most of all about how infrastructure must be built. Its creation has always been a political as well as commercial act – true of his state-backed Holyhead Road, true of the motorways and true of Crossrail or promises of a Northern Powerhouse now. His story shows that the links we find essential today were created through a combination of individual enterprise and state support, often against the objections of the particular owners of property, to the lasting good of the country. You can't leave everything to a cost-benefit analysis.

The economic crash of 2008 refocused Britain – or at least certain of Britain's politicians for an uncertain amount of time – on the need to build modern physical systems to support our lives: roads, railways, digital broadband and sewers. There is a new level of ambition for and interest in infrastructure and it has been prompted in part by a respect for what we have inherited and an understanding that we have stretched that inheritance to its limits.

Things that Telford believed in have come alive again. If that sounds like a morality tale, then it is meant to be. It is not too large a step from the way Telford worked to the High Speed Two (HS2) project (on which I assisted as a political advi-sor) or the debate about how to expand London's airports or whether to charge for roads. Telford would have understood the dilemmas, insisted on innovation and elegant design and known how to work the parliamentary system: indeed the hybrid bill introduced for the high-speed line in late 2013 owes its procedure almost entirely (and often unhelpfully) to the process of parliamentary approval which authorised almost all of Telford's large schemes. Watching MPs unroll maps and parliamentary counsel spell out the antique language of legislation, I sometimes wished for Telford still to be at the other side of the table to understand it all and lift our ambition when it sagged. I noted with envy that the parliamentary process for Telford's great new road to Holyhead – the HS2 of its day – was completed in less than six months, while securing legislation for just the first two phases of HS2 will take around six years, assuming it is secured at all.

Even so, Telford is more than a cypher for what we must do now. He certainly had luck and it helped him: the luck to be born just as Great Britain was making itself; the luck to draw on industry, empire and rich men wanting to invest in technologies and a growing government ready to back them; and most of all the luck to have been born in a remote part of Scotland which nonetheless gave him enough education, and enough companionship from others of similar energy to his own, to go out and change Britain.

But there was more to it than luck. The man was as curious as the things he created. His was not a normal life or character, or an obvious one. A shifting spirit ran through him, like a restless iron shadow. He never settled. He kept on the move, year after year, on foot, by horse, by coach, observing, thinking, testing, designing, chivvying. He didn't have a partner – man or woman – or any sort of close family, or ever seem to need one. His friends away from Eskdale were mostly fellow engineers in the network of builders and managers he cultivated.

He was the shepherd's boy who loved the countryside and yet helped industrialise it; the ambitious man on the make, almost a hustler, who was also a self-published amateur poet; the man who became such close friends with a Poet Laureate that the latter wrote a book about their companionable travels together. He was the man who walked and rode the length of his country, but didn't have a settled home until he was in his late sixties; the man who buried his youthful political views in horror at the French Revolution's consequences, but whose work was radical; the loner who charmed everyone he met; the man who was sometimes funny, often self-observant, and at times – as he aged – intimidating and blunt to the young generation of engineers he trained.

In the 2010 general election I went to Cumbria to see a friend standing for Parliament. One morning we called at a farm in the deep countryside of the Bewcastle wastes, a rolling, open landscape of sheep, thin moorland grass and windblown trees running down towards the valley of the Esk and the Scottish border. The farmer's surname was Telford. It was a renowned surname, I suggested, as we sat drinking tea in his warm kitchen. Mr Telford agreed: he knew all about and was proud of the man who shared his surname and perhaps his ancestry, and who had grown up 250 years before on a farm not all that different to his own, just a few miles north, across the Esk. 'He left, but we're still here,' he said of his namesake. There are still many Telfords, and Telfers, in the Borders and they still celebrate the local shepherd who became the first hero engineer of the modern age.

That chance meeting led to this book. Were the roots of his success to be found in these hills? Where did he learn his trade? Who encouraged him? Then I thought, too, of the Telford structures I had seen near my family's home in Shropshire, a county he made his own as an ambitious young man: not just bridges and canals, but rebuilt castles and churches too. Why, given Telford's remarkable life, the diversity of his work and the fortunate survival of so many of the things he created, was he not more celebrated?

In 2007, the 250th anniversary of his birth, a small cairn was built on the hillside in a valley running off Eskdale, near the site of the hut in which he was born. Telford remembered this place all his life, and wrote of it in his letters and poetry. And because it has changed so little he would recognise it still; and that, too, is part of the magic of his life, a man whose structures developed Britain but hardly altered the place he was born, so that the houses have the same names and the track runs out in the same place, just by a little farm where he was born called Glendinning. Visit it now, at the start of this story, and it is a safe bet that there will be no one there; just a few sheep.

If only people knew what they were missing.

The Early Shepherd

By the time Thomas Telford died on 2 September 1834 none doubted that he deserved his place among the men and women who shaped Britain in the noonday of industry and empire. He was buried in the most important church in the richest city in what was then the dominant nation on earth. His was not the grandest of monuments in Westminster Abbey, nor his funeral the most elaborate or thronged, but the very fact of it was a mark of singular honour. He was the first engineer to be interred there and one of very few, even today. The second, the pioneer of railways, Robert Stephenson, lies beside him. In the Abbey's Chapter House library, a calf-bound notebook records in order the names and locations of contemporary burials. Charles James Fox is listed on the opening page; Robert Browning and Charles Darwin on the last; and immediately before Telford's name comes that of William Wilberforce.

These are the men who made Britain, and that Telford lies in the most elevated company is no surprise. The country has often buried well. But the place he began: that is more unusual. And that is where this story must also begin, beside a different grave, a simple piece of stone in the quiet of a remote church-yard in the Scottish Borders. The journey between them is the life described in this book.

To understand it we must see more than the things which separate these monuments: we must see the steely connection. Trace your fingers through the lettering on the stone in Scotland and you are touching words carved by the occupant of that tomb in Westminster. The former belongs to a father, a forgotten farm labourer; the latter to his son, his only surviving child, the man whose long, industrious life took root in the enterprise of late Georgian Britain.

This place – the starting point in Telford's life – matters more than anything else in this book. The grave can be found in a small churchyard in the mossy borderland where Scotland and England drift together. Even now, it is a hidden place. Few stop at the hamlet of Bentpath, in the parish of Westerkirk, by the waters of the River Esk near the head of Eskdale. But cross a small stone bridge over the river, walk up an overgrown lane, pass through a wicket gate and you'll find a tidy red stone slab set deep in grass and bracken. It is curved at the top, lightly scrolled, with simple crisp-cut words beneath. 'In memory of John Telford who after living thirty-three Years as an Unblamable Shepherd died at Glendinning 12 Nov.r 1757.' And then it goes on to record another Thomas Telford, his life unknown and undescribed, who must have died in infancy before his famous younger brother was even born, and whose name his parents in their sorrow could not discard. 'His son Thomas who died an infant.'

A thousand gravestones in a thousand British churchyards must look much the same. But this is an exceptional spot, beneath open hills with ancient names: Effgill, Mellion Muir, Westerkirker Rigg, on the edge of what were once known as the Debatable Lands. To come here is to touch a past before easy communication brought uniformity. This isn't grand country. Sparse and self-contained, this little patch of border was once part of no man's kingdom and under no man's firm rule, not quite Scotland and certainly not England. There's still a stubborn resistance to category here. The saffron-robed monks from the Buddhist monastery which has found its modern home just over the hills fit in quite happily, as they might not in those British rural scenes we call archetypal. It is another Scotland to the Highlands, too; less famous, less visited, less recognised and perhaps less boastful but as strong in character and shared stories as any of the big mountains and glens to the north, and with families of 'names' – not clans – which hold traditions together with just as firm a clasp.

And in this story Eskdale matters because Eskdale keeps returning. Telford left, but he did not run away. He was no refugee in his later success and his life was marked by a telling loyalty to the place and circumstances of his birth. 'I ever recollect with pride and pleasure my native parish of Westerkirk, where I was born, on the banks of the Esk, in the year 1757,' his autobiography declares, a passage whose charm belies the fact that the surviving manuscript suggests it was drafted by his editor, eager to add personality to the heavy book, after his death.[1]

Many great lives start with a sort of fracturing: a shock in childhood, and perhaps, in later life, a forgetfulness or even a rejection of the starting point. With status can come pretence. But Telford was never embarrassed by his roots. He always declared a connection to the hills and the people amongst whom he grew: a remembering that was out of all proportion to the usual nostalgia for childhood. In an age when to visit Eskdale from London took several days this always-overworked man returned often, sometimes twice a year, until quite close to death. The massive build; the undisguised broad Scots accent (a talking point among his English contemporaries); the hardiness; the bluff face; the mess of hair, large nose and lack of interest in dress; the willingness to work and travel anywhere and in all conditions; the man who could wolf down his meals; the early rising; the rejection of luxury and indifference to honours or wealth: all of these speak of a man of plain sensibilities, who thought his origins not just the reason for his success but, as he became more famous, part of the reputation he wanted and cultivated. Today we might call it branding; we might speak of that must-have of a modern profile: authenticity. Telford's authenticity was authentic. Though it did become a brand and he did cultivate it, it wasn't invented.

You can sense its imprint even now if you visit Westerkirk and walk away from the bubbling Esk and up a side valley towards the tidy farm at Glendinning, where his father was a shepherd and where Telford was born in a simple thatched *sheilding*, no more than a hut. It is as quiet here today as it was then, a few miles from Langholm, the only close town. Not much has changed: there is a cairn in Telford's memory and nearby a small plantation in which it is said that the rubble of his birthplace can be found. The precise spot does not matter. What does is the sense of place. Over the last 250 years the hills of the valley along Meggat Water have been upholstered in trees and the houses rebuilt, but many still carry the same names and as you pass them you pass places the boy would recognise. This is a central part of the lure of Telford's life: that a man so keen on change and construction and who shaped his time as he shapes our own came from a place that escaped many of the forces of the last two centuries. The sounds are still of sheep and running water; and beyond Glendinning the road stops and the valley seems to run out into the sky.

———

Thomas Telford was not born with the name which made him famous. His christening is recorded in the Westerkirk parish register in a broad scrawl: '9th Augst. John Telfer in Glendinning, had a son baptised named Thomas'.[2] Telfer was the simpler spelling of the surname he later used, just as in childhood the formality of Thomas gave way to Tam or Tammy. So it was as Tammy Telfer that the infant began what was to be a simple childhood; tough even, with many of the hallmarks of poverty though few of desperation. The meaning of Telfer, or Telford, is a source of speculation: some say it comes from the old French for 'to cut iron', a nickname for a strong man of iron. If so, it was right for him.

The boy was born in late summer 1757. Months later his father was dead – the cause unknown – and, since the cottage had been tied to his farm work, his mother, Janet Jackson, was without a home and had to leave Glendinning with her infant.

It was a small, private tragedy of the sort which passes unmentioned in so many lives. In the Borders and at this time she was at least lucky in that the Telfords and the Jacksons were part of a breed of intermeshed families who supported each other in hardship and whose local roots ran as deep as any aristocracy. Now, they came to her aid. At Whitsun 1758 she moved with Thomas to the home of a cousin, a cottage a mile or so down the valley.

Today this house, called The Crooks, stands as a solid building, but then it was hardly grander than a shelter and, in its modesty, typical of its time and place. 'We were informed in the valley that about the time of Telford's birth there were only two tea-kettles in the whole parish of Westerkirk,' recalled Samuel Smiles, the first of his biographers and the only one to visit the valley within approximate memory of his lifetime.[3] Given to moralistic fables, Smiles has been much distrusted since, but he captures fairly the simplicity of Telford's early days. He paints a picture of a cheerful and daring boy known as Laughin' Tam by local legend, who 'spent his time. . . in high summer on the hillside amidst the silence of nature. In winter he lived with one or other of the neighbouring farmers. He herded their cows or ran errands, receiving for his recompense his meat, a pair of stockings and five shillings a year for clogs.'

It was a ruddy-cheeked childhood and in later life Telford talked nostalgically about it, though if you walk now in the cold rain and early dark of an Eskdale autumn you cannot but think that there must have been hard days which drove the boy to think of escape.

Mother and child had no choice but to be close; an only son and an only parent thrown together (some have suggested that a stepbrother may have lived with them too) at an age when most families were large. Without a father and without siblings his mother was a dominant force in his early life, although what he really thought of her he never recorded. The suspicion must be that he was always more respectful than adoring. His later correspondence suggests she was a tough and austere character. In his autobiography there is no mention of her, though she lived, supported by him, until her death in 1794. He was dutiful, at least. He wrote letters to her in simple block print, so she, nearly illiterate, could read them, and once he had left the valley he badgered his friends for information about her wellbeing.[4] 'I want to know how my Mother is. . . and what other country news you can furnish, your store has been a long time to replenish.'[5] Such loyalty carried with it an awkwardness on both sides. 'Her habits of economy will prevent her getting plenty of everything, especially as she thinks I have to pay for it which really hurts me more than anything else,' he complained.[6]

Her last years were painful. 'Your kindness in writing and paying so much attention to my mother is doing me the greatest favour that can possibly occur in the course of human life,' he told his closest friend in the valley as she aged. 'All that we can do is to render her situation as comfortable as possible and continue to give relief to her complaints. . . I am very anxious to come down to see her, and shall most certainly as soon as I possibly can,' he concluded. 'I am rather distressed at the thought of coming down as I must see a kind parent in the last stages of decay on whom I can only bestow an affectionate look, and leave her, her mind will not be much consoled by this parting, and the impression left upon mine will be more lasting than pleasant.' She died not long after that honest letter was sent.

These few descriptions of family association stand out in Telford's story because they are exceptional. He grew up short of the close relations who might have created a fulfilling home, and through his life, this stark isolation never changed. He was always solitary. He grew used to it; even proud, as one letter, sent back to Eskdale to mock the rumour that a supposed 'niece and heiress' of his was about to marry, suggests: 'I fear I have no claim to credit from this connection as I never had either brother or sister, and no relation of that name that I ever heard of even within the wide circle of Scotch Cousinship – altho' of this a very fair proportion falls my share.'[7]

With no father and no siblings, the child was forced to look outward. The effect on his character was lasting. His loyalties were always to his friends; to other families into which he was welcomed; and as an adult to the men with whom he worked. He bounced into other people's lives and stayed in them. Had he been less able, he might have been thought a pest. As it was, Telford's companions in the valley were the making of him and their stories are in some part his, too.

———

Travelling at the start of the eighteenth century, Daniel Defoe found that Scotland 'had the most enlightened peasantry in the world' – an impressive contrast to the darkness of the rural English poor.[8] The cherished Scots myth of the 'lad o' pairts', the boy from nowhere such as Telford who could throw his talents at anything and make his way, had meaning. In Scotland, especially the Borders, 'the parish schools had become deeply embedded in the rural community and gave rural education a legal status and permanence lacking in most European countries'.[9]

Visit the Esk – stay in the neat little town of Langholm, with its grey stone houses and whitewash, its cricket ground, woollen mills, woods and fields – and it is easy to think of this as old-fashioned, unchanging Britain; so loyal to tradition that it was part of one of only three constituencies in Scotland not to elect a Nationalist MP in 2015. But not far beneath the surface there was and still is a strong thread of modernity and adventure. When Telford was born, in Edinburgh and Glasgow, not far to the north, the Scottish Enlightenment was flourishing. In the valleys and small towns of his childhood there were also echoes of change, an opening of horizons, an interest in ideas. Perhaps it is more than chance that the first man to walk on the moon traced his roots to the Armstrongs from the Borders.

It was, it is true, also an old-fashioned place, renowned even in Telford's time for its quirky rustics. In 1776 one London paper reported that a surgeon had just died in Langholm: 'his Age was 136 years; in his Practice in the Medical Profession for above 100 Years past'.[10] But the metropolitan sneer in that report misrepresents what was happening in the valley as the young man grew up there. He was not an exception in his energy and later fame. He was typical. 'This parish is very remarkable as being the birthplace of men who have signalled themselves in every department of literature and science,' wrote one early nineteenth-century gazetteer, and it was no exaggeration.[11] 'The lower classes are sober and intelligent. The generality of them are fond of reading; and as they have an ample supply of books, the shepherds in particular have acquired a degree of knowledge and information beyond what might have been expected from their situation in life.' Eskdale was isolated, cold and remote in the 1700s. But it was no backwater. It nourished creativity and ambition and its sons and daughters ventured outwards and upwards in an amazing and exceptional manner.

To understand what made Telford you have to delve into this dynamic Eskdale culture. It was partly Scottish, partly British, partly global: lives given colour through learning, technology, war and empire. As a child and as an adult Telford was part of a remarkable network of mutually supporting friends who inter-married and exchanged surnames so that it is hard down the generations to pull them apart. Their family trees are all a mess and the point is not to untangle them but accept the complex and rare energy which linked them. They supported each other in their successes and failures and older ones brought the younger ones on. It is not to diminish Telford's individual genius or character or sheer diligence to say that it was the connections he made as a boy which unlocked his qualities.

The bond was formed at the parish school in Westerkirk, which Telford attended. Today there is none: it closed in 2000, though a love of learning still breathes within the little library next door, cared for by the last schoolmistress (a library in whose survival Telford played a later and significant part). This school stands out as a breeding ground of ambition. It fed talent, producing a string of achievers out of all proportion to its share of the population or its economic strength.

Even that simple gravestone at Westerkirk shows it. He can have had no direct memory of his father. But when still almost a boy he made an unusual effort to mark his father's life: he carved the words quoted on the grave at the start of this chapter. His friend Charles Pasley (a dashing military officer and engineer, born in Eskdale in 1780) records that Telford 'set up his father's humble monument, the work of his own hands, in Westerkirk churchyard with an inscription, in which he designated him a blameless shepherd, probably from having read of the "blame-less Ethiopians" in Pope's Homer'.[12] Pasley knew Telford well enough in his lifetime to have had a grasp of the truth of the story (though his is the only contemporary source that confirms it).

And what of that reference to Homer in the inscription the young Telford cut? It would have been no great surprise from a waistcoated young Etonian, trained to conjugate dead languages almost since birth and remembering his parents in the blustering family chapel at some frigid Palladian pile. But at the time Telford put up the stone he was still nothing more than a fatherless teenage boy, born in a quiet valley and educated in its parish school. His borrowing from Homer was more than showy, too: the choice of phrase to describe a father he had never known strikes a delicate balance between respect and pity. You do not call your father 'an unblameable shepherd' – you do not make that his lapidary memorial – without some subterranean hope of doing better. The child who cut those words must have determined privately that he was going to escape.

His route was friendship and at this point in Telford's story the remarkable families of Eskdale come to the fore. All were close to him and all were grander. Their surnames – Little, Malcolm, Pasley, Johnstone and Pulteney – run through his life.

Of them all, it is the Littles to whom he was closest. No one ever knew Telford as they did. It is to two Little brothers, Andrew and William, that Telford wrote his longest and most revealing series of letters, full of references to boyhood, his emotions, habits and ideals. After their deaths he wrote to William Little's widow and it is thanks to the care taken by subsequent members of the Little family to preserve a selection of these letters that so much is known of the early decades of Telford's life. The pleasure in reading them is diminished only by the fact that the letters sent by the Little brothers to Telford have not survived among his papers. Whether they were lost, or destroyed, is unknown.

Andrew and William came from an established Eskdale family, distantly related to Telford through his mother. Their father was a surgeon and they attended Westerkirk school alongside Telford. Andrew was a brilliant man, intellectually every bit his equal and one of the boys of their generation who became Eskdale adventurers. He served at sea as a surgeon on Liverpool slaving ships, only to 'lose his sight by lightening on the coast of Africa'.[13] Back home, he became an influential schoolmaster with a small private establishment in the Borders. The later ingenuity of his pupils is evidence not only of his own abilities but of the nature of the training that the pair must have been given as children at the parish school.

Telford's letters to Andrew offer an opening into their young life together: he was the sort of lad who could write to send 'my compliments to Jennie Smith tell her she is a Canterrin sort of lassie'.[14] There is a cocky good cheer to his boast, on first arriving in London, 'that glutton boy John Elliot will have no body to keep him in countenance now; tell him if he'll come here that I'll fight him at Hyde Park Corner'. That is the voice of one very good friend talking to another: and it is as spoken words that they are best understood, since Andrew could not see well enough to study the letters himself. His blindness meant that he must have had his letters read aloud to him, either by his brother (with whom he lived) or by his pupils, whom he employed to read books to him. Did this affect their intimacy? There is no sign of it, though one wonders what Telford might have written had he known the letters would be private. They are beautiful and startlingly personal, as in this outpouring from Telford in 1786 about the bond they had formed as boys in Eskdale:

> we are to toss to and fro in this bustling stage that no sooner
> have we got acquainted with a valuable friend than inevitable
> necessity divided us from that enjoyment and sends us from

him to form other perhaps unwished for connections. . . we reflect with regret on the companions we have left, and are apt sometimes to undervalue early attachments to my dear Andrew are deeply engraved on the young heart susceptible of the most lasting impressions. The traces in a young and generous bosom are not easily defaced – and if I may venture to judge favourably on my own – tis from the warmth and unremitting pleasure that arises in reflecting on the companions of my youth.[15]

Little was not only a sensitive friend: he was a talented and literate teacher. He taught Charles Pasley, a member of the second important Eskdale family in Telford's life. Pasley's skills were indicative of the energy of Eskdale at the time. He was renowned for being able to read the Greek Testament by the age of eight and trumped that four years later by writing a history (translated into Latin in the style of Livy) of the mock battles between the boys on either side of the River Esk, nicknamed the Langholmers and the Mucklemholmers.[16] Classical prose and conflict together: the curious match of focused ambition and wildness which was such a part of Eskdale.

In his mid-Victorian memoirs, Pasley recorded the spirit of the gang of Langholm youths of which he was part. One older boy 'formed a regiment, for so it was called, though only consisting of 40 or 50 boys from 10 to 12 or 13 years of age, of whom I was one, armed with sticks whom he drilled on the kilm-green'.[17] His fighting continued. He knew Nelson, fought with Sir John Moore at Corunna, and was both bayonetted and then shot while leading a raiding party at the battle of Flushing in 1809 (in the year it took him to recover he learnt German). But he also shared with Telford a fascination with engineering. He led the Royal Engineers and, after Telford's death, brought new standards of safety and comfort to the railways as inspector-general of the system. His writing roams the possibilities of his age, from essays on cement to practical hints for carrying out a siege. He liked blowing things up (pioneering the underwater detonation of wrecks), but he corresponded, too, with poets such as William Wordsworth.

Pasley was one of six boys from this generation who became famous as the Eskdale Knights, who as men won titles for their service and ability, not their social status. Another of them was Sir James Little, from the clan which included Andrew, Telford's closest friend. The other four were brothers from the third of the exceptional families in Telford's life in the valley, the Malcolms. 'I was for some years a school fellow of the elder brothers of that distinguished family,' he recalled.[18]

George Malcolm, their father, was one of those clever people who, never quite crystallising their talents, pass on to their children the ingredients of success. Well educated and intended for the Church, he ended up as an impoverished gentleman farmer and land agent instead. He took over his house at Burnfoot in 1761, just down the valley from The Crooks where Telford had recently moved with his mother, and it remained a stronghold of the Malcolm family until 1962. He was, one of his son's biographers recorded, 'a strong-minded, an honest and a pious man – but he was not a prosperous one'. Hard-working, full of schemes (he once tried to make his way in the wine trade), he spurred his children to brilliance. The family produced an admiral, a vice-admiral, a lieutenant-colonel and a major-general: men who were later the guardian of Napoleon on St Helena, an aide to the Duke of Wellington and the imperial governor of Bombay, but whose childhood experiences were all rooted, with Telford, in the valleys that bred them.

George Malcolm's surviving letters to Sir William Pulteney, the owner of the local Westerkirk estate (and later Telford's principal champion), for whom he served a local land agent, are a vivid record of the struggles of the time. They are full of things such as failed harvests, unpaid rents and the devious blocking of spawning salmon from reaching the upper parts of the River Esk by English landowners south of the border. They also record the life of a determined and industrious family. From their house at Burnfoot the Malcolms must have lifted everyone around them. The energy was at times almost terrifying. When John Malcolm left school he was taken to London to apply for a post with the East India Company. 'Why, my little man, what would *you* do if you met Hyder Ali?' asked a director. Aged just twelve, John was blunt: he would 'out with his sword and cut off his heid!'[19] He got the post and rose to be a major-general. As his biographer wrote, he came from 'a healthy and vigorous tribe, who forded the Esk, clomb [*sic*] the steep hill-sides of Douglan and Craig and gambolled in the heather. There was a good parish school in Westerkirk; but better still, there was plenty of fresh air and free scope for exercise, and the boys in early childhood, swimming in the flooded waters of the river, or scampering about the country on rough ponies, learned lessons of independence which were of service to them to the end of their lives.'[20]

Such, too, must have been Telford's childhood in this place, among these people. And as they grew up, the centre of their small world will have been the great house at the bottom of the valley and not far from the school, owned by the Johnstone family. If the Littles excelled through learning and the Pasleys and Malcolms made the most of the opportunities that came their way in military and imperial service, it was this fourth and most illustrious family of the valley, the Johnstones, who opened those opportunities. Theirs was the grandest and richest Eskdale family of all, a pushy (and, by some, disliked) political dynasty.

To find physical proof of the Johnstones' status today, you must return to that small graveyard where Telford's father is buried and which even now contains so many elements of the Eskdale story. Lift your eyes for a moment from the unblame-able shepherd's simple marker and you cannot miss the huge and strange monument that stands hard by. It is an unexpected sight in a place that is otherwise so modest. A pillared Greek Doric mausoleum, it looks a little like a Tardis landed unexpectedly in a foggy field. There is something atheistic, almost occult, about it all. Ringed by carved ox heads – classical bucrania – with carved human skulls on the inside, the mausoleum was designed by the grandest of Edinburgh architects, Robert Adam. It stands as a piece of the Scottish Enlightenment turned to stone, but it also speaks of the wealth, energy and culture which flowed around Telford's birthplace and through his childhood.

Built in 1790 and recently restored, it is a monument to Sir James Johnstone, whom Telford will have known as a boy and whose family house, Westerhall, still lies next to the graveyard. The Johnstones – especially Sir James's brother William – were a force in the Borders and in the House of Commons, and became one, too, in shaping the next, definitive, steps in Telford's life. The bond, over subsequent decades, became quite exceptionally close: not just the usual obligations owed between landowners and tenants, but something more intimate. Indeed, it is impos-sible to overstate the importance of the connection that grew between the family marked by the modest Telford grave and the family marked by the grand mausoleum. It is as if the two monuments and two families were each of a pair. They stand evocative of fast-changing times.

It is Eskdale which made this linkage work. In much of Britain – and certainly most of England – things would have been different. Somewhere else, a family of grandees and a shep-herd from a rough farm next door might not have been bred, taught and buried in such proximity: there would have been a family chapel, a public school, perhaps a peerage, certainly class interest to keep the upper sort apart from the lower. But in Eskdale, by the nature of its past and its isolation, there was a shared community. Boys from different backgrounds were schooled together and played together and as they grew older their families helped each other, too.

At the age of twelve Telford's school days ended and his happy childhood in the valley drew to a close. He was apprenticed to a stonemason at Lochmaben, a small town near Lockerbie not far away to the southwest. It should have been a lucky break, a good position in a place better connected than Eskdale. But it did not work out. Thomas was – or said he was – ill-treated by his master, and ran back home. It must have been a trying moment for his mother and it was to prove the last time in their relationship that she had the better of him. Together with her cousin Thomas Jackson, the land steward on the Johnstone estate, and thanks to his position a man of influence, she found a more congenial master, a stonemason in nearby Langholm, just a little way down the Esk. Telford's journey had begun.

———————

There was no better moment to be born, if you wanted to change the world by building things, than Britain in the second half of the eighteenth century. Everything was in flux: ideas, technologies, identities. Vulgar fortunes; foreign flavours; atheistic thoughts; terraced streets; London; novels; newspapers; MPs; city dealers; the Royal Navy; turnpike roads; iron, canals, coal and steam – all of these and more came together to fuel the new belching, swaggering Great Britain. To its critics – inside and out – the nation was something of a hooligan on a spree.

This was the country which sang 'Rule Britannia' for the first time and those words had bite. London could hit you, or make you, or break you. To find something comparable today you have to look towards Silicon Valley for the push of change and pull of technology; or towards China for the casual brutality, pollution and willingness to give anything a go. If eighteenth-century Britain was less regimented and less populous than China today, life could turn out to be just as thrilling or horrific. You could be born in a peasant village and make a fortune, or end up in a factory – or a brothel, or dead.

We don't always think of the eighteenth century like this now. It tends to be the Victorian century that followed which is defined in our minds by industry and empire. The eighteenth-century industrial revolution, to the extent that the term is still trusted, is seen as just the start of something bigger and more overwhelming to come. To modern eyes, the Georgians are often pictured through the lens of the National Trust tea room: polite and elevated, all titles and country seats; or alternatively, corrupt and foppish, the protectors of rotten boroughs and foolish losers of British North America.

And of course some of these pictures have substance. The central state was small and weak. The monarchy was settled, if underpowered. America was lost. There was no revolution. There was a self-enriching aristocracy. There was great – probably growing – inequality as national wealth outstripped wages: the problem of capital which challenges economists now challenged workers then, too. There was no certainty about social mobility and economic historians suggest society was less, not more fluid than today, even if rapid urbanisation created the chance of disruption. There was a strong, if repeatedly defied, recognition of social hierarchy, and the Capability Brown lifestyle was much desired by those with good ancestry or new money. The rural gentleman and the benevolent squire were popular caricatures even if deference was patchy, and the London mob always stood ready to knock the hats off the gentry or worse. Cities, even London, were small by late Victorian standards, though growing very fast. The countryside was always within reach. George III did like keeping pigs as well as repairing clocks. There were repeated calls for the reformation of manners as well as of Parliament and a fear of the horrors that could be brought about by what was called luxury.

But to see eighteenth-century Britain as one large Stubbs portrait with a few smoky forges tucked away somewhere in Derbyshire and Shropshire is to get the picture wrong. The driving force of the time was not tradition but an extraordinary mass of radical and intellectual energy. It gave momentum to everything, it touched everyone and it reached everywhere. In the face of change the status quo often tottered, pushed by a confident Parliament whose legislative heft and defence of property gave a formal underpinning to the new economy. Fortunes were made, fortunes were invested; and everywhere things were built or fixed up. One artist in particular captured this energy: not Stubbs, but Joseph Wright of Derby with his luminous reflections of industry and science. Our subsequent respect for the work of the former over the latter says more about the Arcadian nation in which Georgians would like to have lived than the messier one in which they did.

Yapping and unguided, Britain was becoming modern very fast: a terrier of a place, smaller and less refined than some of its European rivals, but peculiarly successful and dynamic. This was the land of the good connection, the lucky break and the smart idea, a place for chancers and the clever. The novels and plays of the time – themselves part of a spreading literate culture – are full of novelty and social insurgency: Henry Fielding's Tom Jones, the shameless orphan, or Humphry Clinker, Tobias Smollett's amiable ostler who turns out to be a gentleman's son.

Clinker's story is significant for another reason too: the driving force in the novel is travel, a journey around Britain. The technology of transport was at the forefront of the things building the new Britain. Starting behind much of the rest of Europe – the French state, then as now, oversaw better, straighter, faster and emptier roads – Britain pulled ahead to the point where its achievements became an object of wonder to visitors. The 1750s brought the country's first canals and a mania for laws to turn muddy rural tracks into sometimes less rutted and fitfully profitable turnpikes. Slow stagecoaches gave way to frequent 'fliers', with an increase in speed which could be proportionately more striking than the acceleration brought later by the railways. 'Who would have believed, thirty years ago,' the novelist Richard Graves had one of his characters say in 1779, 'that a young man would come thirty miles in a carriage to dinner, and perhaps return at night? or indeed, who would have said that coaches would go daily between London and Bath, in about 12 hours; which twenty years ago was reckoned a good three days journey?'[21]

Not everyone liked it, of course. As another writer lamented:

Wives staid at home, but now the turnpikes bring,
All to learn vice, buy pins, see the king;
'Tis on the turnpikes that we ought to rail,
The turnpikes, where sin runs up the nail.[22]

For the most part, however, the ability to travel was seen as a positive. It brought ideas, books and newspapers. Better connections spread the sense that all sorts of things were possible. And these connections reached deep into rural Britain and changed it. Nothing stayed still. New ideas were everywhere. The wealth of nations, in all its forms, was gathering pace.

———

This was the world in which Thomas Telford came of age, after being born in a year of impulsive forces even by the

British standards of the time. In 1757 Admiral Byng was shot 'to encourage the others' in the fight against France; Robert Clive captured Calcutta (and with it sustained British influence in India for a century); William Pitt the Elder lost power (briefly); James Watt began experimenting on steam at Glasgow University; and in Edinburgh Adam Smith and David Hume were developing their ideas and exchanging letters. Both men rooted their philosophies and theorising in facts, in data, in observation of the material world; neither anchored their assertions in Authority – of the divine, of the ruling classes, or of the Ancients. 'If we take in our hand any volume; of divinity or school metaphysics, for instance,' wrote Hume in 1748, 'let us ask, Does it contain any abstract reasoning concerning quantity or number? No. Does it contain any experimental reasoning concerning matter of fact and existence? No. Commit it then to the flames: for it can contain nothing but sophistry and illusion.'[23]

Such empiricism shaped the age and shaped Telford. War, intellect, technology and philosophy were being brewed in one great national cauldron. This was true of Scotland in particular and Telford's life went on to exemplify it. The Act of Union in 1707 had left intact the nation's systems of law, local government and education: they remained distinctly Scottish. But as the historian Linda Colley has shown, the creation of Great Britain offered ambitious Scots a bigger, lucrative stage on which to make their name. Today, some Scottish nationalists paint an imagined history of a usurped nation being swamped by Englishness, but to many English at the time it seemed the other way around. Scotland was well placed and keen to take advantage of the opportunities created by union, trade and war and the arrival of law and order.

And though such forces originated very far from Eskdale, change and opportunity shaped Telford's early life nonetheless.

When he sat down to write the first pages of his attempt at autobiography he didn't reflect back to memories of stability – the
usual rhythm of a rustic youth – but instead looked to the things
that were shifting around him as he began work as an apprentice in Langholm. Even in this deeply rural patch of Scotland,
'roads were substituted for the old horse tracks, and wheel
carriages introduced'.[24] A hundred years before his birth, the
borderlands had been properly wild; the home of raiding families, the infamous Reivers, thieves on horseback, taking refuge
in fortified towers: a culture the Tory MP for nearby Penrith,
Rory Stewart, has compared to modern tribal Afghanistan. As
Telford noted, Langholm had 'been exposed to frequent and
destructive inroads'. By the time of his youth this was all but
over; perhaps talked of – and exaggerated – by old men – and
mixed with more recent memories of the Jacobite army which
passed by a dozen years before Telford's birth. By the time
of Telford's own old age both the wild Borders culture and
the Jacobites had become the stuff of comfortable nostalgia
and storytelling in the novels of Sir Walter Scott. His defining novel, *Waverley*, was set 'Sixty Years before this present
1st November, 1805', the year of the last Jacobite rebellion.
Though the novel's climactic scenes are set in the Highlands, its
title came to be associated with the Borders as 'Waverley country' and the railway which later ran through the area became
renowned (and then much missed) as the Waverley line. Part
of it reopened in 2015, rebranded as the Borders Railway, Scott
having passed out of fashion.

Telford warmed his hands before the embers of this nostalgia. 'I am running into prolixity on border details, pardonably,
I hope, in an old man speaking of the scenes of his youth,' his
autobiography states, in another endearing passage seemingly
drafted by his editor. Whoever wrote it, by the time the book
was published the old days had gone.

This changing world matters very much. Born in Eskdale at any other time, Telford's life would surely have followed a different path. A century before, and the boy who was sent to the hills to guard sheep would have stayed there. A century later, and civil engineering was becoming (not least through Telford's own efforts at bringing formal structure to the practice) the bastion of more educated, professional men; and Telford, who started as neither, would not have found it so easy to rise so high or so fast. Born today, Scottish nationalism might have drawn him to Edinburgh rather than London. So Telford hit his moment. Skills learned on the job could sweep him along; connections could carry him upwards; talent and determination could do the rest. Relentlessly, almost every record of his long working life shows a kind of reckless drive, never stopping, never slacking, always pressing on – but never forgetting where he came from either, always drawing comfort, often pride and sometimes opportunity, from his beginnings.

In retrospect he liked to paint this as a stable time, a period of quiet progression, but the unsettled, ambitious side to his character was to the fore from early youth. 'You know while I was in Eskdale... I was a bone of contention,' he wrote, in early middle age, to Andrew Little – throwing the line out without supporting it, assuming that Little would know all too well what it meant.[25] There must have been conflict, but what about and with whom is lost. All that remains is the knowledge of the potent character at the centre of it all.

––––––––––––

'The early part of my life was spent in employment as a mason,' Telford wrote in his *Life*.[26] It was a start which always defined him. 'I ever congratulate myself upon the circumstances which compelled me to begin by working with my own hands, and thus to acquire early experience of the habits and feelings of workmen.' As an older, grander man he must at times have been a bore going on about the lessons he learned young, and frequently disapproving of what he saw as the weaknesses of softer generations. 'It has happened to me more than once, when taking opportunities of being useful to a young man of merit, that I have experienced opposition in taking him from his books and drawings and placing a mallet, chisel and trowel in his hands.'

Telford learnt how to use those tools well in his early days in Langholm, 'the solid knowledge which only experience can bestow', and for the rest of his life it left him with a feel for materials and a tolerant respect for the skills and lot of his workmen. He was practical, at home in the countryside and ready to ride in the rain, jump into the mud, redesign an arch or sketch a way of draining a particularly boggy bit of road. He knew – because he had cut stone himself – when granite was a better choice than sandstone in a wall, what a man could dig in a day and how much a horse could carry. But though his learning was solid not theoretical, he did not scorn planning, calculations or the advice of others; and he was willing (at least until he grew old and stubborn) to change, most of all in his embrace of iron as the most daring of new materials.

He must have been taught this flexible way of thinking from the start. From his early teens he worked as an apprentice and later as a more qualified journeyman for his master, Andrew Thomson in Langholm. He was thrown into building, carving and designing all the things that Eskdale's growing rural economy needed. As Telford admitted, much of the work was modest, building 'dwelling houses for the farmers' and 'bridges, numerous but small. . . over the mountain streams'.[27] But even these structures were being made more solid than ever before and often being built in stone for the first time. The characteristic landscape of the area that survives today was being formed: stout farms and walled tracks; and there was plenty of work thanks to another of those strokes of good timing which came Telford's way. Eskdale, even a few years before his birth, was medieval in much of its character. But by the time he came of age the energetic young Duke of Buccleuch and Sir William Johnstone were busy improving it. Simple huts and pony tracks were being replaced with stronger structures and new roads.

Under Buccleuch's guidance, Langholm, in particular, began to expand. The duke oversaw construction of a new settlement on the west bank of the Esk, laid out on a regular planned grid. New Langholm, as it became known, was a small echo of the grander New Town being built to the north in Edinburgh and the project gave Telford not only employment but his first chance to work on a substantial bridge, linking the new settlement with the old, across the Esk. Its construction was overseen by a local stonemason, Robin Hotson, working with his master, Andrew Thomson. For Thomson's apprentice it must have been a life-changing introduction to the art of civil engineering. It was also an introduction to a lifelong working relationship with his Eskdale friend, Matthew Davidson, born in Langholm two years before him and already an experienced mason.

In design Langholm Bridge is far from radical: three grey rubble arches supported by neat dressed piers – but it was strong and good-looking and, twice widened, still carries traffic on what has been named Thomas Telford Road. Underneath the westernmost arch Telford's reputed mason's mark is still visible: a pair of outward-facing triangles linked with a line and an x – though whenever I visited it the torrent was too great to search without risk of being swept away towards the sea.

Samuel Smiles, Telford's first and most florid biographer, repeats a local story about the building of the bridge.[28] Hotson's contract specified that he was not only to build the bridge but also to maintain it for the first seven years and once, when he was away from town after its completion, his wife Tibby was persuaded that the roaring waters of the Esk were about to wash away the piers. She is said to have run down the street crying: 'We'll all be ruined – we'll all be ruined! Oh where's Tammy Telfer – where's Tammy?' Tammy was sent for and found. 'Oh Tammy, they've been on the brig and they say it is shakin! It'll be doon!' she cried to him. Telford tried to persuade her it would hold firm but she stood with her back pressed against a parapet nonetheless, leaving him doubled up in laughter.

The story suggests that Telford was growing in confidence. No longer a shepherd boy and no longer a mere apprentice, he was becoming a skilled and popular local figure: trusted, says Smiles, 'to do small jobs on his own account such as the hewing of grave-stones and ornamental doorheads'. Today, not much of Telford's known work remains in Langholm but one piece at least is said to be his: a tidy stone wall and blank pillared arch. It is a modest thing but well done of its kind, now moved from its first location to stand in a small garden, backing on to the Reivers Rest Hotel, with a plaque recording its creator.

Within four or five years of starting work Telford was already established. A less ambitious man might have been happy enough with easy success, making things in the town for the rest of his life: there was a call enough for such skills. But Telford was restless. He wanted new horizons, new knowledge and he set out to find them. 'At the age of 23 I considered myself to be a master of the art as practised in the county of Dumfries,' he wrote in his *Life*: a boastful claim, but not perhaps a vain one.[29]

Eskdale Tam

Not quite two years after Telford, on a farm not far to the west, another boy was born into a poor and struggling family. He was to become the definitive Scotsman. There is no certain report that Robert Burns and Thomas Telford ever met, though a letter from Telford implies strongly that they did. They would have scented similarity immediately: the accent, the education interrupted by farm work, the loyalty to their roots matched by a shared desire to escape them – all these they had in common.

Something else would have drawn them together too; indeed did draw them together, for Telford wrote to Burns (it is unknown if he received replies) and they had friends whose paths crossed both lives. Both men were published poets who wrote in the voice of and for the people with whom they grew up.

It is an unexpected conjunction. Today, Telford the engineer is remembered. Telford the amateur poet and keen but private literary critic is unknown. It is hard to imagine now, when arts and science take such separate paths, that an engineer could also pass his time in verse; just as it is hard to imagine Telford as a young apprentice in town for the first time with a bit of money, a little freedom, a new trade and an eye to the main chance but not much education, throwing himself so eagerly on books and writing as a hobby, rather than drink or sport. He did, however, and though his enthusiasm for literature was lifelong it first took flight during his time as an apprentice mason in Langholm in the 1770s.

Poetry took Telford forward and remained an ever-present accompaniment to his life. The cause was not loneliness or boredom. As a youth, he was sociable and not short of friends. But books and verse-writing allowed him to reach out beyond the place and class into which he had been born. It gave him the power of education and expression. At school, he learned to read. But it was outside that he developed his learning to an extent exceptional even for an erudite age. He read books, Smiles suggests, given to him by Elizabeth Pasley, a member of the sprawling border family which was such an influence on his life. 'The joy of the young mason may be imagined when Miss Pasley volunteered to lend him some books from her own library.'[1] Throughout his life, Smiles writes, 'he always had some book with him, which he would snatch a few minutes to read in the intervals of his work'. He adds that one of the first books Telford borrowed from her was *Paradise Lost*, and taking the book to the hillsides with him he 'read, and read and glowred; then read, and read again'.

There is something so idyllic about the image of an eighteenth-century shepherd boy reading Milton's verse in the border hills that the story sounds too perfect to be true. But it might well have happened. He was more than a milky-eyed sop taking respite from a tough trade through popular fiction. He was a scholar. His letters are peppered with references to serious books of all sorts and the things he read shaped the things he wrote – because writing in verse mattered to him as much as reading the work of others and even when at his busiest he made time to write and edit his poems, giving up sleep to do it. At least a dozen of his poems survive; others, mentioned in his letters, have been lost. Many are long; some were published and their composition spans his entire life from the age of twenty-one to seventy-three – the last known piece being written in 1831.

In part this enthusiasm was not unusual in an age when printing was cheap and reading part of everyday entertainment. Novels and epic poems were, if not exactly everywhere or always of good quality, at least a familiar element in many people's lives. It was not uncommon, for instance, for officers in the Royal Navy to write verse – as one, from Eskdale, did, sending them to Telford, who then wrote straight away (as he always did) to Little to describe their contents and judge their merit. 'Eskdale seems to be the soil for Poetasters,' he told his friend, after receiving the verses from the officer, then serving in Antigua, who 'fills all the West Indies with his perfumeries'.[2] The words, he added, were 'composed in a midnight watch. . . he informs me he has finished (nearly) an Epic Poem besides a burlesque Poem in the Miltonian Style; so that he will arrive in England loaded with his works'.

Writing, for Telford too, was an escape from duty, not a calling. As he once told Little, poetry 'is to me what the Fiddle is to others, I apply to it in order to relieve the mind after being much fatigued by business'.[3] Always, work came first – 'I mind houses more than Poems' – and he rated his own writing with a kind of modest smile.[4] In one later letter to Little he told his friend: 'I cannot think yet of sending you the Ode on May, as I have not had time to correct it – but in order to show you my great talents. . . I here subjoin a slight specimin – a sort of tal, al, de, ral before I begin to sing in good earnest.'[5]

Yet for all these professions of modest talent, literature mattered deeply to him. He besieged his friends by letter with compositions, questions and suggestions of what to read. 'Have you read Mr Stewart's book – *The Philosophy of the Human Mind*? Have you read Alison on the principles of taste? What do you think of those two books?'[6] He took care on his travels to visit libraries and spend time with writers. Whenever he visited Bristol, for instance, he told Little, 'I worship at the shrine of

Chatterton fancy yourself in the church where he <u>pretends that</u> he found all his poetic treasures.'[7] He had a critical eye and a neat turn of phrase which might have sustained him as a journalist. In March 1792 he wrote to Little, 'enclosing the Essay on Grandeur. . . the verses are not bad but the Idea is as old as the Hills and as hackneyd as the road to Highgate'.[8] In another letter, sent in 1786, he loaded praise on 'young Master Scott' – tantalisingly, perhaps the sixteen-year-old Walter Scott who met Burns for the only time that year and who had, briefly, been to school in Kelso, forty miles north-west of Langholm.[9] The boy could (it is imaginable but unprovable) have visited Eskdale then too and encountered Telford. 'His perfumaries are wonderful: why Pope was only a fool to him!' he enthused in his letter. 'The luxuriance of his Genius will produce Laurells and Bays so rapidly that the field will be overspent before the corrective hand of Judgement and literature has time to form them in to wreaths.' Whoever Master Scott was, it was quite a review.

But what of Telford's own verse? No claim should be made for his work as significant but it deserves, at least, to be known and studied as a footnote in the development of Scots writing and identity. The fact that it is overlooked today is perhaps a consequence of Telford's success in other fields. Or perhaps the reason is that Telford himself never made any serious claims for his composition.

Like many young Scots of his time, Telford was an avid reader of a magazine produced by Walter Ruddiman, one of the most successful publishers of his time. *The Weekly Magazine, or Edinburgh Amusement* was celebrated and eclectic, publishing essays on things as diverse as the inhabitants of Catalonia, the natural history of the honey bee, mathematical competitions, extracts of parliamentary speeches and a recipe for making potato bread without flour, as well as book reviews and two or three diverse pages of verse sent in by readers. Because it was free of news, it was also free of tax and sold around 1,400 copies a week. Each copy will have been shared widely and its dense pages still speak of the learned, curious and outward-looking culture of Scotland.

Ruddiman published the first edition of the poems of the short-lived but notable poet, Robert Fergusson, later a major influence on Burns. In April 1779 his magazine also published a far lesser work by an anonymous poet from Dumfries (the county in which Eskdale lies).[10] It includes this lumpen passage:

> At ony time when ane has leisure
> Your Magazine affords great pleasure;
> They'll fin't a braw poetic treasure.

The author described himself only as 'Jockie Mein'. He was not – thankfully, given the quality – Telford. But he was the spur to the first publication of Telford's verse in the same journal soon after. Telford clearly knew Jockie Mein's identity (which, it is reasonable to speculate, may have been a school friend's) and his first poem, an *Epistle to Mr Walter Ruddiman* (described by Telford as 'honest Wattie') is a teasingly competitive reply to the piece.[11]

Printed in May 1779, when Telford was only twenty-one and a journeyman stonemason in Langholm, it is a vigorous and boastful piece of writing, but not without merit. The style was his own. The voice was his. And the tone undeniably Scots. It is charming even if it is only an interlude in the life of someone who went on to find greatness doing something completely different. From the start it has a bullish, full-throated local voice:

> Here, honest Wattie, may be seen
> My hearty thanks to Jockie Mein;
> But envy of malicious spleen,
> I do assure ye,

He needa care for critics keen,
Wi' a' their fury.

And as so often with Telford's writing, it goes on to refer to
the self-education that comes from reading. 'And I had got ye'r
Magazine, And glowring owr wi' eager ein,' he writes – before
going on to describe 'how happily ilk shepherd reads, Sic tales,
clad in plain Scottish weeds'. It ends with a typical bit of Telford
bombast:

Lang may ye sing, well may ye phraze,
Ha'e rowth and plenty a' ye'r days;
And sall gar a' our green braes
Ken weel ye'r name;
I'm sure ye still sall ha'e the praise
Of
ESKDALE TAM
Langholm

Two months later, the journal dismissed the poem, in its
regular series of critical responses under the heading 'To *our*
Correspondents'. Telford's verse was, it said, 'a faint (*very faint*)
imitation of the late ingenious FERGUSON'.[12]

As a first review, it was not glowing. It was also unfair. Even
in a literate land and one grounded in the folk songs which
soon formed part of Burns's own tradition, Telford must have
stood out as something of a boy wonder by getting his efforts
published in Edinburgh. Eskdale Tam was the author's spirit,
the Scots voice was real and that description of a happy, reading
shepherd clad in plain tweeds came from life. The author was
gripped by a sort of romantic cheer that was not (as in some
writers) an escape from the remote place he came from or the
rural life he had led, but a celebration of it.

This leads to a question. Did this young man, who was to become one of the leading engineers, toy first with the idea of becoming a poet? It is not idle to wonder. He did more than dabble; effort and persistence was put into work that was not worthless; he seems to have had a minor talent which he might have tried to develop. Writing in 1788 he explained that 'it has ever been and ever shall be my aim to unite those too frequently jarring pursuits, Literature and Business'.[13] There's much that is original in his verse, and even as a beginner he takes poetic risks. Indeed, Eskdale Tam's first piece of writing was thought good enough to be included in a volume of Burns's own verse, published in 1793, without any suggestion that it was not by Burns himself.[14]

In later life he was to go on to cultivate the friendship of poets including the Poet Laureate, Robert Southey. By then, however, he had come to accept that he could never be the equal of such men. If a life in poetry and literature ever beckoned, the realist in him concluded that this was only a daydream, and not to be. But we should not discount the idea that he had once wished it were otherwise. As he made his name in the 1790s and made some money Telford published the most extensive and autobiographical of his poems in a generous bound volume, in large print on wide pages. He cannot have expected it to sell; this was clearly an indulgent folly for friends. But the act suggests he took writing seriously and thought it good enough to preserve.

The piece he published, 'Eskdale. A Descriptive Poem', was written (Telford wrote later, apologetically) 'in early youth, when the situation of the Author gave him little opportunity of being acquainted with English Poetry'.[15] Its simple rhymes and metre are eased by a palpable fondness for the place it describes and by the knowledge on the reader's part that this life was lived, not just observed, by the boy who was to write his tribute to: 'Thy pleasant banks, O Esk! and shady groves'.

It is a (much) lesser Gray's 'Elegy Written in a Country Churchyard' and is very likely to have been influenced by that work, written in the summer of 1750, which Telford will certainly have read. The whole thrust of Gray's 'Elegy' is a wistful tribute to the unfulfilled possibilities of humble people in rural life. Gray speaks of what for good or ill they might have been, but never had the chance to become. It is not fanciful to speculate that his 'A youth to fortune and to fame unknown' might have touched a chord in Telford, who writes of 'the early Shepherd' who 'seeks his flock at morn' and who goes on to describe how:

western Suns with mellow radiance play,
And gild his straw-roof'd cottage with their ray,
Feels Nature's love his throbbing heart employ,
Nor envies towns their artificial joy.

And though the rest of his life was spent building the structures that supported urban life and industry, this dismissal of the 'artificial joy' of towns never left him. His greatest work – the Holyhead Road, the Menai Bridge, his canals and the thousand miles of road he built through Scotland – was set and built in deep countryside, and intended to enhance not obliterate it.

———

In the end, of course, as a peasant Border poet, Burns left him in the dust. Nonetheless the parallels between the pair are worth exploring. Both men had poor, farming fathers; both men worked on farms as a child; both published poems in the 1780s; both came from similar places. Burns was born two years after Telford and not that far away to the west, in Ayrshire.

But it would be wrong to assume that Telford's efforts were a pastiche of Burns.

Though some of his later verse carries a weak (and acknowledged) echo of his far superior contemporary, his first effort was published in *The Weekly Magazine* before anyone had read anything by Burns. At the time it came out, in 1779, Burns was still a lad listening to ballads and writing verse on a windswept, boggy farm and several years away from his own first publication in 1786.

Did they meet? And if so when? Sadly there is no firm record. But on 7 December 1794, less than two years before Burns's death, Telford wrote to Thomas Boyd, the Dumfries builder and architect who erected Burns's farmhouse at Ellisland. He clearly knew Boyd well. The letter gives the strong impression that he was on first-name terms with Burns too (and knew of his drinking). It has a young man's heady quality to it:

> I hope that you informed my friend Robin Burns, that I was very desirous of paying my respects, but tell him that unless he leaves off his baudy songs, that 'he'll get his Fairen – "In Hell they'll roast him like a Herring".' – Tho if he goes on in his old way, not even a <u>she Devil</u> will be able to meet with a Milt in him.
>
> But after you have abused him properly do tell him, that the first time I meet Mr Alison, we will drink his health, for in case we should be consigned to a <u>Neuk</u> in his neighbourhood, it would prove of some consolation to be on decent terms with one another. – Farewell – I am yours very sincerely
> Thos Telford[16]

This is not the sort of letter you write about someone you admire only from afar. But for Telford Burns's importance went wider than friendship. He seems to have acknowledged at once and without rancour his literary superiority. He saw him as someone who could give voice to the ambitions of rural Scotland and had a duty to do so. He also intervened to encourage him. In 1792 he wrote enthusiastically to Andrew Little to describe the despatch of a set of his verses to his now published and much-celebrated Borders contemporary. 'Like a True Poet I am paying you for the stockings with a batch of verses – they have just been sent to R Burns.'[17] A friend, he went on, had encouraged him to send them in the hope:

> . . . that Mr Burns' attention might perhaps by this means be fixed on those happy, innocent and picturesque views of the manners of the Scotch peasantry, which ought to be preserved as a lesson to Mankind, and we are of the opinion that if Mr Burns would exert [h]is wonderful talents in this his peculiar line, he might actually prove the means of preserving that virtuous character as you well know there is no mode of impressing the mind of youth equal to the simple Poem, if I can succe'd in this I shall set myself down as deserving well of the land of Cakes.

This long poem by Telford – one of many sent to Burns by his admirers – was found amid the latter's papers at his death in 1796. It was thought good enough for extracts to be included in the definitive collection of Burns's works published by James Currie in 1800, who found it 'a poetical epistle from Mr Telford, of superior merit'.[18] Currie – who was a friend of Telford's and was born in the same part of Scotland a year before him – annotates the verses to explain their references to Eskdale. Echoing Telford's own explanation of its meaning, Currie noted that the poem was written 'in the versification generally employed by our poet himself. Its object is to recommend to him other subjects of a serious nature.' The words praise Burns but were drawn from Telford's own experiences: he was, indirectly (but consciously) making a modest comparison between himself and his contemporary – both writing in a dialect that was as much Scottish as English.

'Pursue, O Burns! thy happy style,' he begins before going on to praise the writer for expressing the strength of rural identity: 'thy gen'rous flame, Was given to raise thy country's fame.'

The most personal passages – a poetical description of Telford's boyhood – come towards the end. He writes of education and reading, describing himself as 'the virtuous boy' who attended:

> The parish school, its curious site,
> The master who can clear indite,
> And lead him on to count and write,
> Demand thy care;
> Nor pass the ploughman's school at night
> Without a share.

This school made him into:

> The tenty curious lad,
> Who o're the ingle hings his head,
> And begs of neighbours books to read

Telford often returned to Eskdale in his writing, as he does here. 'And thro' the woods, I hear the river's rushing noise, Its roaring floods,' he writes, but this poem touches on something else, too, which he hardly ever refers to elsewhere: religious faith. He describes the shepherd boy's 'holy joy' – 'His guileless soul all naked shown, Before his God.'

This, again, must be self-descriptive: but there is little further evidence that later Telford felt any faith at all. (Once, after studying a book on 'the immense Marble Temples of Upper Egypt' he declared with a farmer's eye that 'it seems very likely that all the Gods of the World have come from there – only the breed changes with the Country where there are brought to, and then they are crossed til their grandfather would not know them'.[19]) He seems, from his writing, to have found greater spiritual power in the forms of nature and landscape than in ceremony. He does not say what shape – if any – 'his God' took, only that in the boy at least, there was some sense of awe. Theists and atheists alike would struggle to claim Thomas Telford for their own. It can only be said that he never became a churchman, nor ever lost a sense of wonder.

The strength of this writing is enhanced by knowledge of the success of the man who wrote it but even without that it is moving in its respect for landscape and childhood. What it lacks is the human depth, the sharp and sometimes wry observation, that marks Burns out. Nothing by Telford comes close to the egalitarianism of 'Is There for Honest Poverty' – 'A Man's a Man for a' that' – or the kind wit of 'To a Mouse'.

And Telford agreed. His aim, in writing this poem and sending it to Burns as what was in effect a piece of eighteenth-century fan mail, was to encourage Burns to write for and be read by people of the sort they both grew up with – and most of all by the young. As he continued:

> Or may be Burns, thy thrilling page
> May a' their virtuous thoughts engage,
> While playful youth and placid age
>   In concert join,
> To bliss the bard, who, gay or sage,
>   Improves the mind.

Some readers may recoil at what to modern ears verges on Scots schmaltz, the tartan and shortbread world of bonny lads and lasses. But this was written well before the nineteenth-century reinvention of Scottish tradition; it is not unreal. Nor is the pride with which Telford ends his poem:

When winter binds the harden'd plains,
Around each hearth, the hoary swains
Shall teach the rising youth thy strains,
   And anxious say
Our blessing with our sons remains,
   And BURNS'S LAY!

He was right about Burns's enduring power – and this, remember, in a piece written in the 1790s, before Burns's death and before the publication of his collected works in 1800 brought him wider fame. Telford was not following fashion. He was helping set it. He saw in Burns from the start the authentic genius that has made him the most celebrated writer in Scotland's history, and he could detect it because it spoke of the background they shared.

For Telford, Burns's work was serious poetry of intense meaning: and to write such a long poem of his own in response (longer, even, than Currie managed to include in his collected volumes of Burns's work) must have taken effort for a man who by this point was busy with his real employment as an engineer and who did not need to write to gain notice.

He kept up his loyalty, writing to Currie in 1800 (when he published the collected works) 'every post brings me encomiums upon Burns and his Editor. Certainly not more than they deserve. . . forgive me for scrawling so I have been in the midst of a general Canal mutiny and [have been] riding in the Sun til I am sick – I shall see you soon.'[20]

By then Burns was dead: a loss Telford mourned in verse in a poem written in August 1796, the most sophisticated of all his pieces. He laments the fact that his hero was left 'in poverty to die' and scorns his move from a farm to a suffocating post in the Department of Excise. Burns was, he writes:

> Unfit to trudge in hampered rules,
> By low details confined:
> But born to triumph o'er the schools,
> And sway the human mind.
>
> Unbless'd that day! when all thy cares
> From Nature's works withheld,
> Exchanged for man's perplexed affairs,
> The pleasures of the field.[21]

Telford cared enough about this poem to have it printed and sent to his friends. He continued to fiddle with it, too: the version he gave to an assistant in 1830 differs in several places from his earlier draft. There is an angry tone to this passage, contrasting the freedom and enduring fame of the poet to the nitpicking of bureaucrats:

> The Muses shall that fatal hour,
> To Lethe's streams consign;
> Which gave the little slaves of pow'r,
> To scoff at worth like thine.
>
> But thy fair name shall rise and spread,
> Thy name be blest by all;
> When to their unremembered bed,
> These slaves of power shall fall.

In this, and other pieces, Telford captured with easy fluency the forms and rhythms of contemporary poetry; and he had something both lyrical and passionate to say. The verse does not read as though he struggled to compose it. What is missing is the occasional strike, the deeply original thought, the breathtaking observation, the insightful thrust, the fresh phrase. Thomas Gray, Alexander Pope and Rabbie Burns, from all of whose very different genres some echo can be heard here, had that genius. Telford must have concluded early that he did not. And instead of envying Burns, he paid tribute to him, and stayed close to poetry, but without pretension, all his life.

3

The Great Mart

The obvious destination for a young Scottish mason on the make was Edinburgh. It was a Hanoverian boom town in the middle of a great reconstruction. The formal New Town with its geometric neo-classical buildings was spreading out to the north of the loch which once bordered the cramped, steep streets of the Old Town.

Telford arrived in the city in 1780, having walked the eighty-odd miles from Langholm. What he saw must have been astonishing for a man who had hardly visited a large town. Then, even more than now, the divide between old and new must have stood out: it was, he recalled, 'a large city of romantic appearance, one-half consisting of very regular stone buildings which have sprung up within the memory of man, the other half forming the old city of Edinburgh'.[1] John Ainsley's map, published in that year, shows the New Town standing to the north of the clustered streets of the Old Town like some alien attachment or a mad, never-to-be-built proposal from a mathematical planner.[2] On the one side nothing is regular; on the other, across the straight-sided canal, which was all that remained of the old loch and which not long after was drained, eventually to make way for Waverley Station, are the ghostly outlines of half-built Princes Street, Queen Street, George Street and Saint Andrew's Square. Building work had filled in only the westernmost part. It was a place in need of good workmen.

Yet there is a mystery. What exactly Telford did in Edinburgh is unclear and why he never seemed to speak of it in later life is uncertain. In the year he spent in the city he is often said to have played a journeyman's part in carving and cutting the pale brown sandstone which still gives the New Town its flavour. But while he looked back proudly in later life at his other early experiences he seemed almost embarrassed by his time in the city, brushing it aside in his *Life*. He never suggested that he had helped build the New Town. Was Telford's time in the city a failure? Or was it too close to Eskdale for him to make the break from his background that he wanted? Either way, to him Edinburgh was more a place to learn than to stay. 'The splendid improvements,' he wrote, 'opened to me a new and extensive field for observation, where architecture is appropriated to the purposes of magnitude as well as utility.'[3] In the Old Town he explored 'the rude features of the ancient Pictish Castle, and... the lofty tower-like dwellings, crowded along a narrow ridge under the protection of the Castle', as well as work by Inigo Jones; 'varieties of Gothic architecture' in the ruins of Holyrood Palace; the crypt of Roslyn Chapel to the south of the city; and on his return to Eskdale in 1781, 'the justly celebrated Abbey of Melrose'.

He is brusque, however, in his description of his scuttling departure. 'Having acquired a general knowledge of drawing, and particularly of its application to architecture, and having studied all that was to be seen in Edinburgh', he left for home. It is curious, at least, that a young man with the right skills in the middle of the construction of one of the finest pieces of urban design in British history should have left so soon.

Still only twenty-three, he was back in Eskdale. The return must have been bittersweet. He was in the place he loved and back to see his mother. But his professional life in the valley was already over and he knew it. Eskdale now could only ever be a stop on the journey. The next destination was obvious. He knew where real opportunity lay in Great Britain. Everyone did. It lay to the south, in England, in London. Work and money were obvious draws. But something else pulled him there, too. Patriotism. Telford was not reluctant to go: he was a Unionist who saw no conflict between his Scottish identity and his British pride.

In 1784, when, in England, he came to write his long poem *Eskdale*, he celebrated Great Britain with unabashed and unforced nationalism. 'Awaked at length, BRITANNIA rea'd her head, And feudal Power and Superstition fled,' he wrote. It was a place where 'One equal law the hostile nations bound' and as a result a bright new nation 'BRITAIN rose the envy of the world'.

The message of the poem is simple. To get on, you had to leave not just Eskdale but Scotland too, as his childhood friends, all described in the verse, did as well. His words burst with pride at the building of a British identity. They are also both an adoration and termination of his time in Eskdale. He knew he would not come back there to live. 'Ambition's wreath, and Fortune's gain', he wrote, are not enough. 'Young hearts' should be taught 'thy simple charms to prize, To love their native hills, and bless their native skies'.

As he wrote that, he will have known that he was leaving his native hills and native skies behind him.

Telford left Scotland for London in early 1782. He was tramping a path set by many of his countrymen. One in particular was an example to him. The architect, developer and all-round man-on-the-make Robert Adam had set out southwards in 1755, first to educate himself amid Italy's classical ruins and then to sell to the English, as expensive good taste, a version of what he had learnt. Like Telford three decades later, Adam found the nation of his birth constraining. 'What a pity it is that such a genius [as myself] should be thrown away on Scotland where scarce will ever happen an opportunity of putting one noble thought in execution,' he wrote to his sister.[4] Telford described the motives for his own journey in a similar way, though a touch less dismissively. 'In the year 1782,' he recorded, 'after having acquired the rudiments of my profession, I considered that my native country afforded few opportunities of exercising it to any extent, and therefore judged it advisable (like many of my countrymen) to proceed southward, where industry might find more employment, and be better rewarded.'[5]

Yet even for such self-assured young men, London must have been an arresting destination: 'this great and monstrous thing', Daniel Defoe had called it.[6] When Telford reached it for the first time at the age of twenty-four he will have been well aware of its reputation for swallowing migrants. Jerry White, a recent historian of eighteenth-century London, quotes from the diary of Samuel Kevan, who was, like Telford, a stonemason from southern Scotland and who arrived in the city three years earlier. 'Here was a soft young lad with out friend or advisor Master of no trade, and without Tools. . . I betook myself to the most earnest request for assistance and direction in that trying & forlorn condition in such a manner, as I never had before, nor durst since,' Kevan recorded of his first, difficult, day in the city.[7]

For new and old residents alike, London was a crowded, filthy, risky and rude place. 'If towns were to be called after the first words which greeted a traveller on arrival, London would be called Damn It!,' one eighteenth-century traveller noted with secret delight.[8] It was a great jumble of modernity and medievalism; a global boomtown run fitfully by the rules of the past or, more properly, not run on any clear authority at all since the engine of its growth was the absence of any single source of order; and because of this all sorts of things, good and bad, became possible. In 1780 the city had exploded in the Gordon Riots – 'London's Bastille Day', White calls it – and the memory of those shocking hours, when an anti-Catholic mob ripped through the city, still worried many when Telford arrived two years later. They feared that the city might explode into rage again. Even if it did not, London was still a place of disguise, chance, commerce and crime. It was where people mixed, identities shifted and authority was tested, and that of course was its appeal.

The other part, to Scots, was just the opposite: familiarity. London may have been a strange place to the newly arrived, but a Scot would at least recognise some of the accents. The Union of 1707 had not just pushed English power north; it had pulled Scottish skills and influence south and Scots were everywhere in the city, strong especially in the professions like medicine and law; and in architecture and building, too. It was this last which drew Telford on his winter journey south. He did it not on foot as a tradesman, which might have been usual for someone of his position, but riding a good horse down the muddy turnpikes (at least if Samuel Smiles's account is to be trusted). The horse was lent him by the grandest of Eskdale men, Sir James Johnstone, who wanted it taken south. Telford dressed for the expedition in a pair of buckskin breeches belonging to his cousin, Thomas Jackson. For years afterwards the latter was said to have enjoyed telling the story of his relative's journey, always ending 'but Tam forgot to send me back my breeks!'[9]

Telford's own record of the trip is less colourful, and blunter: 'I made my way to London, the great mart for talents and ingenuity.'[10] If his early weeks in the capital were a shock, he did not admit it; his arrival eased, perhaps, by Sir John Pasley, a London merchant and uncle of the Pasleys Telford knew in Eskdale. His first surviving letter back home, sent to Andrew Little from London on 12 February 1782, makes a cheery read, the tone perhaps intended to reassure friends that he had not made a mistake in leaving them.[11] He concluded by sending his 'compliments to all the Folk of the north'. The cock of the walk in little Langholm did not want anyone to think he might be beaten by the challenge of his new home, though the place was by many orders of magnitude more daunting than Edinburgh, the only real city he had ever seen before. After bucolic Eskdale, with its childhood games and poetry, it was another world.

Rarely bothered by self-doubt, Telford had arrived with a plan. His starting point was Adam, by now one of the most prominent of the many Scots architects in the city, established as the supplier of a reliable modern style to the capital's rich. If cold glass, vast basements and security gates became the dreary must-have for the international oligarchs who flooded the capital two and a half centuries later, then it was Adam's carved reliefs, Corinthian columns and hard white stone which defined the later years of the eighteenth century for men who were just as powerful, just as rich and often just as unscrupulous. Along with his brothers, James and William, Adam worked hard, worked fast, and worked for money. The Adam style was soon copied everywhere.

By the time Telford sought his help, Adam had had his ups and downs and Telford was lucky, as he was so often, to come calling during one of the ups. Adam had pulled off the spectacular if financially risky Adelphi development by the Thames in the 1770s, funding it, when all else failed, by lottery. By the 1780s he had thrown his resources behind a second, even larger scheme: Portland Place. Intended to be the widest road in the capital, it was to be lined with mock palaces to attract an aristocratic crowd. The uncertainties of the American War of Independence had knocked confidence in London's speculative builders, Adam among them, and his work at Portland Place was cut back. But the end of the war in 1783 lifted the mood.

Though Britain had lost America, it was gaining power and riches, not least from India. Construction in London exploded like gunpowder. Then – as in London today – not everyone thought this a good thing. 'The town is so extended, that the breed of chairs is almost lost, for Hercules and Atlas could not carry anybody from one end of this enormous capital to the other,' lamented Horace Walpole.[12] For a young mason looking for work this growth was only good news. Skills learnt during his Langholm apprenticeship were valuable in a city full of building sites.

Yet even by the standards of an age when the power and obligations of connection were a recognised route to advancement, when the pestering letter and the anxious wait to secure a few words could make a career, what Telford did next was audacious. He secured meetings not only with Adam but with the other leading architect of the time, Sir William Chambers. To imagine something similar today you have to picture a newly arrived eastern European worker managing to call on both Richard Rogers and Norman Foster in order to ask for their advice on bettering himself. Not for Telford the anxious tramping of streets in search of work or the long years of drudgery hoping for a break.

How did he pull this off? Smiles suggests he obtained access through letters of introduction from the Pasley family, as well rooted in the capital as they were in Eskdale. He may equally have arrived with a letter of introduction from the Johnstones: Adam, after all, was employed a few years later to design their family mausoleum that still stands next to Telford's father's gravestone in Westerkirk, and had earlier built Pulteney Bridge in Bath for the family. Whatever the route in, seeing these men was a coup for the young Scot. Doubly so, because an introduction to Adam was hardly a safe route to the support of Chambers. Although the latter had been influenced by Adam early in his career (and later they shared a royal appointment to the post of architect of the Office of Works) the pair of almost exact contemporaries were rivals. Comparing Chambers with Adam, Telford recalled that 'the former [was] haughty and reserved and the latter affable and communicative'. He added that 'a similar distinction of character pervaded their works, Sir William's being stiff and formal, those of Mr Adam playful and gay'.

He did not warm to Chambers; Adam, he says, 'left the most favourable impression'. But his first work in the city was as a mason on Chambers's monumental, near-thirty-year reconstruction of Somerset House. Long afterwards, reputedly, he pointed out his handiwork on the south-west corner of the building whenever he crossed Waterloo Bridge.

Today Somerset House is caught between the traffic-packed Strand and the Embankment. In the 1780s it stood beside the river, inspired by Adam's earlier Adelphi development. The project was typical of Chambers's work: more austere and less ornamented than anything by Adam; influenced by Palladio and recent French design; and led by the state, not a private developer. Chambers had developed a strong line in designing public buildings, not least as the architect who began turning Whitehall into the street of government offices which exists today; and the rebuilding of the decayed remains of the Tudor palace of Somerset House was the largest of them all. It was intended to provide new office accommodation fit for a professionalised army of tax-gatherers, part of a trend of schemes for public improvements in the old core of the cities of London and Westminster, which were looking increasingly decrepit, suffering by comparison with new build in areas such as Bloomsbury and Mayfair, on what would today be called greenfield sites.

For Telford his first job at Somerset House was simply a way in. He intended to achieve much more than a mason's position. He knew his skills were valuable and he was determined from the beginning to build a reputation and a business of his own – not just fall into someone's employ. He always thought ahead. Now this lifelong instinct and the confidence which drove it was serving him conspicuously. 'I acquired much practical information, both in the useful and ornamental branches of architecture; and in the course of my two years' residence in London, I had an opportunity of examining the numerous public buildings of the metropolis of Great Britain,' he wrote.[13]

His meetings with Adam and Chambers encouraged him. 'The interviews with both convinced me that my surest plan was to endeavour to advance, if by slower degrees, yet by independent conduct.'

That account, written in his old age, is supported strikingly by a powerful letter he sent Little from London in July 1783.[14] As a puffing-up of youthful plumage the display could hardly be bolder or more extravagant. It comes closer than any other piece of Telford's writing to exposing the toughness that lay not far beneath his amiable exterior; and though refreshingly frank as an exercise in self-examination (and a telling reminder of his intimacy with Little) some may find something a little distasteful in its spirit. The letter is worth quoting at length. Still attached to the crumbling red wax with which he sealed it, it opens with a nod to the place of friendship above work. 'The meeting of long absent friends is a pleasure to be equalled by few other of our enjoyments below,' he tells Little. But he is not writing to exchange pleasantries and soon gets down to the serious business, his plan for advancement:

'Tis impossible for me to inform you of much concerning myself at present I am laying schemes of a pretty extensive kind if they succeed, for you know my disposition is not to be satisfied unless placed in some conspicuous point of view my innate vanity is too apt to say when looking on the common drudges here as well as other places.

Reading this, remember that Telford had not long before counted among the common drudges; that he was writing even now as nothing more than a young stonemason in a city he hardly knew; and that he had left a small community in which he had already achieved some 'conspicuous point of view' for a city in which, for every Scottish migrant who made his name as Adam had, thousands sank into invisibility. His words tell us not only of his state of mind as a young man, but also of the internal fire which irradiated him at every point in his career.

In short, he had come to London to find fame. And not only fame. He wanted power, too. There is a sweeping arrogance to the next section of this letter.

'Born to command ten thousand slaves like you – This is too much, but at the same time it is too true, for I find the workmen here to be more ignorant than they are in Eskdale shores, not a Mt David among them,' he continued to Little (referring to Matthew Davidson, a friend of them both). Perhaps some of this contempt for English workmen was justified but much of it must have been self-reassurance; Telford was working among these men, but wanted to assert his superiority to them.

What, however, stands out most strongly is an itch to break through the ceiling which trapped him beneath inadequate but socially elevated masters. He goes on to say so in this long passage in radical vein, written just six years before the French Revolution:

There is only one man I have found any intimacy with, he has been 6 years at Somerset House and is esteemed the finest workman in London & consequently in England... the Master he works under looks upon him as the principal support of his business but I'll tear away that pillar if my scheme succeeds, and let the Old beef head and his puppy of an ignorant Clerk try their dexterity at their leisure – This Man whose Name is Mr Hatton, may be half a doz: years older than myself at least, he is honesty and good nature itself and has been working all this time a Common Journeyman contented with a few shillings pr week more than the rest; indeed he is ador'd by both his Master and his fellow workmen, but I believe your uneasy friend has kindled in his Brest that he never felt before: now the Whole is this I propose that if he and I could get a fair footing that we shall carry on business for ourselves there is nothing done in Stone or Marble that we cannot do in the Completest manner; there is nothing but just having a Name here, to make a fortune all my wits are at work to lay down an extensive foundation.

What is remarkable about all this is the evident spirit: an almost indignant impatience to escape working for anyone else, an indignation he feels on Hatton's behalf as well as his own. Yet with that impatience, there is also calculation: he took care to win support from the powerful wherever necessary, and to anticipate their objections and meet them. As he went on to tell Little: 'Mr Adam is uncommonly kind and says that if we try that scheme he'll do all in his power to recommend us.' That power was, of course, extensive, even if Adam may have been drawn not just by Telford's persuasiveness but by the prospect of luring good workmen away from the great project being overseen by his rival at Somerset House.

In the end Telford got nowhere with his scheme to set up in business for himself. As he wrote to Little, 'you'll think Somerset House the finest opportunity, [but] there are circumstances against beginners'. He knew that if he deserted Chambers he would not be welcomed back on his own terms. Chief among the challenges, he explained to Little, was the fact that contractors were not paid for the first two years, 'and then only one year's the other always lying dead as security for the work being done in a proper manner this is an insuperable Bar otherwise it would flatter my vanity much to have it said that I had a hand in that Noble work'.

Vanity, ambition, manipulation and the exploitation of contacts: all these are as much a part of Telford's story as skill, genius or vision, and it is perhaps to his credit that he did not hide them, at least in this long letter sent when he was twenty-five.

He concludes with familiar praise for the place he loved more than any other, Eskdale: he asks Little to 'give my very best' to a friend who was returning from London, and to 'tell him that I envy him the pleasure that he must enjoy amongst the hills of the Esk this Charming Season'.

But just before this comes something significant. He mentions the man who was to provide his planned escape from the life of a journeyman labourer: 'Mr W Pulteney'. 'Sometimes [I am] with him twice or thrice a day,' he adds, with revealing pride. Pulteney was born a Johnstone, in Eskdale – but more than that, he was very rich, very determined and he was to do more than just find work for Telford. He was to pull apart the bars of the young man's perfectly tolerable cage and launch the bird skyward.

He first took flight not in London, but on a project back in Eskdale. In 1783 Sir James Johnstone decided to improve the family seat at Westerhall, and asked Telford to design the alterations. Presumably Johnstone, who must have been familiar with Telford's precocity since childhood, trusted the young man to send his designs up from London. Telford got Andrew Little's help in securing prices from local builders and elements of what was done survive in the house still.

More significant than the practical work, however, was what resulted from it: Telford's valuable, lasting and close proximity to William Pulteney, Sir James's younger brother (he had changed his surname on marriage). 'Mr Pulteney and I have made 100 alterations to Westerhall,' he boasted to Little. These must have been heartening moments for the younger man: to be treated with respect by someone as powerful as Pulteney, a Member of Parliament, was a mark of distinction.

Not yet thirty, Telford was stepping up from his first career, as a mason, to his second, as an architect and project manager: a rise in status that was coming his way faster than even he can have hoped. Work at Westerhall went well enough for him to be trusted soon after with a second project by Pulteney: improvements to the rectory at Sudborough in Northamptonshire, where the MP was patron of the living. This small task turned out to be transformative. At Sudborough he found himself welcomed into one of the families which gave shape to the rest of his life; a family that in some ways became a substitute for one of his own. The new incumbent in the village was a Scottish intellectual, Archibald Alison. As an educated writer – and the author of the respected *Essays on the Nature and Principles of Taste*, a work of serious philosophy which Telford later rated highly – he was much more than a village parson and he took Telford under his wing. His son (also named Archibald and later a notable Scottish lawyer and historian) recalled in his memoirs:

Sir William Pulteney sent down a young Scotch mason who had recently been in his employment in London. My father according to his usual custom, took a great interest in these improvements, and frequently entered into conversation with the head mason. He was so much struck by the decided turn of his mind, and of the vigorous, clear expressions which he made use of on every subject, that he asked him to dinner. This was the first time the young Scotchman had been in such society, and it had a material effect on his future destiny, for it led him to think that he might take his place among the cultivated ranks of the country. 'Archy,' he has often said to me, 'your father was the first man to treat me like a gentleman.' It laid the foundations of a friendship which continued uninterrupted for above forty years and was the source of unmixed gratification to both parties.[15]

That memoir was written after Telford had gained his fame, and after Sir Archibald had gained his baronetcy. The friendship the memoir describes between the families, however, was real, deep and lasting. Telford continued to call on Alison at home, first in Northamptonshire, then in Shropshire and from 1800 in Edinburgh, until old age prevented him travelling.

The next steps in his career were more military than ecclesiastical. In 1784 Telford won a significant commission to oversee the erection of a series of grand buildings at Portsmouth Dockyard, among them a chapel and a house designed by Samuel Wyatt, the latter intended to be formal enough to receive King George III. It was a big step up and another sign that his promises and his promise were being trusted. The Royal Navy was one of the best funded and most prominent products of Britain's rising status as a world power and to be asked to work for it in Portsmouth was a mark of acclaim. Perhaps his flourishing relationship with Pulteney secured him the position; or perhaps he earned it on his own account through contacts with Wyatt, who had worked on Somerset House (and who earlier had been Robert Adam's clerk of works shaping the austere Palladianism of Kedleston Hall in Derbyshire). Either way, the post was part of the web of connections that Telford was skilled at cultivating.

Samuel Wyatt and his brother James, who became the most celebrated country house architect of his age, were both close to the
great engineers of the early steam age, Matthew Boulton and
James Watt; the latter was soon signed up as another of Telford's
permanent allies.

He was plugging himself in to a network of powerful and
ingenious men who were driving the industrial revolution
forward. He was an outsider no more – and nowhere shows it
more obviously than the energetic and boastful correspondence
he kept up from Portsmouth with Andrew Little. In July 1784
(a year after proposing his unrealised plan to set up as mason in
London), he was describing his new life by the sea.[16] He began,
as so often, with a reference to his birthplace and to the lack of
letters from Little: 'I think that. . . all my Eskdale friends have
forgot me,' writing fondly in particular of 'Miss Pasley', the lady
who had first encouraged his interest in reading in Langholm.
Then he moved quickly to his preferred topic, his skill and success
in securing advancement in his own career. 'Commissioners and
officers here. . . would sooner go by my advice than my Master's
which is a dangerous point, to keep their good graces & his both
however I shall manage it.'

Here, too, the presence of Scots in England following the Act
of Union came to his aid. He mentions the 'Surgeon to the Yard
Mr Ramsey Ker from the Banks of the Tweed'. 'Mr Ker is my
friend, so that if anything happens there is a great probability in
the present posture of Affairs that I have a chance by and by to be
employed as Principal Surveyor, for all the officers here would
prefer it – but these things must come in course. I shall endeavour
to secure a proper Basis by performing the duty of my present
station to the satisfaction of all my superiors.'[17]

If he had not been so able, such cockiness might have made
Telford insufferable. If it had not been mixed with modest self-
mockery it might have strained even Little's patience. One

speculates that it must have made him enemies as well as friends. He retained for the rest of his life an ability to flatter powerful people and win their support, matched by an uncommon understanding of the lot of the workmen on his schemes. To rivals of equal status, he was less understanding – and among them, less popular.

In Portsmouth, though, all went well. Work on the house and chapel progressed. Writing to Little almost two years later, in February 1786, he comes across as more confident than ever.[18] He describes the rapid urbanisation of Britain in playful horror, aping (or mocking) the genuine outrage of many contemporary writers (and later the Romantic poets to whom he eventually became close). 'The rage for building has not subsided yet I find. . . our poor, destitute descendants may behold those hill sides formed into streets and canals cut from sea to sea: Dreadful Thought!' Was there a smile on his face – perhaps even then an awareness of his future career – as he wrote?

'And now to your questions,' he continued – Little had been badgering him for a description of what he did and how he lived – 'but first I must go and inspect what's going on <u>under</u> my direction this afternoon.' The emphasis on <u>under</u> says a lot: Telford was enjoying his new power to command, not be commanded.

Now then – <u>you</u> ask me what I do all Winter – I rise in the morning at 7 o'clock and will continue to get up earlier til it comes to 5. I then set seriously to work to make out Accts write on business, or draw til breakfast which is at 9 – next go to breakfast and get into the yard about 10. Then all the Officers are in their Offices and if they have anything to say to me or me to them, we do it. This and going round amongst the several Works brings My dinner time which is about 2 Clock; an hour and a half serves this.

Telford – who grew plumper by the year – was rarely shy of his meals and in Portsmouth he seems to have had four a day. As his description of his routine says, 'I go to Tea til Six – then I come back to my room and write, draw or read til half an hour after nine, then comes supper and Bed Time'.

If this life sounds a little lonely as well as driven – doing the work of a clerk, inking out neat architectural drawings and writing accounts, as well as managing men – then perhaps it was, but these were habits and a pace that he never varied until close to death. They go some way towards accounting both for the huge amount of work Telford accomplished and his failure to build a family life. In his letter to Little, he admits to – and is almost apologetic for – this obsessiveness:

> this is my ordinary round unless when I dine or spend an Evening with a friend but friends of this sort I do not make many. For I am very particular in this respect – nay nice to a degree – my business requires a great deal of drawing and writing – and this I always take care to keep under [control] by having everything ready – then as knowledge is my most ardent pursuit a thousand things occur that would pass unnoticed by good easy people who are contented with trudging on the beaten path but I am not contented unless I can reason on every particular. I am now very deep in Chemistry – the manner of making Mortar led me to enquire into the Nature of Lime so in pursuit of this, having look'd in some books on Chemistry I perceived that the field was boundless.

He was, as this shows, the very model of a self-taught man: but unlike some he had a respect for proper learning. As he went on to tell Little: 'I am determined to study with unwearied attention until I attain some general knowledge of Chemistry.' To that end, he wrote: 'I have a Noble Room in the New House which is as sacred as the Sanctorum of a heathen Temple – I have plenty of Fire and Candle allowed me and often my dear Andrew I do wish to have you along with me and to participate in the mental Feast.'

In these letters his affection for Little, never matched with anyone else, shines through. The pair had a shared past in Eskdale and if Little's own voyaging had been cut short by the blindness he suffered as a surgeon at sea, then at least he could participate at a distance in Telford's great and growing adventure. 'You [would] find Andw that I am not Idle but much in the old Way of stirring the Lake about one to prevent it from stagnating.'

Stirring the lake – a neat phrase – was what Telford was all about. His respect for established order was not matched by any respect for mere custom. He always wanted to innovate. But he did so by flattering those in authority rather than by confronting them. As he suggested to Little: 'I [am] happy in being a great favourite of the Commiss's and I take care to be so far master of all the Business as none shall be able to eclipse in that point.' Adding: 'The Man that will neglect his Business and suffer himself to be led by the nose deserves to be P-x'd.! [poxed].'

'I had rather be said to possess one grain of good sense than shine the finest Puppet in Christendom if I eclipse it shall be minds of men and not their d–d Bodies,' he exclaimed and then, in one of those narrative leaps only possibly in truly personal correspondence, set out to show that he was becoming if not fully middle class, then at least much more than a tradesman: 'After all this I am powdered every day, and have a clean shirt 3 times a week.' Another sign of his social development was his declaration that: 'I take great delight in Free Masonry.' He was initiated into the Lodge of Antiquity on 17 December 1784 and became a founder member of the Phoenix Lodge in May 1786, designing its Lodge Room in the George Inn in the city.

Scrawled, at the end of this extensive letter, on a separate piece of paper, is an affectionate postscript: 'I wish Andrew that you saw me at the present instant surrounded by Books, Drawings, – Compasses, Pencils and pens great is the confusion but it pleases my taste and <u>that's enough</u>.' He signed it off: 'Farewell I am Thos Telford.'

———

He was living in a dockside town notorious for its drinking, whoring sailors out on the spree, a place of impressed men, rum and ships of the line like HMS *Victory*. But Telford refused to have any part in this. As he wrote to Little, 'Portsmouth is what you have said of it, but what's that to me.' The eighteenth-century Royal Navy is remembered today for its raucousness, but beneath the swaggering sea shanties lay a reliance on good planning and good order to keep them afloat and it was in planning and order that Telford intended to make his name. He was a quiet revolutionary caught up in the shifting spirit of the decade before war with France gave radical ideas a dangerous name.

Writing to Little in July 1786, he pondered the forces which had pulled them apart.[19] 'We must be separated we were all born in that situation which enforces an activity to procure a decent subsistence and born where there was no field to act in.' Telford, Little and their Eskdale companions had been 'torn asunder divided and scattered over the face of the Earth – we range in various regions for a scanty supply – while chance gives to others ease, affluence and abundance – the horn of plenty is planted by their cradle and trembling thousands affectedly pretend to hail their conspicuous birth'.

These are angry, dangerous words. They are ones an older, more settled, Telford would have distanced himself from: words his own rich patrons – the Johnstones, the Pulteneys, even the Adam brothers – would have found both insulting and alarming. The man who on the one hand took such care to ensure his superiors praised his hard work and willingness to take on their tasks – the man who sought favours wherever he could find them – was on the other privately scathing about the social order on which their power to help him rested. It is clear from this passage that self-advancement was more than just an eagerness for his chosen trade; it was fed by a deeper hunger to break away from the station in life to which he was born and a certainty that his talents deserved better. Nor did he simply seek advancement within the existing order: his later lack of interest in securing honours or wealth for himself is presaged in this letter in a short passage taken, says Telford, from 'a poem published or at least reviewed last month', by the popular, prototype-Romantic poet William Cowper. 'I have not seen the reviewers so warm in praise of any late production,' he adds.

> But yet I would not have a slave to till my ground
> To carry me, to toss me while I sleep
> And tremble when I wake, for all the wealth
> That sinews bought and sold have ever earned.

The thoughts behind this are radical, a defence of something close to equality. They are certainly not ones his slave-owning patron Sir William Pulteney would have endorsed. Perhaps that is why, having quoted the lines, Telford suddenly pulled himself up short. 'I believe this enchanting, distressing theme has turned my brain,' he told Little.

But he did not sound all that apologetic.

————

Work on the new chapel and commissioner's house kept Telford in Portsmouth for over two years; time he used to train himself as an engineer. 'I had an opportunity of observing the various operations necessary in the foundations and construction of graving docks, wharf-walls and similar works which afterwards became my chief occupation,' he wrote; turgid words which underplay the drama of shaping a growing naval dockyard.[20] But with the project well advanced, he needed a new job. Writing to Little in 1786 he reported that 'this new house and chapel will be finish'd in the latter end of the year after which (for anything I yet know) I will return to Town'.

By that he meant London. In the four years since he had left Eskdale he had built a name for himself, built alliances with men who could help him and built a life outside Scotland. He was rightly proud of this. The only thing missing was companion-ship. He kept in touch with his old friends when he could – or all but one: Miss Pasley, the only woman other than his mother to whom he was ever close, did not reply. 'I had wrote Miss P so often without being able to procure the least return: that I began to judge her silence a kind of hint not to pester any more with long epistles totally devoid of any occurrence that could engage her attention.' There's more than a little melancholy in that reflection. Telford must have made an impression on Miss Pasley as a youth of no account. It dismayed him that her inter-est had wandered.

And this is a suitable moment to draw attention to the way this man came across, on meeting. Thomas Telford appears to have had a rare facility not only in his trade, but also at inspiring – often on first encounter – extraordinary admi-ration for and confidence in him as a person. There was something about this young tradesman that caused others, on making his acquaintance, to believe in him and want to give him a chance. What was it? Later in his career his reputation helped. At this early stage, however, his fame lay ahead, yet he already appears to have struck others as special.

Why? Was it his wit? Was it personal magnetism? Was it sheer self-belief? In part it must have been his physical presence, the thickset masculinity and Scots accent which made him a commanding part of any crowd. Few of the witnesses whose accounts we must rely upon were types much given to describing character with the depth and colour a novelist might attempt, so we can only guess. The effect Telford had upon those who met him was unusual, powerful, immediate, positive and critical to his advancement.

On that hung everything about the success which was about to come his way in an English county far from Portsmouth and far from Eskdale too.

4

Young Pulteney

The fin-backed ridge of the Wrekin hill is part of many Shropshire views: rounded, tree-covered and distinctive. Telford recommended visitors to climb it to see the land. It sits between the plains of the north of the county and twisted, rough country to the south, where pulsating steep-sided valleys confront Wales like battlements. Shropshire is still a thickly hedged, rural county, full of strong stone farms and market towns. It can feel that not much ever changes here; that these really are the blue remembered hills which A. E. Housman celebrated in verse and Ivor Gurney and Ralph Vaughan Williams in song. But that impression is incomplete. There is another Shropshire: an urban land of industry and innovation, a place where minerals have been ripped from the ground and forged in inventive processes whose relics still draw visitors from all over the world. This was the Shropshire which sells itself today as 'The Birthplace of Industry', and that claim is not overblown. It is also the Shropshire in which Thomas Telford made his name.

Today, the Wrekin stands above the new town named Telford in his honour. It is built above a warren of defunct coal pits and the work of creating it involved 'civil engineering works which would have impressed the Pharaohs of Egypt', claimed one promoter.[1] But there is little good design.

What is of interest in Telford lies not in the middle but on the fringes, and comes from the past. A couple of miles to the south lies Ironbridge Gorge, cut deep by the River Severn. Neat workers' cottages now cluster above sleepy wooded banks, but in the 1700s this was one of the industrial hearts of Europe with a succession of villages along the river, including Coalport and Coalbrookdale, with its ironworks. An energetic museum trust has opened ten fine museums here and cares for many other sites, preserving things that Telford, too, came to wonder at, including the innovative iron bridge itself – five years old when he arrived – which gives the place its name. It was in Ironbridge Gorge that he first saw the technological possibilities of modern transport and industry. These works still stand out as audacious: things such as the narrow tar tunnel which runs under the hill into old coal mines, its walls seeping into bituminous pools; or the Hay Inclined Plane, which dragged canal boats up the valley's side in iron tubs running on steep rails.

The inclined plane opened during Telford's time in Shropshire, but he had no part in it. To find his mark, travel across the town to a field grazed by sheep and planted with turnips and mangelwurzels. Here, also under the shadow of the Wrekin, sits a curious metal structure, supported by two traditional stone abutments. These are the remains of the Longdon-on-Tern aqueduct, Telford's first radical success. The canal once carried by the structure has gone and its remains have been bulldozed. Only the aqueduct, now drained, has survived, its thick iron plates and girders polished by age. It is short and low; spectacular only to those who know something of its history. But it was part of the making of Telford and the story of how he came to work on its design and the story of how he got to Shropshire and established himself there are also the stories of the success he made of his life.

It was in Shropshire that Telford moved from mason to architect; from architect to engineer; from stone to iron; from youth to middle age; from hope to success. Many of the best projects he worked on can still be found in the county or just over its borders. Here, more than anywhere other than Eskdale, his name is still cherished. Yet he arrived in Shropshire by chance and with no expectation of what was to follow. And paradoxically he was drawn not by the new canals and furnaces of Coalbrookdale but by a relic of the county's medieval past: the riverside remains of Shrewsbury Castle. Its owner had invited him to work on their reconstruction and he had leapt to oblige. Telford was on the rise but he had not yet risen so far that he was free to turn a big job down.

Telford's arrival in Shropshire in late 1786 marked the moment he fell completely into William Pulteney's orbit and he did not escape it fully until the latter's death twenty years later. The pull of his gravity was intense, and for the young man it must have been an unsettling attraction. In his heart Telford was no respecter of persons. He was careless of others' prestige or status and resolved to make his way in the world on unadorned merit. He liked others to imagine – and, later, wanted history to remember – a mind focused on work, design and execution. But to rise in his profession he recognised early that he needed patronage. He sought it, used it, revelled in it and made his way by it.

The two men were brought together in Shropshire by their shared roots in the Borders. Pulteney had been born beside the Esk, near Westerkirk, thirty years before Telford. He may have known him as a boy. Back then, Pulteney was nothing more than a third son in the Johnstone family who owned the big house in the valley, Westerhall. Born with no expectation of inheriting the family estate, Pulteney had embarked upon the typical life of a Scotsman of his class and education in the middle of the eighteenth century.

First, he left Eskdale to train as a lawyer in Edinburgh, where he mixed with rising members of the Scottish Enlightenment. Then he moved to London where – in 1760 – he married Frances

Pulteney, a relative of the immensely rich Earl of Bath. The couple were unexceptional, wealthy enough to play a part in a growing city but not distinguished in any particular way. Then they struck an eighteenth-century goldmine. First Frances's cousin (the established Pulteney family heir) died. A year later the Earl of Bath died, too. He entailed his huge estates to an elderly bachelor, who in turn and (by then) to no one's great surprise, died, in 1767. As an only child and the surviving heir Frances Pulteney inherited the lot and, through his wife's wealth, William Johnstone was rocketed from the ranks of the minor gentry to the top table in British national life. His wife and later his daughter, a peer in her own right, retained much control over the money. But William Pulteney – as he became, changing his name from Johnstone immediately on inheriting the money (he became Sir William in 1797 when he inherited the Westerhall estate too) was now a man of fabulous means and national importance.

He was also a man of peculiar, austere character; intellectual, even: when he altered his name in 1767 he called upon 'my friend' David Hume, the philosopher, to attest on his behalf.[2] Pulteney did not lapse into the luxuries that can come with wealth and he used what he had well, in property developments and investments in land in Britain and America. He was an improver and activist, most of all in the fast-growing city of Bath, and he enriched himself beyond his inheritance. He was also a man of odd habits and calculated speculations; judgemental, unsociable. Not for him a life of silk drapes and crystal decanters. He had no stately home, no interest in fox hunting or dancing; no title, few friends; he left no will at his death; he was a man so frugal in his dress and diet that it was said he lived on bread and milk. For all that, he was believed (probably correctly) to be the richest commoner in England; and for all his hunger for status (he once threatened his daughter with the unfortunate prospect of marrying the unmarriable William Pitt, if the latter made her the Marquess of Bath) Pulteney remained an agitated outsider, a Scot in the provinces, not a swaggerer in the drawing rooms of London. His politics verged, increasingly, towards the Tory side of Whiggery; 'this country enjoys a degree of liberty which may excite the envy of the whole world', he wrote in 1779, proposing that America be brought back into 'a new and magnificent empire on the pillars of freedom'.[3] But as MP for Shrewsbury he played only a marginal role in the machinations of Pitt's government.

Perhaps only someone as peripatetic as Telford, with his own Eskdale roots, and restlessness and lack of a family life, could have connected so closely and so quickly with a man such as this. In his own manner of living and lack of interest in his own status, reward or even property, Telford echoed Pulteney consciously. Their relationship developed rapidly beyond its practical origins into a lasting, dynamic intimacy of which Telford, at least, was endlessly proud. His letters are full of boyish praise for the merits of his patron and the favour he earned in return. 'His good opinion,' he once wrote, 'has always been a great satisfaction to me, and the more so as it has neither been obtained nor preserved, by deceit, cringing, or flattery; on the contrary I believe, I am the only man who speaks out fairly to him, and who contradicts him the most. We quarrel like tinkers.'[4] 'His manner is remarkably engaging,' he wrote on another occasion.[5] 'How comes it to be so? Why by that plain simplicity and natural ease which ought to be the study of all men, the moment that is departed from there is something which takes place that is disgusting.'

Not everyone could see this. Telford knew that others found Pulteney odd. 'Mr Pulteney is a man who does not court popularity, he is distant cautious and reserved to mankind in general but I believe to the few in whom he can confide there is no man more open.' His character, he told Little, was matched only by their old Eskdale acquaintance Mrs Pasley. 'They both are convinced they live only for the good of mankind and they both despise riches, were I to say so to the world, I should be laughed at but I would laugh at the world.'[6] He was anxious when Little and Pulteney met and delighted to hear both that the encounter went well and that Little pressed a poem by Telford onto his patron. 'It was rather better than if they had come from me,' he added; a certain reserve remained.[7]

These bursts of self-exposure in Telford's letters are telling. They show the emotional supports to his public façade. They also show his awkwardness about the real nature of his adoration for his patron. He wanted more than a cosy place for life under someone else. He wanted to rise up, to equal if not surpass him, and even as that began to happen he must have been aware of the compromises that this required. The young man who protested more than once in letters to Little that he did not intend to be a slave to anyone found himself drawing support from a patron who not only owned extensive sugar estates in the West Indies (there is still a Westerhall rum from Grenada) but who in 1798 shipped 144 slaves to Westerhall from the island at an annual rent for their labour of £1,375, though by then slavery in Scotland had formally been ended.[8] Telford was too intelligent to avoid seeing the contradictions, but the stain never – in his mind – detracted from his patron's merits. 'His ideas of simplicity are perfectly similar to what I have always conceived in my own mind,' he told Little. 'I have myself for about half a year taken to drinking water only, I avoid all sweets, and never eat any nick nacks. I have sowens and milk every night for my supper [a Scottish dish made using the inner husks of oats after milling]. In short Andrew I am a queer creature, and not ashamed of being thought singular.'[9]

In this, Telford was doing more than just admiring Pulteney. He was seeking to copy his patron's manner. He began aping his diet. He even began to take his name. 'The world says I am young Pulteney, I wish I was,' he wrote to Little a year after arriving in the town.[10] Young Pulteney: a title that throws a spark into the darker corners of Telford's mind, lighting, briefly, both the ambition and the jealousy which powered him. He would not be content until he made the title solid.

Telford's work for Pulteney in Shrewsbury began modestly, reconstructing part of Shrewsbury Castle for his patron to use as lodgings on his intermittent visits to the town. By now he was starting to be established as an architect, a man who had done his apprenticeship, studied design in Edinburgh and London, won the approval of Robert Adam and proved himself managing public works for the navy but failed to set himself up as a contractor as he had hoped, at Somerset House. He was not at this point in any sense an engineer or associated with the things, canals and roads, for which he is remembered.

He needed opportunity and Pulteney could provide it. Both men chose well. The young Scotsman hit Shropshire with astonishing force. His initial task at Shrewsbury Castle was small – Telford refaced some exterior stonework, fitted a staircase and created a series of rooms in the Gothic Revival style. They were, wrote Katherine Plymley, the daughter of one of his new Shrewsbury friends, 'in true taste'.[11] Most of the interior of Telford's castle was swept away in the 1920s, but one room remains and another of his works is still in the grounds, an octagonal summerhouse, named Laura's Tower after Pulteney's daughter. Telford housed himself in his patron's property in Shrewsbury, too. 'Mr Pulteney has been gone a fortnight and I am again commanding officer in this renowned fortress,' he wrote to Little in early 1787, scrawling his address boastfully as 'Castle Salop'.[12]

He wanted to be more than a provider of fripperies to the rich, however. The castle was just a way in. His appointment – with Pulteney's backing – as county surveyor in 1787 furnished a bigger canvas. Formally the post gave him responsibility for bridges and the maintenance of official buildings. Informally it was a foot in the door to so much that was going on in Shropshire, and especially Shrewsbury, a town that was trying to escape the degraded remains of its medieval past. There were plenty of builders on hand in the county but until Telford arrived no honest man of independent judgement, no outsider who could be trusted by all sides to oversee contracts and rise above petty corruption and feuding. Pulteney was part of this conflict, as a local MP who led a political faction opposed to Lord Clive (the county's lord lieutenant and the famous Clive of India's son). But with a mix of charm and toughness Telford wormed his way around this. Soon, he was working on an extraordinary number of public and private projects. He was a man who knew how to make himself very useful, very fast.

In January 1787 he reported to Little that he had 'been very much hurried'. It was an understatement. Already he had drawn up '5 or 6 plans for different alterations at and above the castle'; had 'made a Survey and Plan of part of the High Street' (persuading the town authorities to vote to sweep away half of it); had 'drawn a plan for a private Gentleman to form a kind of square in another street'; had taken charge of plans for a new public bridge; had proposed 'another plan for a Country house for another Gentlemen which will cost about a <u>thousand</u>'; and had taken on the county's drainer of lands – 'I have weighed him in the balance and found him wanting – wanting of what? Why of <u>Abilities</u> and <u>Integrity</u> – and these unluckily happen to be little requisites not easily to be disposed of.'

All this from a man of only twenty-nine with no links to this part of England, who had been in the county only a few months, who had no money or social standing and had done nothing of real substance anywhere else. With good cause, Telford (always self-aware in his confessional letters to Little) viewed himself as an uncontainable, perpetually disruptive force:

. . . it may be self-confidence, but I have often observed that there has always been a bustle where I was – how that came about I know not – but these are the simple facts, you know while I was in Eskdale. . . I was a bone of contention, when I came to London I had nearly caused a contention at Somerset house – at Portsmo' the navy B[oard]d and Ad[miralty] were engaged, and since my arrival here, there has been one continual sense of contention, where I have always been the prominent feature; in short I am here now exactly as in Eskdale.[13]

Exactly the same as in Eskdale; by saying that Telford was admitting to more than huge energy (though that was present). He was also showing another side to his character: his ability to inveigle himself into the affections of people who could be useful to him, to make himself indispensable, to attach his name to projects involving others, too, to turn their actions to his greater advantage and glory, without their ever realising it. It was a talent which played well alongside his more outwardly obvious skill in design and it lifted him above others whose native ability in engineering may have been almost as great but who did not know how to deploy it with such cunning. No passage in Telford's letters to Little shows this more than the following admission:

. . . you will not be surprised if some projects should enter my brain in order to extend the scene of action, the Gentlemen having once interested themselves, and publikly declared their opinion, it becomes their own act, and rebounds to their honour, to render their Surveyor respectable: and on my own part it requires a continual something to keep the spirit awake. This has hitherto luckily been the case and I think there's a chance for its increasing, for you know it is impossible for me to be at ease – but this is all between ourselves for we must let others only into the facts as they happen, and not the train of causes that produce them.

As a soliloquy of calculated candour and ambition, it could almost come from the lips of Iago. It is a stage whisper to his confidant Andrew; it shows that Telford was playing with the grandees of Shrewsbury and winning and perhaps even playing with Pulteney's trust, too. Certainly Pulteney would have been horrified to read it. But Telford makes, and makes honestly, an important point regarding major contracts: that the client acquires a gently appreciating interest in the contractor's perceived success.

By now – his manipulation carefully hidden to all but Andrew Little – Telford was sliding himself happily into Shrewsbury life: literate, hyperactive, ingenuous, a sort of county Figaro in his ever-present willingness to solve problems. He rebuilt private houses in the town and at least one unidentified country house outside it. He was widely consulted. One such incident became famous. In May 1788 Pulteney asked him to examine St Chad's church, Shrewsbury, where the roof was leaking. He reported back that the churchwardens should be much more concerned about cracks in the walls, since the whole building 'was in the most shattered condition'.[14] He told Little what happened next:

Popular Clamour overcame this report – These fractures were said to have been there time immemorial – and it was said by even sensible people that Professional men always wish'd to carve out employment for themselves: and that the whole might be done at small expense which they proceeded to do – and I gave myself no disturbance when lo: & behold on the Morning of the 9th inst the very parts I had pointed out, gave way – and down tumbled the mighty mass – forming a very remarkable, magnificent Ruin, while astonished and surprised the inhabitants were roused from their delirium – <u>tho they have not yet recovered from the shock</u>.

His advice had been ignored and a local builder had begun hacking at the foundations to the tower. When the clock struck four, the bell's vibrations caused it to collapse. The episode was perfect marketing for the county's new architect and he was always proud of the story – retelling in his *Life* his warning, as he left the vestry, 'that if they wished to discuss anything besides the alarming state of repair of the church they had better adjourn to some other place where there was no danger of it falling on their heads'.[15] It was the start of what proved a lifelong sideline as an ecclesiastical architect. A second church, St Mary's in Shrewsbury, was the subject of patient efforts by him to secure its structure over a decade. He propped up its medieval roof and renewed its Gothic appearance, placing a new pulpit in the church in 1788 which – he was pleased to see – 'carried off more applause than the sermon'. Though Telford's characteristic architectural style involved the square lines, sharp angles, overhanging eves and wide sash windows characteristic of north-west England and the Borders at the time, to the extent that he was at times in his simplicity almost a putative modernist, he was no careless obliterator of the past and respected (more than his twentieth-century successors as town planners) elements of Shrewsbury's medieval core. He sketched the rotten remains of the fortified gatehouses which stood at the ends of the Welsh Bridge, before it was replaced in the early 1790s.

By now, Telford was shaping Shrewsbury. But it, in turn, was shaping him. It was unkind and untypical, in telling the story of St Chad's, for him to imply that he was in a town of dunderheads. It was anything but. In the late eighteenth century Shrewsbury was a place of unusual improvement and intellectual energy, part of what Trinder has suggested might be called the Shropshire Enlightenment, based on 'a commitment to scientific enquiry, an optimistic view of the consequences of industrial development, investment in canals and a determination to eradicate slavery'.[16]

This drew on – and some of its members joined – more celebrated groupings in larger, better-connected towns such as Derby and Birmingham, whose famous Lunar Society brought together leading industrialists and inventors. Telford was too junior to be part of that creative Midlands world. But he was in touch with it at one remove.

He ascended carefully. At first he mixed with modest men. He joined the Salopian masonic lodge when it was founded in 1788, listing his employment as 'surveyor' in its membership book, alongside wire-workers, carriers, dyers, innkeepers, a surgeon, and a self-described 'gentleman farmer'. Those business-like trades suggest Telford was one of the middling sort, not the town's elite. But he was ascending fast: and in time that claim to be a surveyor was replaced among those who knew him by the more elevated title of 'architect'.

In February 1788, he reported to Little, 'I had a visit from the celebrated John Howard Esq – I say I for he was on his tour to Jails and Infirmaries and they being under my direction, of course was the cause of being thus distinguished I accompanied him to the Infirmary and Jail.'[17] By this time Howard (the famous prison reformer) was near the end of his life and on the last of his tours of inspection, during which he estimated that he had travelled more than 42,000 miles. Telford responded to Howard's moral cause with passion: 'You will easily conceive how I enjoyed the conversation of this truly good man – and with what attention I would court his good opinion,' he wrote; 'I consider him as the guardian Angel of the miserable and distressed, travelling over the world merely for the sake of doing good, shunning the society of men and afraid of being taken notice of.'

He admired both Howard's application and his restlessness: traits he shared. 'He assures me he was born a domestic man, and that he hates travelling, that he never sees his country house but he says within himself: "O might I but rest in this spot and never more go three miles from home I should be happy",' Telford told Little. 'But he is now entangled in so extensive a plan that he says he is doubtful that he shall ever be able to accomplish it – he goes abroad soon. He never dines, he says he is old and has no time to lose.'

Howard's foreboding was right: the great reformer died by the Black Sea less than two years later, from typhus, which he had caught nursing another victim of the disease. But his visit to Shrewsbury had a lasting effect. Even before he arrived, the town had begun planning a new prison, along lines set out by Howard. On inspecting the plans in person, he pointed out a series of flaws. The site was next to the castle where Telford already had work under way, and he was asked by the town's magistrates to take over work on the prison and improve the designs. Today his gatehouse stands at the front of a larger Victorian structure which (by the time it closed in 2013) was once again a source of scandal for its overcrowded, insanitary conditions and a source of woe to the prison reform trust that carries Howard's name.

Telford also persisted with Howard's proposal that prisoners be given useful work to do. 'I am in a hurry today and yesterday and to day have been to the quarter sessions and you know that I must have a finger in every Pye – I had several things in hand, but the principal was about setting convicts to labour,' he told Little. 'They have a Dress which I had made of White and Brown cloth counterchanged in each Garment so that they are Pyebald,' – adding that 'they each have a light-weight chain about one leg'.

This crew could be difficult. Felons 'are a troublesome family. I have a great deal of plague from them,' he complained. But he set them to work, their tasks quite possibly including an unexpected project, the excavation of the ruins at Wroxeter, a few miles from Shrewsbury and the site of the Roman city of Viroconium, once the fourth largest in Britain.

Telford's interest in ruins was not new: he had visited them in Scotland and adapted Shrewsbury Castle with some respect for its antiquity. But his work at Wroxeter stands out in the history of early archaeology because it was conducted with a proper respect for recording the order in which things were found and because it involved more than a destructive raid to find valuable objects. The site was well known: Telford records that famers were in the habit of farrowing the ground to find coins. 'They know where there are ruins underneath by the corn being scorched in dry weather,' he told Little. But on 5 June 1788 workers looking for stone to repair a blacksmith's shop came across some exceptional remains, 'a Bath plastered with Roman Cement, a floor paved with large Tyles, and some Pillars'. For Shrewsbury's little crowd of intellectuals, brought up on classical texts, the destruction of such a sight was too much to contemplate. 'The literati were roused,' he recorded – one of them being the Revd Francis Leighton, of Shrewsbury, whose summary of the work (attributing it to 'Mr Telford, the able architect') was later published by the Society of Antiquaries in London.

With Pulteney's backing (and funding) he began his first excavation 'in order that men of learning might satisfy their curiosity'. The bath house he unearthed contained 'a dressing room, a Cold Bath, a hot Bath, a large Sudatorium, four Tessalated floors and places with pillars of Tyles'. He recorded his findings in an accurate, coloured, cross-sectional illustration – and then, with the always-present energy of a polymath, read up on Romano-British antiquities before illustrating the site in two watercolours still held by the Society of Antiquaries in London (drawings, one modern archaeologist notes, which 'provide an archaeological record as detailed as any produced by the leading antiquaries of the day').[18] Living Thomas Telford's life must have been exhausting.

He now found time for friends, too, and even for games. He enjoyed the company of literate men and women, chief among them Archibald Alison, whom he had first met when he repaired his home in Northamptonshire. Alison was, by now, living in Shropshire and another beneficiary of Pulteney's patronage. 'Never was a household more rejoiced than was ours with his arrival. No sooner was his well-known white horse seen passing the door than the whole family rushed down with tumultuous joy to receive him,' Alison's son recalled in his memoirs.[19]

By common consent, lessons, work, and occupation of every kind, were abandoned; and the whole period of his sojourn, which seldom exceeded two days, was one continued scene of rejoicing – games and sports of every kind, both within and without doors, in all of which he took an active part, succeeded each other without intermission, till, exhausted by joy, the whole children were sent to an early bed. My father and he then sat down and spent half the night in discussing the vast projects for the internal amelioration of the country, which he had already conceived, and a great part of which he lived to carry into execution. Never was a more simple heart united to a more powerful understanding – he was a lamb in play with us, but a giant in council with men. In our games he and I – the oldest and youngest of the party – were always on the same side. 'Mr Telford,' said I frequently, 'I've got a plan, here is my opinion.' 'Lord bless the boy!' exclaimed he, laughing, 'let us hear his opinion.'

These were happy times for Telford: perhaps the happiest of his life. He was, if not rooted, then at least resident in a single place in a way that was never the case again until his old age. With the Alisons, he became part of something close to a family, adored by the children for his indulgence. He once left a hammer and nails in the nursery to occupy them, to the shock of their parents who returned home soon after to find the nails driven into every wall in the room.

He was also getting richer, busy with his work and able to enjoy company and entertainment. In 1788, at the newly opened Shrewsbury theatre, he recorded that 'a Mrs Jordan from London has played 6 nights'.[20] She was already famous and went on to become more so, as the mistress of William, Duke of Clarence, later King William IV. Telford was wowed by her too. 'She is certainly one of the finest Comic Actresses of this Age, at least I never saw any thing in that way that could come in any degree of competition,' he told Little. 'Every word, every look is nature – and she is thus irresistibly charming from her unaffected simplicity of manner.'

Less promising were other entertainments in the town. 'Shrewsbury Races are begun but I care nothing about them, I shan't go near them,' he told Little determinedly. Music was no better. 'I have a Ticket sent me for the Concert so that I shall go and have a peep in there,' he told his friend. It did not go well, he added in a postscript later that night: 'You know my Organs are not well tuned – for sound. If they will let me hear the words I will tell them what I think of the matter – now I have been at the concert I might as well have stayed at home – it was all very fine, I have no doubt, but I would not give a song of Jock Stewart for the whole – my corse organs are not susceptible of those enchanting, sweet delightful sensations.' He was unpersuadable: 'it is certainly a defect but it is certainly a fact – I have no enjoyment for fine Music. I sit down and am as attentive as any mortal can be, nay endeavour to interest my feelings but all in vain. I feel no emotion unless an inclination to sleep be reckoned upon.' The poetry of theatre was preferable – and perhaps the celebrity of it, too. 'One look, one sentence from Mrs Jordan has more effect on me than all the fiddlers in England.'

Hateful or not, the concerts and theatre were a sign of a town with a lively intellectual life. The most famous of the Shrewsbury families which took part in it were the Darwins: Charles was born in the town in 1809, the son of a surgeon friend of Telford and grandson of Erasmus Darwin, among many things a natural philosopher, surgeon abolitionist and promoter of canal building. Telford knew the family well, writing in 1792 to tell Little that he had been reading 'a very wonderful and masterly performance' by Erasmus Darwin called 'The Loves of the Plants', which portrayed the sexual organs of plants as though they were a bride and groom (and which implied the theory of evolution half a century before his grandson). This piece of writing, accomplished in the form of a poem, has been described by the historian of the Lunar Society, Jenny Uglow, as 'one of the most extraordinary – some would say bizarre – works in English literature. Arching between two eras, it was a final exuberant flowering of Enlightenment experiment and optimism.'[21] It impressed Telford, too: 'In the text, which is beautiful poetry, he has contrived to introduce a sketch of almost every thing which is in heaven above or the Earth below, or the Waters under the Earth – and in the notes which are very extensive and valuable – he has expatiated at large upon every subject mentioned in the Text, and has given to the world a multitude of most eccentric and amusing Theories that were ever engendered by the human Conception. From the formation of Planetary Systems to the humble Daisy, nothing escapes his attention.'

Natural science, however, was not at the forefront of Telford's mind when he read Erasmus Darwin's work. The poem was published in 1789 and reissued in a longer form in 1791. Telford commented on it in 1792. These were years of radical hopes and terrors, as first the revolution in France took hold and then Britain was dragged into a war which lasted for much of the rest of his life. Telford was at first swept along and then – later – repelled by what took place. This was not his normal way of thinking. He never much cared for politics or current affairs, as he told Little with typical steadiness two years before the French Revolution broke out. 'With regard to politics you shall have it all your own way – for I freely confess that I know nothing of the matter, I am therefore unfit to give an opinion for I am so totally enveloped in my own business that I very seldom read a paper. . . however I intend to stretch a point, and read one now and then.'[22] By 1792 that had been cast aside amid the excitement from France. There are echoes of the conversations he must have had with his Shropshire friends in his letters to Little. He compared Thomas Paine's writing with that of Sir James Mackintosh; both had been provoked by Edmund Burke's prescient denunciation of the Revolution to produce bestselling ripostes (Mackintosh, unlike Paine, eventually recanting and turning in favour of Burke). Of the two writers, concluded Telford, Paine was 'better calculated to work on the general mass of mankind' and 'fully rouse the attention, and fill the whole soul with indignation against the whole system of despotism and all its appendages'.

But it was Darwin's poem which filled him with even more delight. He singled out to Little a passage 'expressive of the Author's sentiments concerning Civil Liberty' – a section about electricity, lightning and Benjamin Franklin. This, he suggested, was 'proof that we have the good and wise of every county – the powerful advocates of Civil Liberty and their Voluntary Pens are of more avail than all the mercenary swords of expiring Despotism'. Off on a tangent, he continued, excitedly: 'The French have summoned the Prussians to evacuate Verdun, which they have wisely obeyed – I believe I told you that Spires [Speyer] and Worms are in the hands of the sons of Liberty – and Coblentz is quivering at the approaching Tempest – I mean the Nobles and Priests the people are embracing their deliverers.' Even in isolated Shrewsbury, far from any Channel port, the Revolution was thrilling – and at that moment, seemingly commendable, too. 'You must have heard that Paine is appointed one of the Committee for revising the French constitution,' he added.

He was caught up in a national frenzy which led William Wordsworth to feel that 'bliss was it in that dawn to be alive'. That was in France, of course. But for a vigorous few months Telford was persuaded that Britain, too, was on the skids:

I am convinced that the situation of great Britain, tho' perhaps not quite so alarming as he represents, is yet such that nothing short of some signal revolution can prevent her from sinking into Bankruptcy, Slavery and Insignificance our pecuniary embarrassments are so perplexing, our statesmen are so fond of power the pernicious system of continental connections is so prevalent, while the influence of distant colonization is necessary to the support of the corrupt administration, that perhaps little change is to be expected from the ordinary course of Government. It would require the united abilities, integrity and independence of a hundred <u>Pulteneys</u> to purge the bloated mass and restore it to vigour and that alas! can never be expected.[23]

To a modern reader, taught to look back on the age of Nelson, the industrial revolution and the brink of a century of British global dominance as a high point in our national story, it is salutary to be reminded that thinking people at the time could be persuaded that Britain was going to the dogs or, as Telford put it, 'it must make every sincere lover of his country grieve to reflect with what rapid progress we are hastening towards inevitable Ruin'. Politicians, he concluded, were 'obstinate, corrupt and vain'.

In penning that he either forgot that his patron, William Pulteney, was a Member of Parliament, or was confessing to Little a private scepticism that he dare not share more publicly. Either way, he overreached himself with his next move. He sent, under Pulteney's postal frank, a copy of Tom Paine's free-thinking *Rights of Man* to Little, who had it read to him by his pupils and then allowed it to circulate in Eskdale. Pulteney's previously apolitical sidekick was encouraging revolt in his master's backyard. The reaction in Eskdale was predictable. The valley staged its own little aftershock of revolution. 'It appears that Mr Paine's Pamphlet has had its full effect on the Langholm Patriots, tho' you do not tell us who were so warm in the cause of Liberty,' he told Little.[24] And Pulteney was furious. It was not what he expected from his county surveyor. When he found out that Telford had sent Paine's work without his permission, he came close to sending him out of the county into disgrace.

Perhaps that was a lesson to Telford. Or perhaps his radicalism never was very deep. Or perhaps – like many others – he
was appalled by the course of the Revolution as the Terror took
hold. Either way, he was a man who derived much from climbing up the settled order and one who had much to lose from its
collapse. From now on, as the news from France darkened, first
caution and then outright hostility to radicalism overtook him.
As he wrote to Little, 'steady resolution founded upon thoro'
examination and firm conviction is much to be preferred to
the violent ebullations of enraged passions'. This is the 'steady
on' voice of an engineer coming to the fore; the emerging and
increasingly Tory voice of a man whose associations by the end
of his life would all lie on the reactionary side. The mob was
getting closer to his door and he knew that it was no respecter
of intellect or enterprise. In 1791 crowds in Birmingham set
about one of the geniuses of the Lunar Society, Joseph Priestley,
a man Telford respected deeply. 'The vile dregs of mankind
immediately arrived, and in a little while the Drs. Phiosplical
[sic] apparatus his valuable manuscripts, and every part of his
household fell a sacrifice to their implacable fury,' he protested
to Little. He ended his letter with a jibe at those members of the
Anglican clergy who delighted in the attacks on dissenters such
as Priestley. 'May the Lord mend their hearts and lessen their
incomes.' But as reports of the horrors of the French Revolution
spread, he sided with its opponents and Pitt's spying, taxing
state. So did many others.

Telford knew which side he was on: that of order, property
and authority. 'We may seriously rejoice at the welfare of our
neighbours and anxiously desire a reformation at home, but we
should carefully avoid any tendency to excess or anarchy for
they are too frequently the attendants of popular reformations,'
he told Little. When Birmingham rioted, Shrewsbury held
firm: 'there were some slight alarms in the town but nothing

happened, the watchmen were called out for several nights but we have now discontinued them', he reported. A new sort of Britain was emerging: militaristic, big spending, at times authoritarian. The Age of Enlightenment was at an end. From now on Telford left politics 'to the care of those active citizens who spend their time in discussing what they rarely understand'.[25]

Something Like Bonaparte

The 1790s were years of fears and disasters. Of hunger, bad harvests and rebellions. Of one spring so dismal that the birds failed to sing and another so sodden that floods swept the land. And most of all, of war. On 21 January 1793 the French king, Louis XVI, was executed. On 12 February, Pitt told the House of Commons that Britain would fight the Republic. It was, he said, an ideological war against an enemy 'that aims the total ruin and freedom of Great Britain'.[1] It was also war which went on, in different forms, for more than two decades and which, through the schemes for infrastructure it demanded and the industry it supported, shaped the rest of Thomas Telford's life. Just as the Second World War propelled technology, accelerating the creation of things such as computers and jet engines, so the Napoleonic Wars forced the pace and opened minds. Necessity encouraged daring.

It was a useful time to be an engineer, but not necessarily a happy one. If, in this period, a sort of frenzy overtook Telford – a maze of work and travel which eventually expanded beyond Shropshire to take in almost every part of the United Kingdom – then the causes lie partly in the strange times in which he lived. Britain was regulated, drilled and brought together as never before. 'The wars were like permanent bad weather,' writes Jenny Uglow in her luminous account of what it was like to live through them. 'So all surrounding that people stopped referring to them and merely said "in these times". They affected everyone, sometimes directly, and sometimes almost without their knowing it, and in the process the underlying structures of British society ground against each other and slowly shifted, like the invisible movement of tectonic plates.'

In his Shropshire vicarage Archibald Alison, Telford's learned friend, sat distraught at the consequences of the revolution in which he had placed so much hope. 'He received daily the *Chronique de Paris* from France; and so extreme did his anguish become during the Reign of Terror, that I have repeatedly heard him say that as long as Robespierre's power lasted, he never shut his eyes 'til four in the morning,' his son recalled.[2] The Alison children may not have shared their father's anxieties but nor they did escape the war's cultural consequences. 'In our games with Mr Telford, we had been divided into two parties – the French and the English; and. . . there were regular surprises, combats and prisoners taken on both sides.'

French ways and the French war became the target of grievances of every kind. In 1796, in Shrewsbury, Telford's patron, by now titled as Sir William Pulteney, faced the most fractious and corrupt election of his parliamentary career. To the tune of a popular song, 'As I Was A Driving My Wagon One Day', his opponents marched through the town crying:

A Freeman I am, and a Freeman I'll be,
For all in this Kingdom thank Heaven are Free;
No system of terror here governs our votes;
No one third, nor two thirds are cramm'd down our throats.[3]

Pulteney, who won the election, was in fact no friend of the new French constitution, which under the Directory replaced a third of the parliament every year and required two-thirds to come from the old Convention. But substance hardly mattered. What counted was the general unease, which is why, though Telford's correspondence with Little continued its familiar rhythm in these years, there was a new and fretful undercurrent to it. 'Our bankruptcies are still increasing,' he wrote in 1793.[4] Two years later, as the harvest failed completely and the more benevolent Shropshire employers paid from their own pockets to feed the poor, he feared insurrection. 'I look forward with some apprehension to the ensuing winter and spring. . . Now that war has desolated all the Corn Countries of Europe. . . the film is removed from our Eyes, and all that can be done is to economise, and make our scanty portion suffice.'[5] By 1798 national ruin seemed close at hand: 'we are subscribing most liberally here to oppose these terrible Frenchmen, who by the bye, seem to want other worlds to subdue – not considering how difficult it will prove to govern what they already possess'.[6]

Telford could not be a bystander in this conflict. But nor did he become an enthusiast for combat. France, in his view, had gone mad, but like many early admirers of the Revolution he hoped for a return of sanity. 'I have always disapproved of our meddling with them at all with regard to their own affairs, but I am equally averse to them settling ours,' he wrote. His affection for Paine and radical politics faded fast, but he saw the jingoism into which many others lapsed as an improper substitute, constraining the sort of free thinking which supported progress and commerce. 'Honest John Bull with his national bluntness, likes to trudge on in the trammels of system,' he complained.

Telford, however, did not trudge on. Nor did he economise. And he broke through the trammels of system. Despite the war, the decade was the making of him. At first in Shropshire and then darting around Britain, his life became a mass of sketched designs, scribbled calculations, arbitration, estimation and consultation. 'I am toss'd about like a Tennis Ball, the other day I was in London, since that I have been in Liverpool, and in a few days I expect to be in Bristol – such is my fortune – and to tell you a bit of a secret, I truly believe that it suits my disposition,' he told Little in July 1799.[7] A few months later he boasted of a schedule which took him in just a few days from Liverpool to Shropshire, to Chester, to Liverpool again and then to Worcestershire, managing canals, settling disputes, expanding ports. 'So you see what sort of life I have of it. It is something like Bonaparte when he was in Italy – fighting battles at 50 or 100 miles distance every other day.'[8]

Bonaparte, by then, was the national bogeyman and even by the standards of Telford's self-aggrandisement, it was taking things to the limit to compare himself to the commander of French forces, a man whose armies had in that year alone roamed into Egypt and Syria, who had just declared war on Austria and who – in the month the letter was written – had staged the coup of 18 Brumaire to establish himself as First Consul of France. But Little will have known what he was getting at. Telford saw himself as a general fighting a war at home every bit as hard and as significant as the battles abroad; and he saw himself as equal in character.

And that attitude never varied. '<u>I shall Sir, set out to join my armies in the north</u>', he boasted in 1813 to Charles Pasley, with his glorious military record.[9] 'My uninterrupted 30 Years Warfare seems still distant from termination,' he wrote to him again, three months after the Battle of Waterloo, describing the physical links he was building between England, Scotland, Wales and Ireland, 'this is my mode of subjugating Kingdoms'.[10]

In his glassy, polished *Life*, written in old age, Telford despatches his time in Shropshire smoothly. He glides from architecture to canal and bridge design in a handful of pages, as if his progress was easy. This, of course, is the view of a man looking back at the end of a successful life and a man who also, as the canal historian Charles Hadfield has argued, took care to edit difficult passages and other people out of his story as he prepared the volume he intended to define his reputation after his death.

The reality of this time in Shropshire was different. He made his own luck. He fought hard. He searched out new tasks and discarded old ones with clinical intent, as he evolved from county architect to civil engineer. The way he presented himself to the world as he did so disguised a character of boiling energy and not a little insecurity. In his private writing he described himself as 'the principal actor' – a telling term. He was driven by an obsessive need to be at the centre of the swirl. 'In the operations of the business of such an extensive populous Rich tract of Country, you will easily perceive there must be frequent and sometimes violent agitations in often clashing interests to contend with or reconcile, and in my present situation the ostensible and active portion of this Office generally falls to the share of your humble Servant,' he told Little. This was a boast, not a complaint.

He was always manoeuvring like this. 'I have no objection to be engaged with this man of war,' he told Little, when Lord Clive, the greatest landowner in the county and an opponent of William Pulteney, got in his way.[11] Faced with Clive, lesser men would have faltered. Telford secured success. At the quarter sessions in Shrewsbury, he boasted, he carried every point 'tho I was obliged to speak a good deal, and even bully a little'. The battles in his life must have seemed endless. 'I was in hopes that I would have enjoyed Eskdale for a week or ten days, but that is now over for the present,' he told Little in November 1795.[12] 'There is (between one thing and another) scarce a week, but what I have to attend some public meeting, and you know, that . . . it has cost me some labour and attention, to bring all this about.' This passage is followed by one of those characteristic explosions of condemnatory private passion which Telford shared with Little and no one else. 'He who neglects his duty, deserves and will be neglected by Society, <u>he may draw nutrition propagate and rot</u> but he is unfit for the grand operation of human society.'

This harsh and demanding injunction to his fellow men gives us a glimpse into the inner thoughts of a man who was not generally regarded by those who met him as ill-natured or unkind. Telford would have been resented and scorned for such bossy censoriousness, were it not for the fact that he seems to have hidden this entirely from everyone apart from Little. Those who met him at this time encountered a cheerful and generous man, on the brink of a prosperous middle age, whose rise from nowhere offered the sort of good-luck story that played well during a difficult war. His success seemed to justify the British belief in their national liberality. Plainly he was something of a secret moralist, too.

Pictures of him from this period show a bluff man in a heavy coat with a thick mop of hair falling over his broad forehead and wide cheeks. You can imagine him out in the rain, or helping fix a loose shoe on a horse, or lighting a fire. He could see how something ought to be bolted together; could sketch out a rough calculation; could draw a structure with a practised eye and pen. People enjoyed his company, loud-voiced and curious. They respected his dedication and his knack of getting to the practical heart of a problem, working it through until he had found a solution. But they also respected his serious, bookish intelligence. His youthful delight in poetry, philosophy and classical learning did not fade. Whereas today the intellectual and the practical bent have tended to diverge, so that an academic is not expected to be also a builder, this was not the case with Telford or those he mixed with. His mind could turn easily from design to learning and back again. He continued to spend time with Shrewsbury intellectuals such as the physician and naturalist Robert Darwin, 'a very clever man' of whom Telford could boast 'I am intimately acquainted' – well enough that he later sought his advice on the best way to inoculate against the pox using serum from a cow.[13] On Bonfire Night 1793 he dined with a typically literate and progressive group: among them Archibald Alison, Archdeacon Joseph Plymley and his sister Katherine. 'I have not often spent so pleasant an evening,' she recorded in her diary.[14]

The Plymleys were at the heart of the Shropshire Enlightenment, opposed to slavery and fascinated by science. But they were also, like Telford, unafraid of getting mud on their boots. Later in life, Katherine explored North Wales alone on her pony, while over many years Archdeacon Plymley tramped through every Shropshire parish, however remote, recording the conditions in a detailed diary. He was fascinated by schemes for the improvement of roads and farms (though he confessed that 'I am very little of a practical farmer'[15]). Later, he enlisted Telford to contribute a chapter to a printed account of the economic potential of the county. Like Alison, he became a close friend and promoter of the rising engineer. And in this world, by turns bookish and solidly practical, Telford blossomed. As Katherine continued in her account of the evening:

Mr Telford is a man I highly respect. Born of poor parents. . . brought up a common working mason, he has by uncommon genius and unwearied industry raised himself to be an excellent Architect and a most intelligent, enlightened man. His knowledge is general, his conversation very animated, his look full of intelligence and vivacity. He is eminently chearful and the broad scotch accent that he has retained rather becomes him. He has been settled in Shrewsbury for some few years. . . and has been engaged in many public and private buildings, he is the architect for the new church at Madley and has just received a very advantageous appointment. . . But praise of a higher kind belongs to him, what he procures by his merit and industry he bestows most benevolently and liberally: frugal in his own expenses, he can do more for others and what he does he does chearfully.

She was, perhaps, smitten with this big, jovial, engaging Scotsman, so at ease in the company of his literate friends. She offers no sign of spotting the turbulence which drove him. But though incomplete, the picture Katherine Plymley paints of him is not untypical. Her description of his benevolence rings true. He often gave gifts and help to those he thought needed and deserved them and in his time in Shropshire none benefited so much as the two people to whom he was closest. 'To set my mother and you above the fear of want has always been my first object, altho' I have never told you so before,' he had told Little in a letter sent a few years earlier. 'I have at present the opportunity of earning money. You have not and I therefore insist on the privilege of sharing a little. . . this is not ostentatious because I don't wish anybody to know of it besides yourself.' The end of this letter is particularly revealing. 'If I don't do something set me down in your own mind for a rascal, til you tell me otherwise yet remember I am not rich yet.'

The 'yet' stands out. The ambition may have been concealed from public view but it was never far away. He was on the march and he knew it.

If one secret of Telford's success was his brilliance at what today would be called project management then a huge component of this lay in the quality of the people he chose to work with. His projects became theirs – and theirs his, since it was usually Telford who ended up with much of the acclaim. He trusted them, trained them, travelled with them and depended on their skills. In many cases, the bonds lasted for life; or even longer, since as he grew older he often worked with the sons of fathers he had known earlier in his career. He liked to command and became dependent on something close to an army of men. Boastful though it was, his self-comparison to Napoleon in his letter to Little was not wholly absurd and perhaps a little tongue-in-cheek. And if Napoleon was lucky in his generals, then Telford, in turn, was fortunate in his choice of friends.

Three, in particular, stand out: men whose working associations with each other and with Telford lasted for life. From now on this gang would travel together from project to project, an anchor which held firm in the most challenging of times, and their families integrated until they were at times almost one. When, in the early 1790s, two of these friends lost children in quick succession the infants were buried in the same plot in Shrewsbury. Later, their surviving sons were often sent to live with Telford, to learn his trade; and in adulthood some continued to work with him on schemes begun by their fathers. It was a web of relationships which supported everyone involved and from which it was hard to escape.

One of these men was John Simpson, 'a treasure of talents and integrity' who crops up many times in Telford's story.[16] Born in Midlothian in 1755, he was a stonemason and project manager of 'diligence, accuracy and irreproachable integrity'. His death in 1815 was a practical and emotional blow not just to Telford but to others in this tight working partnership. 'I mourn him deeply indeed,' wrote the second of them, Matthew Davidson.[17] He was, his obituary in the *Shrewsbury Chronicle* recorded, 'a man of the strictest integrity, generosity and benevolence; a warm and steady friend, a most affectionate husband and indulgent parent'.[18]

Davidson was even closer to Telford. The pair had known each other since infancy. Born in Eskdale, educated at Westerkirk school, and trained as an apprentice mason in Langholm, he was an intimate friend of Telford and a fundamental part of his life. Their correspondence was lengthy and varied, by turns formal, with Davidson taking a humble and informative tone when discussing work with a man who was by now substantially superior in status, but then bursting out into jovial equality when the letter would be seen by each other and no one else. As Telford teased in one letter home: 'Davidson is now at my elbow, and is perplexing me with questions about. . . the public house Landlords of Langholm.'[19]

He was a man of unusual manner, odd enthusiasms and self-taught learning: another product of that powerful Eskdale culture. One of Telford's later friends, the Poet Laureate Robert Southey, described him (after his early death) as 'a strange, cynical humourist' – which is an unkind way of saying that he was sarcastic, outspoken and practical – perhaps even a manic depressive.[20] Said to have resembled Dr Johnson in both looks and manner, Davidson was so well read he was known as 'the Walking Library'. He emerges as one of the fundamental heroes in Telford's story, though those who did not understand him could not see it. He never stopped rubbing people up the wrong way. While working in the Highlands – a part of Scotland whose landscape and people he purported to despise – he told someone complaining about the weather that the rain would not hurt the heather crop. To an artist, drawing the landscape, he said that it was the first time he knew what hills were good for, and of Inverness he claimed that if justice were done there would be nobody left alive there after two decades apart from the Provost and the hangman.

He was infamous, too, for his obsession with the health bene-
fits of cold water, once throwing two buckets of sea water over
a maid to cure her of fever (it worked) and swimming regularly
in the icy Moray Firth. 'I have no hesitation in saying that a man
may live healthy many years, even in the noxious atmosphere of
London, solely by means of temperance and the daily use of that
elegant Classical medicine, Bathing in cold water,' he declared –
which did not prevent Telford, who had no truck with such
folly, outliving him by a quarter of a century. 'I keep reading
at night as usual,' he wrote to his son. 'Thucydides engages me
much. He thinks and writes so much like an Englishman.' Note
'an Englishman'. Davidson was Scots but he went even further
than Telford in building a life away from his roots, marrying
a Welsh woman and always claiming to regret his move to the
Highlands, where he died in 1819. 'He would not admit a seat
in Heaven if there was a Scotchman admitted into it,' one of his
colleagues complained.[21]

John Simpson came to Shropshire to work on the rebuilding
of the collapsed St Chad's church in 1790; Matthew Davidson
arrived not long after to lead the construction of Montford
Bridge over the Severn on the main road to North Wales (a road
which Telford would, two decades later, also reconstruct). It
was a characteristic project for the county surveyor and the first
of forty-two bridges that Telford would eventually build in the
county. It was done well enough that the bridge still stands, the
construction of its three tidy arches in red sandstone overseen
by Davidson who, Telford says in one of his infrequent tributes
to anyone else in his *Life*, '(as ever afterwards) well performed
his duty'.[22]

If Simpson and Davidson provided familiarity – a Scots accent
and a way with stone that Telford knew too – then the third
of the trio who worked with Telford from this period brought
something different. He was, for a start, not Scots but English,

from Shropshire. William Hazledine was a millwright's son who was, as Telford came to know him, venturing into iron production. His business in Shrewsbury was modest compared with those of the great men whose foundries lay along the Severn Gorge, such as William Reynolds, John 'Iron Mad' Wilkinson and Abraham Darby III. Telford knew many of them, too: he called Wilkinson the 'Iron God' and learnt from Reynolds's innovation in promoting canals, towpaths along the Severn and scientific study at his home in Coalport, where he kept a collection of fossils, a library and a laboratory.[23] He became a good friend too of Charles Bage, a structural engineer who designed the first building with a cast-iron frame, the Ditherington Flax Mill, which still stands in Shrewsbury. But it was with Hazledine (who cast the columns and beams for the mill) that Telford was to build his most durable and creative relationship as a civil engineer. Without this, he might have remained nothing more than a competent and ambitious county surveyor, the now-unremembered originator of good stone structures.

Telford's description of him – 'the Arch conjuror himself Merlin Hazledine' – has often been quoted in support of his wondrous powers. Research by Andrew Pattison, a former Shrewsbury doctor, has changed the story a little.[24] He points out that when Telford used this term in a letter to Davidson in 1796, Hazledine was yet to make anything in iron for his friend and Telford was simply referring to his surprise at encountering him in London while 'I was conjuring about a Spring for the Coffee house Door'. The circumstances behind the story hardly matter. Hazledine was to prove a wizard and the title stuck.

One can imagine their first meeting, either as fellow members of Shrewsbury's masonic lodge in the late 1780s or, more probably, a little before. Hazledine was 'a man of few words, with a bluff manner and unpolished Shropshire accent', according to Pattison. He was also someone 'of immense energy, throwing himself into a bewildering array of different branches of work'. In that, he was something of a second Telford. Both had worked their way up as apprentices and were now making their careful way into middle-class lives. Both knew what it was to labour for a living. If Hazledine's start in life was a little more secure, then it was hardly more comfortable. His father's notebook records the terms of his brother's indenture: 'if he loses any time or does disobey his father or his mother, he shall upon such offence have his arse rubbed with a brick quite raw'. Similar stern conditions must have applied to Hazledine's apprenticeship too.

The pair shared an untutored love of books and Burns's poetry. But it was the revolutionary possibilities emerging from the blast furnaces and forges of industrial Shropshire which most excited them. Together, Hazledine and Telford tested the possibilities of iron, the radical material of the age. Like concrete in the twentieth century, it seemed to presage not just a change in the manner of construction but perhaps the way people could live. Even Thomas Paine had been drawn to the thrill of building with it. With the backing of Benjamin Franklin, he designed a prototype iron bridge in the 1780s and in 1790 he created what became a peripatetic one. Cast in Rotherham, intended for use in America, it was first erected as an exhibit in London (winning praise from George Washington and Thomas Jefferson). Some of its wrought-iron strapping was then reused in Wearmouth, Sunderland (without Paine's involvement) in 1796. His life as an engineer was short. Telford's was just beginning. But for a moment the paths of their lives touched.

It might seem curious that a material so familiar and ancient could – after over 3,000 years of constant human use – become transformative in the hands of men like Hazledine. But something important had changed. Before the eighteenth century, iron production was a modest business and a local one; simple, charcoal-fuelled forges scattered near any source of ore made the material for essential objects. Anything big and structural was made of wood or stone. Then, in 1709, at Coalbrookdale in the Severn Gorge, the first Abraham Darby changed the economics of production by using coking coal rather than charcoal to smelt the ore. Though the transformative significance of that single event has been disputed and it took many years for the new technology to replace the old, it was part of a move to efficient production and distribution that unlocked a second iron age, in which Telford was to play a predominant part. At first the material used was cast iron, strong in compression, with a high carbon content, but liable to crack under tension. Later, after Telford's time, this gave way to wrought iron, hammered out by hand or steam presses, and much better able to cope with the stresses of civil engineering. Iron, in turn, gave way to steel by the end of the nineteenth century, but in its time it was the emblem of advancement.

Shropshire was not alone in developing its use. There were great furnaces in South Wales and a large and competitive iron industry exporting from Sweden. But there was nowhere like the Severn Gorge, 'the revered heart of the industry', for testing new ideas.[25] In 1761 one Shropshire man had himself buried in an iron tomb; soon after Shropshire produced both the first iron rails on horse-drawn railways and before the end of the century the first iron boat floated (in front of an amazed crowd) on the Severn.

By the time Telford arrived the county also had its new showpiece, a bridge built from cast-iron parts by John Wilkinson and Abraham Darby III and opened in 1781. It was part of a memorable and much-visited scene. Tourists, and industrial spies posing as tourists, came to stay at the new Tontine Hotel to be awed by the gorge, which blended a fast-flowing river, sylvan walks and a Doric rotunda with the raw hell of iron-making. 'The approach to Coalbrookdale appeared to be a veritable descent to the infernal regions. A dense column of smoke rose from the earth; volumes of steam were ejected from fire engines; a blacker cloud issued from a tower in which was a forge; and smoke arose from a mountain of burning coals which burst out into turbid flames,' wrote Carlo Castone della Torre di Renzionico Comasco, an Italian aristocrat, in 1787.[26] As for the iron bridge itself, 'it appeared as a gate of mystery'. Telford was never bored by this scene: in the 1830s, aged almost eighty, he insisted on taking a visitor travelling from Wolverhampton to Shrewsbury via the gorge, to see its iron industry, then not quite yet in decline.

When the French war broke out in 1792, many of the leading families in Shropshire's iron industry found themselves in a quandary. On the one hand, war offered work. Ironmasters in Wales grew rich on casting cannon and shot. On the other, many (although far from all) of the gorge's leading families, among them the Darbys and the Reynolds, were Quakers, 'their lives of tough commercial sense were tempered by calm meditation and determined philanthropy', writes Uglow.[27] The restraints of wartime forced them to adapt the base metal to new tasks. 'Coalbrookdale was thus a curious place, a powerhouse of invention and of spiritual illumination, as well as of furnaces, iron and steel.' This led to collieries and canals, to roads, steam pumps, iron bridges and tunnels – to wealth, to trade, to experimentation – and to the acceleration of Telford's already rocketing career.

In early 1793, Telford was still turning his hand to the tasks of a county surveyor, which involved buildings as much as bridges. He did not see himself as an engineer and on the face of it showed little sign of becoming one. His ambition, at this point, was to break through as a successful architect. 'He seems to have had no premonition that his architectural career was drawing to a close,' says James Lawson in the first full study of Telford's time in the county.[28] Two remarkable buildings speak of his talent. In Bridgnorth between 1792 and 1796, he replaced an old chapel with a tidy neo-classical church, St Mary Magdalene, which still stands high on its rocky promontory above the River Severn, next to the remarkable civil war ruins of the town's castle, which lean more steeply than the tower at Pisa. 'Its only merit is simplicity and uniformity,' Telford said of the building, whose construction was overseen by Davidson and whose site was not his first choice but forced upon him when a landowner at the end of the high street refused to sell. Pevsner is kinder, calling it 'a remarkable design, of great gravity inside and out, apparently inspired by engravings of recent churches in France'.[29] An undated letter now in the library of the Institution of Civil Engineers shows that Telford had thought hard about its design and was ambitious for his work. 'This Plan had been formed under the persuasion of having in some measure corrected what appeared to be inconvenient and faulty, in the general construction of churches.' A year later, following the success at Bridgnorth, he designed a second church at Madley, near Ironbridge, which he called a 'very peculiar construction', octagonal outside but oblong within.

What stands out from these buildings is their author's surefootedness in deploying the tasteful, educated styles of his age. He was by now a proper architect, not a jobbing structural surveyor. And he worked hard to improve his skill. The library register at Shrewsbury School records that at the start of the Bridgnorth project Telford borrowed the second volume of Montfaucon's *Antiquities* and the first volume of Stuart and Revett's *The Antiquities of Athens*, studying serious works which underpinned Britain's rising interest in classical archaeology and styles. The exterior of his Bridgnorth church is Tuscan, the interior Ionic, the clocktower Doric and the smooth rustication of the portico wall drawn directly from Stuart and Revett's plate of the Stoa at Athens.

This is not what is normally expected of an engineer and it does not suggest a man who wanted to give up buildings for great bridges. Nor was this the limit of his learning. Continental war meant that travel in Europe was off-limits, even if he had been able to afford it. But Telford was determined to educate himself properly and in the early spring of 1793, aged thirty-five, he set off on an architectural tour of England, visiting Bath – a powerbase of William Pulteney Oxford and London. It was no pleasure ride. He studied the things he saw, took notes and remembered well. He also, of course, made sure to describe the expedition to Andrew Little.

It began with a journey south from Shropshire, through Worcestershire and Gloucestershire – 'I never saw a place that pleased me so much by its external appearance.'[30] Then 'the scene changed most completely for on the side of a rough Hill we stopt to water the Horses at a Paltry Alehouse, full of drunken Blackguards bellowing "Church & King" – with most tremendous vociferation'. Even on the Somerset border, the war could not be escaped. Telford tells the tale of what happened next with verve:

To thicken the plot there happened at that very moment to arrive a poor ragged German Jew, whom the whole of the discerning loyalists immediately accused of being a Frenchman come to take away their liberties, it was in vain that he remonstrated that he was only a poor German who cut the corns and that he only wanted a little bread and cheese – the enraged Landlord (a great brawny fellow) swore he should have nothing in his house, and that as he was a constable he would carry him before the Parson of the Parish (a Justice) who he was sure would send him to Jail – however partly by my interference and more by the following happy thought of the Landlord, things were softened a good deal, the Landlord all at once with his formidable knife – sliced off above a pound of raw Bacon from a ham which hung over our head and swore that if the foreigner chose to escape prison he should swallow the bacon raw as it was, the poor Jew was at first in a worse plight than ever – he declared that he was de Jew and durst not – but the Landlord was so pleased by this new Idea of his that he totally forgot the Church and the King and the Parson and all and as I furnished poor little Moses with something to pay for his bread and cheese; and by the time the coach started I may venture to assure you they were all perfectly reconciled.

Soon he was in Bath, a city built from 'the greatest plenty of beautiful material at the cheapest rate in the world'. He was won over by the work of John Wood, the architect of the Royal Crescent, but not by that of his successors. 'Since his time, altho' the rage for Building has been unbounded, yet there has none inherited even a portion of his genius' – not choosing to add that many of the buildings he disliked were being put up by Pulteney. He inspected Roman remains – including some being demolished – and travelled on to London, on a road that was still the haunt of highwaymen. 'We escaped safe, tho' the collectors had been doing their duty on Hounslow Heath.'

In London – where he stayed with 'an honest Scotchman' – he studied omnivorously, examining in the libraries of the Antiquarian Society and the British Museum works by Palladio and Inigo Jones; reading 'most of the editions of Vitruvius' and 'a book on Chinese Architecture'; looking at 'several Models of Indian things sent from the East' and 'some fine Plates of Egyptian Antiquities'; and pictures of Shakespeare's characters in a gallery. He praised pictures he saw by Joshua Reynolds, who 'was not content to delineate merely the cold features of a likeness, he has with the true spirit of a philosopher and a poet, always seized some characteristic expression'.

It was a rapid and conscious period of self-education, a typically Telfordian accelerated learning course, 'so that with the information I was before in possession of I now have a tolerable good notion of Architecture'. His visit to Oxford added to it, with 'a profusion of paintings, sculptures and Architecture'. He saw pictures by Raphael, Rubens and Titian at Christ Church, early works by Inigo Jones and Christopher Wren and spent time with a mathematician who was publishing the works of Archimedes. Back in Shropshire he persisted in the summer of 1793 with architectural study – 'my favourite pursuit' – 'it will be probably only slowly, as I must attend to my practical employment but. . . if I keep my health and have no unforeseen hindrances it will not be forgotten'.

First, though, he had to complete his tour, travelling through Birmingham, by then an expanding industrial city 'famous for <u>Buttons</u> and <u>Buckles</u> and <u>Locks</u> and ignorance and <u>barbarism</u>'. 'Its nick nack hardware is proof of the first and its Locks and Bars are evidence of the last of the Assertions,' he wrote, 'and the disposition of the inhabitants confirms the theory.' Birmingham was where a mob had destroyed Joseph Priestley's house not long before and it was infamous for selling slave chains to plantations. It was never a place he liked.

The winter of 1794 was awful. On 28 November, drying out in front of a blazing fire at the Bush Inn, Carlisle, with a long journey to Manchester ahead of him, Telford complained that he had been soaked. 'It has been a day in which heaven and Earth

& "may I couple Hell" seemed to mingle in horrid conflict – the pouring of unintermitting Torrents.'[31] He had only just got across the River Esk before its banks overflowed. In Shropshire, the weather was just as bad. 'This season has been severe beyond all precedent,' he told Little the following spring.[32] 'The storm of Frost and Snow kept accumulating for two months, after which a very hasty thaw caused a greater inundation than has ever been known before in England.' The great flood of 1795 devastated much of the West Midlands, with Shropshire particularly badly hit. Dozens of bridges were damaged. In the Severn Gorge, where the river rose twenty feet above its usual level, houses were washed away and at least one forge was inundated. Shrewsbury was cut off. Nothing so severe had been seen in memory – or has been seen since.

Yet for Telford all this destruction was an opportunity. It was at this point that his focus shifted firmly and finally from architecture to engineering. As the water drained from the woods and fields, he found himself in the middle of reconstruction projects in which he was to achieve brilliant success. The first was at Longdon-on-Tern, where the torrent had smashed the half-built stone aqueduct being built on the new Shrewsbury canal. Within weeks of the flood he was appointed engineer of the project (his predecessor had recently died) and tasked with rebuilding it. This is the structure which sits today in a field near the town of Telford, the first of its kind. Yet as with so many of the schemes which made him famous, there is something of a gap between the role he allowed others to believe he had played and the probable reality of what happened.

'I have just recommended an Iron Aqueduct,' he told Little on 18 March 1795, 'it is approved and will be executed under my direction, upon a principle entirely new, and which I am endeavouring to establish with regard to the application of Iron.' This is an overstatement – or a lie. He was far from the only champion of iron in Shropshire, or even the one who came up with the idea of using iron at Longdon. That honour, he conceded in an essay he wrote shortly afterwards, went to Thomas Eyton, chairman of the canal committee – although 'the principles of construction, and the manner in which it should be executed were referred to Mr Reynolds and the writer of this article'.[33] Nor was the aqueduct quite the first in Britain to be built of iron: a small structure in Derby opened a month before. But there is no doubt he took a leading part in bringing the structure at Longdon about – and no doubt, either, that he wanted the credit. 'I proceeded with confidence of ultimate success,' he wrote in his *Life*, 'although the undertaking was unprecedented and generally considered hazardous.'[34] The result was the first full-scale, navigable, iron aqueduct on a canal and one that did not, as its critics feared, leak or shatter in the first frost. If nothing more it proved what greater things might be possible.

The chance came soon enough. At nearby Buildwas the great flood had ripped down a medieval stone bridge over the Severn and as county surveyor Telford was responsible for its replacement. He could have settled for a wooden structure, as he did in construction at Cressage between 1799 and 1801, or stone as he did on other Severn crossings after the flood, in Bridgnorth and Bewdley, where he designed a notably elegant, broad-arched design in local stone a world away from the futuristic engineering in iron he attempted at Buildwas. But it had not been missed that in the floods the nearby iron bridge had 'firmly stood and dauntless braved the storm' (as one newspaper put it) even as stone structures had collapsed.[35] That famous bridge had been built from iron parts cast to form traditional masonry joints. In his autobiography Telford pays tribute to the ironmasters who created it – he later sought out the original papers to study the design – but adds that 'they had not disengaged their ideas from the usual masonry arch, the form of which in iron is not graceful'. Nor was the original iron bridge strong enough, he pointed out, to hold off the pressure from the abutments, which forced the arch up in the middle: a problem which has persisted in the two centuries since. It still stands, though much repaired. Famous for using iron, it missed same of iron's potential.

At Buildwas he was determined to improve upon it with an elegant and efficient design which drew on a Swiss practice developed for timber crossings. He threw one very flat arch across the river, carrying the roadway and taking the force of the abutments, and a second sharper one rising up from beneath to support it. Despite having a bigger span than the first iron bridge just upstream, it used under half as much metal and it lasted over a century, before it was replaced not because the iron failed but because the abutments on the unstable river bank shifted too far. Had it lasted only a few years more, into the age in which roads came back into use and its author's achievements were respected, Buildwas Bridge might be famous now as the definitive creation which marked Telford's coming of age as an engineer. Instead attention has focused on another, much larger project, intended to serve Shropshire, but which in the event hardly did, which occupied ever more of his energies in the second half of the 1790s. It was a project that drew together the strands of his time in the county: his place in the affections of powerful patrons; his skill in turning people and meetings to his advantage; his charm; his quick mind; his learning; his determination; his interest in industry; his enthusiasm for iron; and most of all the efforts and energies of his friends, Simpson, Davidson and Hazledine. That project was the Ellesmere Canal. Together they were to build a miracle.

The Stream in the Sky

An ethereal flash of darkness and bright water shot high above forest, Pontcysyllte Aqueduct is today as terrific in the sense Telford would have understood the word as it was in November 1805 when it opened. Contrasts of scale add to the fear it can create, the impression of height and length sharpened by the small things: the finger's width of iron protection lying between each passing craft and the airy abyss beneath and around; the shallow lip above the trough so that a swimmer in the channel doing an energetic butterfly stroke would send surges over the side; or the absence of railings on the water's edge which would make it possible, if one did not pay attention, to roll off the top of that snuggest of things, a narrowboat, mug of tea in hand, and drop 126 feet onto the rocks and rapids of the River Dee below.

There is something unsettling about a transit whose purpose is to carry water soaring over water, seeming to offend the natural order of upstream and downstream, gravity and flow. Like the Pont du Gard in France, Pontcysyllte creates awe out of all proportion to its actual utility, almost as though generating wonder were its primary purpose. The structure is a piece of mathematical precision over a hundred feet high and just over a thousand feet long, carrying a channel of water twice as wide as it is deep. It stands on eighteen tapered stone piers supporting nineteen iron spans, each made up of four iron ribs, each of which was in turn cast in three huge pieces. It is also pure in form, perhaps the most brilliant single emblem of the genius which powered Britain's industrial revolution. There is no ornamentation, beyond the broad iron panels which hide the ribs of the arches and make it appear solid, not a trace of Gothic or neo-classic, despite its creator Telford's architectural understanding. Its attraction lies in its lightness and sharp edges; it does not compete with the deep and surviving beauty of its rural setting, but frames it. One might even argue that it is one of the first great structures of the modernist school of architecture, a machine for movement, made out of the newest of materials.

Today, the structure (pronounced '*pont-cuss-ull-teh*') is busy with tourists drawn to what since 2009 has been a World Heritage Site. They come to see the spectacle and to celebrate Telford and to celebrate his colleagues, who are now, as they were not before, given just billing for their work here too, as in the historian Peter Wakelin's expert and lavish recent guide. On a summer's afternoon there seem to be more foreign voices than British ones; parents and children edging along the footway; scared dogs; a New Zealand family in a hired narrowboat struggling to reverse after meeting another coming the other way and finding that among the things Telford and his colleagues designed for the aqueduct, room to pass was not among them. On the TripAdvisor website visitors warn each other of the terrors of crossing, just as 150 years before the traveller George Borrow's guide in *Wild Wales* told him 'it gives me the *pendro*, Sir, to look down', and Borrow himself admitted to feeling dizzy.[1] These days there are large signs, insisted on by health and safety, warning parents that their children could fall between the widely spaced railings by the narrow footway, though there is no record of this ever having happened.

What visitors are not told so loudly is that Pontcysyllte Aqueduct has right from the start served as the showpiece it is today, long lacking a purpose as significant as its potency to impress. True, it carries 11 million gallons of fresh water each day from Wales to Cheshire – a recent use. But the channel which brings this flow from the serene Horseshoe Falls, west of Llangollen, was a late addition and never intended to be wide enough for much commerce. The main line of the Ellesmere Canal, as it was intended to be built, edges out from Shropshire, dares to cross Pontcysyllte, enters a tight basin at Trevor – and surrenders immediately outside the Telford Inn, a tidy black and white house with overhanging eves designed by its namesake (and built by one friend, John Simpson, for a second, Matthew Davidson, to use). Onwards from here there is a missing link through the hills where the canal should have continued – but never did.

The investors who funded the Pontcysyllte and the engineers and workmen who built it delighted in the glory of the structure they created. But they also created what today appears to be a bridge to nowhere. True, the Ellesmere Canal Company may not have seen things this way at the time. It was well aware that the vast majority of the Denbighshire coalfield lay north of the Dee. The company and its successors made healthy profits from a steady flow of traffic of coal and lime, which arrived on a light industrial railway connecting Trevor Basin on its upstream side. But the aqueduct was intended to be part of a grand through-route to Chester, not the distant end of the line. As such it resembles one of those bejewelled masterpieces created by apprentices to prove the skills they have learned rather than for pure utility. And this was certainly Telford's masterpiece – chosen in old age above any other of his creations as the background to his official portrait in the Institution of Civil Engineers.

People came to see it from the outset: poets, painters and scientists. To Charles Dupin, the French mathematician, who visited in 1816, 'the canal, enclosed in its iron envelope, hung, like something enchanted, on its high slender pillars, a supreme work of architecture, elegant and unadorned'.[2] To the American writer, Washington Irving, it was 'that stupendous work'. To Sir Walter Scott it was the 'stream in the sky', and to the Romantic poets it was emblematic of an age in which the natural world could be enhanced and embellished by man. Percy Bysshe Shelley came to call it 'the wonderful bridge of junctions' – an interpretation of its Welsh meaning, 'the bridge to join', referring to the older and lower road crossing which can still be seen below. Robert Southey described (in tottering verse no better or worse than the engineer himself might have managed) how:

> Telford who o'er the vale of Cambrian Dee
> Aloft in air at giddy height upborne
> Carried his Navigable road.[3]

William Wordsworth came too and in his poem 'Steamboats, Viaducts, and Railways' declared:

> In your harsh features, Nature doth embrace
> Her lawful offspring in Man's art; and Time,
> Pleased with your triumphs o'er his brother Space,
> Accepts from your bold hands the proffered crown
> Of hope, and smiles on you with cheer sublime.[4]

It remains sublime today.

———

On 10 September 1792 canal mania hit Shropshire. Ellesmere, a little town to the north of the county, was swamped. A committee formed of the great and the good, including the mayors of Liverpool, Chester and Shrewsbury, had gathered at the Royal

Oak inn to invite subscriptions to fund a planned canal linking the Rivers Mersey, Dee and Severn. The scene was frantic: a 'paroxysm of commercial ardour', recalled Rowland Hunt, the chairman.[5] 'The memorable tenth of September can never be forgotten by the writer, who had the honour to be left to defend the hill near the town. . . from the excessive intrusion of too ardent speculation: – the books were opened about noon, and ere sun set a million of money was confided to the care of the Committee.'

It was a day (reported the *Chester Courant*) in which normal order was abandoned. 'Shrewsbury, about 16 miles from Ellesmere, was so crowded on the nights before and after the meeting that many people found very great difficulty in getting accommodated: several gentlemen being obliged to take care of their own horses, cook their own victuals, and sleep two or three to a bed.'[6] By the end, 1,234 subscribers had offered £967,700 between them, which even on a simple adjustment for inflation would represent some £100 million today.[7] By any standards this was a gamble: and it was also more than the project was believed to need. Applications were scaled back to just £246,500.

What did these investors – the nobility, gentry and rising middling types from solid shire counties such as Shropshire, Cheshire and Leicestershire – get for their money? At this point, little more than shares in an inky line on a map. Or not even that, since while the Ellesmere Canal's commercial purpose – to link the rising port of Liverpool with industry in Denbighshire and Shropshire and the then-navigable River Severn south to Bristol – was agreed, its actual course was anything but. The investors had funded an idea. Somebody now needed to turn it into a canal. The committee had decided it needed 'an Engineer of approved Character and Experience' to lay out a route and turned at first not to Telford, who had no experience of canal building and was seen in Shropshire principally as an architect, but to the older and experienced William Jessop, 'the first engineer of the kingdom'.[8]

Modest, practical, often overlooked, and in the 1790s seriously overworked, Jessop quickly recommended a route which would cut south from Chester, to Wrexham, Ruabon, Chirk and eventually to Shrewsbury. From the very start he also planned to build an aqueduct of unprecedented height, although at this point not an iron one, at Pontcysyllte.

Elsewhere in Europe, early canals and rail lines were planned, sometimes ruthlessly, by the state. In Britain there was a mad, joyous, local scramble, backed by conflicting powers secured by Acts of Parliament, which led the route of the Ellesmere Canal to become tangled very quickly, until it became less a single canal than a spider's web of projected or completed waterways running throughout north Shropshire and Cheshire and linking to others which in the 1840s merged to become known as the Shropshire Union Canal. It was a scramble Telford was hungry to join. He will have seen the growing excitement in Shropshire caused by proposals for canals, and contrasted it with the mundane parts of his work as county surveyor, which 'give a great deal of unpleasant labour for very little profit in that they are like the calls of a Country Surgeon'.[9] He had spent the first part of 1793 on his architectural tour of Bath and London, trying to educate himself upwards and out. He returned to Shropshire to plot a second and more effective escape: a new career as a canal engineer.

The manner in which he muscled his way into the project is one of the controversies of his life. According to the story he told at the time (and always stuck to) his involvement in the autumn of 1793 came as a happy surprise. 'The committee of management. . . were pleased to propose my undertaking the conduct of this extensive and complicated work,' he said in his *Life*.[10] He made it sound as if he was the obvious candidate, which given his inexperience he was not; and as if he was in full charge, which given Jessop's continuing senior position as engineer he was not, either.

On 23 September he was appointed to the significant but still subordinate post of 'General Agent, Surveyor, Engineer, Architect and Overlooker of the Works and Clerk to this Committee and the Sub-Committees'. 'My literary project [most probably the revision and publication of his early poem on Eskdale] is at present at a stand and may not unlikely be retarded for some time to come as I was last Monday appointed the Sole Agent, Architect and engineer to the Canal, which is to join the Mersey, the Dee and the Severn,' he told Little within days of the news.[11] 'I had no idea of any such thing until an application was made to me from some of the leading gentlemen and I was appointed at their meeting, tho' many others had made much interest for the place.' He took pains to sound as if he had agreed to the appointment with reluctance. 'This is a great and laborious undertaking, but the line which it opens is vast and noble and coming in this honourable way I thought it too great an opportunity to be neglected.' So far, so decent.

But this story is untrue. Even his boast to Little – that he was the sole engineer – was misleading. Jessop remained the chief engineer for the rest of the decade and was shocked when he found out what had gone on. 'I have your letter and feel myself under some difficulty in answering it,' he told the canal committee on 2 October 1793; 'I am uninformed of the motives which have directed the resolves of the Committee.'[12]

'I am quite unacquainted with Mr Telford and with his Character,' Jessop continued coldly. 'I think as you do,' he added, 'that no one Man can properly undertake the actual direction of the whole so extensive a concern' – but Jessop wanted them to choose someone other than Telford, most probably William Turner, a junior colleague who had surveyed a line for the company even before Jessop's appointment and had already been asked to begin designing the crossing at Pontcysyllte.

The stage was set for a showdown between Jessop and Telford. But it was an anti-climax. The general meeting in Shrewsbury on 30 October, where Telford's appointment was to be confirmed, was a big affair, attended by many of the most powerful men in the county. But not by Jessop. 'If the Committee should consult me on this question I should tell them so,' Jessop wrote haughtily, 'but I cannot be at the next General Meeting' (a reluctance to travel and to turn up when required being one of his characteristics). In this letter to the committee he had also revealed just how dilatory he had been in giving the Ellesmere scheme his attention. 'From the little acquaintance I have had with you,' he began his letter, writing from distant Newark. The committee was well aware that their big name chief engineer was deep in canal mania elsewhere. Telford, by contrast, was on hand to help, respected locally, and eager to start. He was also, unlike his rival, prepared to fight his corner.

When the day came he played his hand brilliantly. If there had been any truth to his claim that he had not sought the job and took it reluctantly when his provisional appointment was agreed the month before, now he cast that aside and fought for the post like a lion. 'They had endeavoured to raise a party at the general meeting but we were too powerful for [the] opposition,' he told Little four days later, after he had won.[13] 'I had the declared support of the great John Wilkinson King of the Iron Masters. . . I travelled with him in his Carriage to the Meeting and found him much disposed to be friendly.' Arriving in a grand coach, sitting next to the richest and most famous industrialist in the county was a piece of theatre which flattened Jessop. And the prize was well worth having: the canal was, he boasted to Little, 'the greatest Work, I believe, that is now in hand in the Kingdom'.

He was to be paid £500 a year – at least £51,000 in current prices – out of which he had to fund his clerks.[14] It was the greatest position he had held so far but it was not – quite – the top job. His duties were 'to attend meetings, to make Reports to superintend the cutting forming and making of this Canal and taking up and seeing to the due observance of the Levels thereof to make the Drawings and to submit such Drawings to the Consideration and correction of Mr William Jessop'. Jessop was still to have the final say. But he would have to rub along with his whirlwind of a subordinate. And somehow, he did.

Telford and his allies in Shropshire were jubilant. 'Davidson is canal mad,' he told Little.[15] He dropped most of his duties of county surveyor as quickly as he could (retaining a role only in 'Public Buildings and Houses of importance') – though in doing so he took care not to offend Pulteney and his daughter, who was by now a wealthy, titled woman in her own right. 'I have the pleasure to say that they are not disposed to quit me,' he wrote – which in his exuberance overlooks the fact that if anything he was quitting them.

He also knew his enemies were ready, should he fail. 'Contentions, jealousies, and prejudices are stationed like gloomy sentinels from one extremity of the line to the other,' he told Little. 'But as I have heard my mother say that an honest man might look the Devil in the face without being afraid, we must just trudge on in the old way.' That old way – Telford's way – was ever upwards. He had eyed up the prize, planned his move and nabbed it. As he told his friend Archdeacon Plymley afterwards, 'the profession of a Civil Engineer has opened a much wider field, and to me a much more agreeable pursuit'.[16] 'Feeling in myself a stronger disposition for executing works of importance and magnitude than for details of house architecture, I did not hesitate to accept their offer and from that time directed my attention solely to Civil Engineering,' he added in his *Life*.[17]

As for Jessop, Telford saw in his disgruntled superior a man who could be charmed into becoming a tutor: a new source of advice in his perpetual quest for education. Over time he succeeded in this. In his *Life* he suggested with typical breeziness 'as most of the difficulties which occur in Canal making must be overcome by means of masonry and carpentry, my previous occupations had so far given me confidence'. Jessop was mentioned only as 'an experienced engineer, on whose advice I never failed to set a proper value' – though Telford does not specify how high that value really was.

This complex episode matters because it has tainted Telford's reputation. Charles Hadfield, the great postwar historian of Britain's canal system and a champion of Jessop, thought his treatment of the older engineer dishonourable. In the last of his many books, *Thomas Telford's Temptation*, published in 1993, he charged him with a lifelong effort to write Jessop out of the story, 'a gentle but persistent process of character erosion'.[18] It was a controversial attack. Telford, he argued, had at the opening of the aqueduct in 1805 ensured that he alone was named as engineer on the card circulated to guests, and revised early drafts of his *Life* to write Jessop out of the story. 'He could not bear to feel that others might be his equals, even his superiors,' Hadfield wrote, painting a picture of him as an 'old, lonely man near death who had yielded long ago to an impoverished egoism'.

Hadfield's evidence of the care Telford and his executors took to edit his autobiography is convincing. But by suggesting that Telford sought out and destroyed Jessop's missing private papers to cover up the truth of the Pontcysyllte's creation (for

Springtime in Eskdale, James McIntosh Patrick, 1934.

Thomas Telford never forgot his birthplace and returned to it often. The valley in which he grew up, running along Meggat Water, above Eskdale in the Borders, has hardly changed. This picture of his childhood cottage, The Crooks, was painted to mark the centenary of his death.

In 1780 Telford arrived in Edinburgh as a stonemason in search of experience. John Ainslie's map of the same year shows the ghostly outline of the partly built New Town. The young man was impressed by the grandeur but did not stay in the city long enough to shape it.

Like many Scots at the time, Telford saw London as a place in which to make his name. He used his connections to find work as a mason on Somerset House, Sir William Chambers's new Thames-side palace for tax officials, while he plotted his way to greater status and success.

Telford's fond, intimate and confessional correspondence with his childhood friend Andrew Little flowed back and forth to Eskdale for two decades.

In this letter, sent from Portsmouth Dockyard on 1 February 1786, he thanks 'My dear Andrew' for the latest gossip from home. 'Geo. Little's marriage is certainly a strange piece of business,' he writes, before going on to describe the way in which even Eskdale was changing. 'The rage for building has not subsided yet I find ... our poor, destitute descendants may behold those hillsides formed into streets ... and canals cut from sea to sea: Dreadful Thought!'

'Now then – you ask me what I do all Winter,' he continues on the following pages. 'Knowledge is my most ardent pursuit a thousand things occur that would pass unnoticed by good easy people who are contented with trudging on the beaten path but I am not contented unless I can reason on every particular. I am now very deep in Chemistry – the manner of making Mortar led me to enquire into the Nature of Lime so in pursuit of this, having look'd in some books on Chemistry I perceived that the field was boundless.'

'Let my mother know that I am well,' he ends. 'Write soon I am always Dear Andrew – Yours while Thos. Telford.'

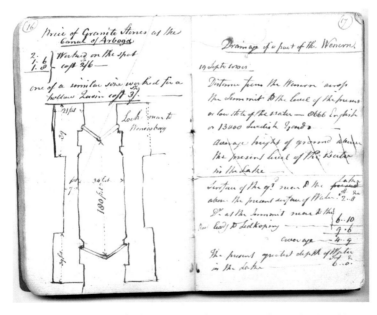

Always on the move, Telford was never without a pen and a notebook and his jottings capture the outpourings of a curious and flexible mind. In the example at the top he calculates that a horse could pull three loaded waggons running on a level trackway. At the bottom he estimates the cost of granite and a design for lock gates on a project in Sweden.

Thomas Telford, Sir Henry Raeburn, probably painted 1801 or 1802.

Telford had his portrait painted several times, most successfully in Edinburgh as he returned, fit and tanned, from an expedition through the Highlands. Already in his mid-forties, he still looked every bit the cocky young man on the make.

Coalbrookdale by Night, Philip James de Louterbourg, 1801.

The English county of Shropshire was where Telford's career took off. The fire and furnaces of Coalbrookdale, shown in this famous painting, encouraged him to start building in iron.

An architect before he became an engineer, Telford built several churches in Shropshire, including t[ ] one high on a crag by the ruins of Bridgnorth Castle. It still brings classical purity to the handsome market town.

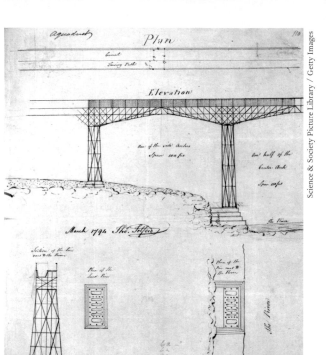

The 1790s saw a race to build in iron, with Telford only one of the engineers involved. This drawing for an iron aqueduct, dated 1794, carries his signature and is the earliest of its type but is disputed as proof that he led the way.

By 1800 his hopes for iron had grown to include this proposed arch at London Bridge, on a massive scale that would still astonish today. Political support fell away when the Napoleonic wars resumed but this print of the bridge that never was became popular nonetheless.

The aqueduct at Pontcysyllte is seen as Telford's masterpiece, though he was not its only creator. Water, iron, forest and landscape come together in poetic perfection. To Sir Walter Scott it was 'the stream in the sky'. Over 200 years later it still astounds.

which, other than the absence of Jessop's documents, there is no proof) is to overstate things. Telford was a self-promoter. He pushed his way onto the project. But he was not mad.

And more than that, there is the evidence of the entry on Jessop published in the *Edinburgh Encyclopaedia* in 1817, by a writer who reveals that he had known his subject for more than twenty years. He was a man 'totally free of all envy and jealousy', according to the anonymous author, 'free from all pomp and mysticism. . . persons of merit never failed in obtaining his friendship and encouragement'.[19] The piece goes on to refer to his relationship with Telford and assert that the younger engineer 'embraced every opportunity of acknowledging, in the warmest manner, the advantages and satisfaction which he derived from the able, upright and liberal conduct of his enlightened colleague and friend'. It blends generous tribute and faint praise, about Jessop's timidity in taking on work and his reluctance to promote himself in print: traits Telford, of course, never shared. Telford must be the author of this entry (as the historian Roland Paxton asserts in a comprehensive bibliography published as part of his entry on him in the *Biographical Dictionary of Civil Engineers*). The voice comes across as his and at least one reference in the piece was repeated in a later article in which he is named directly as the author. The existence and tone of this essay is, in itself, reason to doubt Hadfield's view that Telford wanted to write Jessop entirely out of history, although it is striking that it does not mention the Ellesmere Canal and suggests only that they 'acted jointly' on projects. A fairer account might suggest that he found the older man infuriating but that he respected him, too, even if he was reluctant to say so.

On today's canal both men are remembered. The large motorised barge that carries tourists over the aqueduct is called the *Thomas Telford*; the little craft, pulled by a pair of horses,

Taffy and Hercules, stamping out of their stables in Llangollen, that makes the shorter journey to the Horseshoe Falls, is the *William Jessop*.

They built the boring bit of the canal first: the easy nine-mile section from the Mersey, at a harbour to which the canal gave the name Ellesmere Port, to the Dee at Chester. Construction began in November 1793 and by 1796 it was open and making good money. But the interesting challenges lay further south on the hilly route along the Welsh border, with its dramatic planned crossings over the Dee at Pontcysyllte and the Ceiriog near Chirk. The story of these years is one of the diminishing of Jessop's role and the rise of Telford's authority, as the project proceeded from conception to completion, though the pace at which this happened is in part a matter for conjecture.

The minutes of the canal committee show that Jessop was nominally in command of the strategy, if not the detailed design, until after the turn of the century. But the extent to which Jessop ceded effective control or Telford grabbed it, firing ideas by letter to Jessop and the committee, is unknown. Credit is best shared; but if Jessop held the final authority, then it was Telford and his allies who were more often on the spot and got things done.

Certainly, Telford bowed to his superior at the start. In January 1794 a meeting attended by both men agreed to build a three-arched stone-masonry crossing at Pontcysyllte, the one that had already been designed by Turner, 'with such Alterations therin as Mr. Jessop shall communicate to Mr. Thomas Telford'.[20] By this point Jessop had retreated from his initial idea of a high crossing, proposing a less satisfactory lower one with locks climbing up on either side. At a second meeting the same month he delayed a decision to advertise for a contractor for the structure 'until such time as Mr Telford shall have consulted Mr Jessop upon the subject'. And by March Telford had drawn up his plans 'which have been settled and approved by Mr. Jessop' and set out to find a contractor to build it. None of this suggests Telford had the upper hand or yet claimed to do so.

Things were well placed. The two engineers were co-operating. A mason was employed to start quarrying stone at Pontcysyllte and on 21 February 1794 Matthew Davidson was appointed Inspector of Works at the site. Then in 1795 – the year of floods, in which Telford took on his second role as a canal engineer, building the small aqueduct in Shropshire at Longdon – these plans were ripped up and replaced. Stone gave way to iron in the engineers' thinking, and a low crossing to a high level one. It was a crucial choice. The metallic skeleton of the structure which stands at Pontcysyllte began to emerge. Was this bold switch Telford's doing or Jessop's? Historians have debated the matter, with Hadfield taking Jessop's side. There is evidence to support both arguments.

Fighting Telford's corner is a drawing, with his signature on it, dated March 1794 that appears to show an iron aqueduct supported by trestles. L. T. C. Rolt, in his biography of Telford, suggests that this was intended for Pontcysyllte and takes this to show that using iron here was Telford's idea.[21] So, more definitively, does the engineering historian Roland Paxton in a paper published to mark Telford's 250th anniversary. He suggests Telford discussed the drawing with both William Reynolds, the ironmaster, and Jessop, and that it shows a high-level iron crossing of the Dee at Pontcysyllte.[22] But it is jumping the gun to suggest that he had settled on iron as a definite answer by 1794. The dimensions of Telford's drawing for an iron bridge – if that is what it is – are all wrong for the Welsh site. The drawing may be a copy of a design from somewhere else, by someone else.

There is no record in the canal company minutes of a proposal to use iron at this point and anyway, even if it was being considered by Telford, no one had told the mason, who was still busy cutting blocks for the intended lower stone structure on the site.

Instead supporters of Jessop have stepped forward to claim the honours. He was already an authority on the structural use of iron and was a founding partner in the Butterley Ironworks in Derbyshire, which built an iron aqueduct on the Derby Canal that opened a month before Telford's larger one at Longdon.[23] On 14 July 1795 Jessop wrote to the canal committee with a radical proposal. 'I must now recommend to the Committee to make this saving by adopting an Iron Aqueduct at the full height originally intended which on correcting the Levels appears to be 125 feet above the surface of the water of the River Dee. . . the arches or rather openings of the Aqueduct may be seven of 50 feet each the remainder may be raised by an embankment.'[24] A fortnight later he wrote to Telford: 'looking forward to the time when we shall be laying the Iron Trough on the Piers'.[25] That suggests the pair were working in harness; if Jessop was advocating iron, then so was Telford in what was a year of engineering revolution, building his iron bridge at Buildwas as well as the crossing at Longdon. He had already befriended ironmasters and he surely influenced the decision to switch to iron at Pontcysyllte.

Jessop and Telford deserve joint honours. But in his *Life* Telford did not award them. 'I had about that time carried the Shrewsbury canal by a cast-iron trough. . . and finding this practicable, it occurred to me, as there was hard sandstone adjacent to Pont-y-cysylte, that no very serious difficulty could occur in building a number of square pillars, of sufficient dimensions to support a cast-iron trough, with ribs under it for the canal,' he wrote.[26] Jessop's role was left out entirely. This was at least a little unjust.

Deciding to build the aqueduct from iron, of course, was one thing. Building it was quite another. The attraction in principle was clear. Iron would allow a lighter structure than the massive stone crossings, which had previously been built to carry canals across rivers and which needed thick, unstable layers of heavy 'puddled' clay to keep the water in. If it could be made to work, the savings would be huge. And Pontcysyllte was not the only place iron was to be tried. Both engineers proposed a similar structure at the other major aqueduct on the route, a few miles south.

This now stands at Chirk, a scenic spot where the proposed route had met much opposition from the Myddelton family at Chirk Castle on the grounds of what Telford called at the time 'imaginary injuries to imaginary pleasure grounds',[27] although in his *Life* he admitted it was 'a landscape seldom surpassed'.[28] As with Pontcysyllte, the archive suggests that it was Jessop who led the way. 'If instead of an embankment of Earth, which would shut up the view of the valley, it be crossed by an Iron Aqueduct I should hope the objection might be removed,' he told the committee, 'instead of an obstruction it would be a romantic feature in the view and avoid damage to much of the Meadows and Plantations.'[29] Telford's letters to Davidson at the time show he was taking advice from the senior engineer. 'Mr Jessop is in London but we have hitherto been so busy that I have not had time to consult him as to the Chirk Aqueduct,' he wrote.[30]

Of the two great structures this proved easiest and was built first. It does not look like an iron aqueduct today: the canal is carried on elegant stone arches into a narrow tunnel whose towpath (the first to run underground) still makes for an uneasy walk without a torch. No sooner had he endorsed it than Telford retreated from the idea of an all-iron trough, perhaps following complaints about its proposed appearance by Sir Richard

Myddelton at Chirk Castle, a force in the canal company. He may have been influenced, too, by signs of distortion and potential collapse in his iron structure at Longdon. In the end he used metal only for the base (it was added to the sides, too, much later, where it can still be seen).

In February 1796, Telford put his new plan to Jessop and by March he was able to send Davidson detailed instructions for its construction. The innovative hybrid structure opened in late 1801 and was much praised (including by Telford) for its fine proportions. He used a picture of Chirk alongside Pontcysyllte on the personal seal he applied to his letters. It was the tallest navigable aqueduct ever to have been built.

Meanwhile work at the even higher crossing at Pontcysyllte was advancing. This was by now, more obviously than ever, Telford's project being built by Telford's team. In November his ally John Simpson was appointed alongside (and soon in place of) the previous mason. He joined Matthew Davidson, who was overseeing the works. It was Davidson who drew plans for the aqueduct, sending them on to Telford who presented them to the committee of the canal company. This keen team drove themselves forward with a manic good cheer. '"Live a Thousand Years!!!" ' Telford wrote at the end of one letter to Davidson.[31]

The aqueduct's foundation stone was laid on 25 July 1795 and the following year Telford had an iron plaque for it cast, praising the 'Nobility and Gentry' who had 'united their efforts with the great commercial interests of this Country'. It was a patriotic statement, composed against a backdrop of war, presenting Great Britain as a nation where 'the equity of the laws, and the security of property, promoted the general welfare of the Nation' – hardly needing to add the unspoken coda that things had been arranged rather less well in Jacobin France. Telford, newly recovered from an injury which he called 'the reign of my broken shin', after a blow on his leg reopened an old injury, had worked hard to get the plaque completed. 'Yesterday afternoon,' he wrote, 'John Simpson pull'd up his Breeches and mounted me on the outside of the Mare and off we set to examine. . . the inscription plate.'[32]

It was a gang of mates, from which only William Hazledine was missing; and he arrived soon after, securing the contract to provide iron for Chirk and later Pontcysyllte.

By October 1796 Telford had drawn up a 'Geometrical Drawing of Pontcysyllte' and his letters issue a stream of instructions about putting this into effect.[33] 'Besides the Canal between Liverpool and Chester we are now working upon 20 miles in Shropshire,' he reported to Little in August 1797. This included 'Davidson's famous Aqueduct and which I can assure is already recknd. amongst the Wonders of Wales, for your old acquaintance now thinks nothing of having three Carriages at his Door at a time.'

In the summer of 1797 he was confident enough to spend a week away on holiday 'with one companion'. It was 'a hasty excursion through the most Mountainous parts of North Wales – we passed the mountains of Cader Idris; Snowdon, and Pennamman [Plinlimmon]'.[34] The weather was kind: 'Snowdon we saw for an hour without one cloud, which was a singular piece of good fortune as he may often be visited 20 times in vain.' This was Telford's first visit to these mountains, which he came to know well in the following decades when he built the road to Holyhead through them. 'Upon the whole we were highly delighted, and every part of our Journey fully gratified our expectations. There were parts which very much resembled the lofty green hills and woody vales of Eskdale.' He seems in exceptionally good spirits in this letter, visiting Llanrwst Bridge, possibly by Inigo Jones, in a valley he decided was home to 'the most beautiful race of people I ever beheld. And I am much astonished that this has never struck the Welsh tourists.' He returned via Davidson's house at Pontcysyllte, where the good news was that the pillars were going up rapidly.

The bad news, however, was that there was by now much confusion about what sort of structure they would support. Indeed, the whole Ellesmere Canal scheme was descending into a mass of variations and new Acts of Parliament – which had led Telford to make his first appearance before a Lords committee the year before. The canal was only as useful as the trade it could carry and there were doubts about what that would be. As Telford mused to Little in March 1798, commerce was 'like the operation of a great Machine, to draw money out of the hands of a numerous proprietary to make an expensive canal, and then to make it return into their pockets by a business upon that canal'.[35] War was driving up costs everywhere. Pressure was growing to abandon the link onwards from Pontcysyllte to Chester and send traffic via the rival Chester Canal instead. The trouble was, without the connection north, trade across the aqueduct would be limited.

In late 1797, work had stopped both to allow resources to be concentrated on Chirk and to reconsider the use of iron after problems with it at Longdon. 'The canal affairs have required a good deal of exertion tho' we are on the whole doing well,' Telford told Little the following spring. But the truth was that the whole Pontcysyllte project was being questioned. On 24 January 1800 Jessop wrote to the canal committee with a dramatic proposal: that a horse-drawn railway not only be built to carry coal from collieries near Ruabon, but might be 'continued over the columns intended for the Aqueduct'. This idea embraced the coming technology of rail transport, though the horse-powered wagons would have carried far less than canal boats. The alternative was 'the Aqueduct should be completed to communicate with the Railway on the north side of the piers' – which is in fact what happened, once the water extension north had been abandoned definitively in 1800. Pontcysyllte almost ended up as a spectacular early tramway. Had it done so, Davidson's pillars might well have evolved further in subsequent decades to carry the Great Western mainline railway which now runs nearby. And there would be no great stream in the sky.

Perhaps in that case Telford would have become famous as a railway pioneer. He was certainly starting to ask whether the enthusiasm for canals had overreached itself. In 1803 Joseph Plymley published his *General View of the Agriculture of Shropshire*, to which his friend, 'an engineer and architect, and to whose general merit I am happy in this opportunity of bearing testimony', contributed a chapter on canals.[36] The bulk of it was written in 1797 and offers a contemporary account of building the Ellesmere network ('rather than one it is in fact a system of canals') and the planned aqueduct. In this section Telford is full of enthusiasm for his project. 'The advantages which the county is likely to derive from this canal, are various and extensive; but the most important are the improvements in agriculture, which are expected to follow from the cheap and expeditious conveyance of coal, lime, and slates, and the easy communication which will be opened for markets of consumption for the produce of the land.'

However, a postscript written three years later shows that his thoughts were moving on. 'Since 1797, when the above account of the inland navigation of the county of Salop was made out, another mode of conveyance has been frequently adopted in this county to a considerable extent; I mean that of forming roads with iron rails along them,' he wrote.[37] 'This useful contrivance. . . may be constructed in a much more expeditious manner than navigable canals.' Railways, he points out, are useful 'in countries where surfaces are rugged, or where it is difficult to obtain water for lockage, where the weight of the articles of produce is great in comparison with their bulk, and when they are mostly to be conveyed from a higher to a lower level'.

What these rail lines lacked at this point were reliable steam engines to pull the wagons, though they too were coming, and being developed in Shropshire, not least by John Hazledine, William's brother.

So even as Pontcysyllte Aqueduct was reaching a climax its creators were questioning its design and aware that the technology and the commercial logic were changing. Canals remained the best way to carry large loads and they continued to be built and to make money for three more decades. But though he threw himself heartily into building them, Telford was never an obsessive advocate of canals.

His life was changing as well. Jessop remained the canal's principal engineer until its completion but after 1801 seems to have been less involved. Telford too was giving it less of his time: he was becoming a significant national player as the impact of war forced the government to take a new and strategic interest in infrastructure. Confident in the abilities of Hazledine, Davidson and Simpson, he spent ever more time away from North Wales on what he called his 'Long Circuits' – tours which took him as a consultant, advocate and problem-solver to many parts of the country.

He seemed to have enjoyed this existence. Although Shrewsbury remained the closest thing he had to a permanent address, he was more often away from it than present. In 1798, apologising to Little for not answering his last two letters, he admitted 'the fact is, that unless a person answers letters immediately on receiving them, neglects generally creep in – and especially with such a wandering being as myself, who is scarcely permitted to remain in one place, unless detained by business which occupies the time completely'.[38]

His circuits took him to Scotland, where soon he was to find his next major employment; and he was busy with ideas of every sort: joining a consortium to promote a patented boring machine (and asking Jessop to join because 'he is going everywhere and is everywhere known and respected').[39] Two years later he was busy finishing 'a paper on Mills for the Board of Agriculture; this said paper has grown and grown, yea until it is waxed into a large quarto volume illustrated by upwards of 30 plates'.[40] The manuscript, in Telford's hand, with tidy hand-coloured illustrations of the workings of windmills and watermills drawn by an assistant, survives in the Institution of Civil Engineers, full of proposals to use the waste water of canals and dam every river in England, managing the flow on the example 'now practised under the regulations of the greatest and wisest nation in the East. . . China'.[41] His dealings with Jessop were clearly close enough for him to have sent the text to 'my friend Mr Jessop, whose talents and character are too well known, to need any encomium here' – and he includes Jessop's teacherly reply: 'I have had much satisfaction in perusing your essay on Mills, I think it generally correct, and the arrangement well conceived.'[42] The tone is of a headmaster encouraging a promising pupil: if the pair had fallen out then they had made up again. In his sweeping conclusion, Telford praises the Board of Agriculture for its plan of 'collecting, arranging and circulating. . . a great system of useful knowledge, which has hitherto remained in obscurity and disorder'. Whether he was really interested in mills or simply wanted to flatter powerful people connected with the board – which was worried by the prospect of riots caused by a shortage of flour for bread – is uncertain. But there is no doubting his enthusiasm for 'a liberal system of instruction' in Britain, 'in which the whole of its inhabitants, as the children of one family, will participate'.

At the same time, Telford was building links with other engineers. 'I have just got intimately acquainted with Mr Watt & his son,' he wrote in February 1799. 'He is the steam engine Man from Glasgow – he is great and good, so is his son.' This letter is notable for something else, too. It is the first to be addressed from what became famous as Telford's London lodgings, the

Salopian Coffee House, Charing Cross. This stood at the top of Whitehall, near Admiralty House, a convenient address which was evolving from its eighteenth-century guise as a meeting place where men came to play chess, to a more serious nineteenth-century hotel, where they came to stay while doing business. The location was a sign of Telford's developing contacts and involvement at Westminster, where he was soon at ease, and increasingly respected, as a witness before legislative inquiries and committees. From 1800, the Salopian, where a set of rooms were kept ready for his use and where he could entertain, became a sort of home to him and he remained there for more than two decades, his celebrity a fixture of the business.

Back in North Wales, rock was being cut, molten iron poured and mud excavated by an army of men. It must have stunned the locals in what had until then been a bucolic and isolated corner of borderland; a bit like the arrival of heavy machinery to build a motorway through a farm. As Rowland Hunt said in his oration at the opening, 'wherever the Spirit of Commerce has touched the mountains on the whole borders of Wales, they begin to smoke'.[43] And good progress was at last being made on the aqueduct. In 1800 tenders were let for the completion of the pillars at Pontcysyllte. Jessop recommended that their upper sections be hollow, not only to save weight but to allow workmen to build safely from the inside. The growing height of the structure brought home a challenge that he had already identified back in 1795: 'I see the men giddy and terrified in laying stones with such an immense depth underneath them.'[44] And once the pillars were up there was the unprecedented problem of hoisting the iron supports on top.

Simpson, in charge of the stone, with Davidson in charge of the site, and Hazledine casting the iron at his new foundry established at Plas Kynaston on the north side of the valley, and erecting the pieces in situ, made a brilliant team. To them as much as Telford and Jessop should go the credit for solving these problems. The pillars were built up in turn, from south to north supported by temporary piers; and when a new level was reached they were joined by gangways which carried a construction tramroad used to move the blocks of stone. It was all done with simple technology, no safety ropes and no experience of building anything so high. Unusual care was paid to protecting the workforce: around 500 men when construction was at its peak in 1804, building the aqueduct and the canal running beyond it to the Horseshoe Falls, which supplied water upstream from Llangollen. The workers would have been a mix of specialist craftsmen and locals eager for a wage and perhaps a chance to learn useful skills. 'One man only fell during the whole of the operations in building the piers, and affixing the ironwork upon their summit,' recorded Telford, 'and this took place from carelessness on his part.'[45] On other, less well-run projects, deaths were commonplace and unremarked upon.

By 1801 five of the pillars had passed a hundred feet and the committee had decided it would press on with using them for an aqueduct, not a railway. Hazledine signed the contract for the ironwork in 1802 and, with Telford, built wooden models to test the design for the trough. This included a footway cantilevered out over part of the water channel so that there was room for the current to flow past even as boats crossed, avoiding surges which might have sent a torrent over the edge. Casting the thousands of iron parts needed for the structure as well as the bolts to hold it together was in itself a vast and difficult job, with no room for error: each piece had to fit precisely onto its neighbour. The parts were numbered and brought together in order from Plas Kynaston, just beyond what is now the head of the canal at Trevor Basin. First the ribs of the iron arches were lifted into place, then the trough was bolted over them and finally the towpath (originally wooden, not the present iron) and handrail were fitted. Construction became so routine and quick (and affordable, Telford stressed, costing only £47,018 in total, including the embankments) that by 1805 it was done.

Pontcysyllte Aqueduct was opened officially on Tuesday 26 November in a ceremony which Telford helped devise and whose spectacle and hyperbole were never again equalled in his career. The tone owed something to the cocky mood of a nation which was back at war and winning an empire after the breakdown of the short-lived Peace of Amiens and whose greatest admiral had beaten the French weeks before at Trafalgar.

Like that sea victory, Pontcysyllte was something to shout about. The trough was filled with water early in the day, and by midday clouds had broken away to winter sunshine, 'adding, by its lustre, to the beautiful sight of various carriages, horsemen, and persons, descending, by every road, path, and approach, leading towards that great work', reported *The Gentleman's Magazine*.[46] Thousands of people filled the valley, to hear the explosion of sixteen rounds of artillery (fired from guns captured at the Battle of Seringapatam in 1799) boom back and forth between the hills. The first barge carried the leading investors on the canal committee. Telford crossed in the second, carrying two Union Jacks, followed by a third with a band playing 'God Save the King, and other loyal airs'. Behind that came representatives of the trades that had built the structure, waving a huge banner with the fulsome slogan: 'Success to the Iron Trade of Great Britain of which the Pontcysyllte Aqueduct is a specimen', written across it.

No one was to be in any doubt about what made the structure special, or about its commercial purpose. A horse drew up on the iron railway leading to the basin at the north end and coal was shifted from five wagons into the symbolic, first-loaded barge to cross.

The workmen were fed on roast mutton, beef and ale at Hazledine's foundry; the gentry attended an equally convivial meal at the inn at Ellesmere, the place where, thirteen years before, subscribers had crowded to invest in the canal. At each seat, they found a card listing the dimensions of the structure and naming its creators. 'Thomas Telford was the Engineer', it read – followed by the names of Davidson, Simpson, Hazledine and William Davies, who built the embankment.

There was no mention of Jessop. He did not attend the dinner or the opening day. Was this exclusion intentional? Jessop and Telford were by this time working together on other projects. But it is hard not to speculate that the older and still more celebrated engineer might, had he known about it, have been a little hurt to see so many tributes to a man to whom he had taught the craft of canal building. In his oration at the opening, Hunt, who had seen the canal through from start to finish as chairman, took care to praise both men in a speech whose sweep, balance and generosity it would be hard to better. He praised: 'our General Agent, Mr Telford; who, with the advice and judgment of our eminent and much respected Engineer Mr Jessop, invented and, with unabating diligence, carried the whole into execution'. That seems fair, alongside tributes to 'Mr Hazledine, the spirited founder of the Duct itself'; as well as 'Mr Simpson, the accurate mason, who erected the pillars; the well-computed labours of Mr Davies, who constructed the mound or tunnels adjacent; and the careful and enlightened inspection of Mr Davison [sic], who overlooked the whole'.

'Thus will Pontcysyllte stand for ages,' he concluded, 'a Memorial of the active loyalty of those who were studiously industrious in its creation, and will record the eminence of Patriotism, as well as Progress of the Arts, both of which have been so successfully patronised by our beloved Sovereign.'

It stands there today and does just that.

The Appearance of Plenty

If he was lonely, he hid it well. But he must have missed them. By the time Pontcysyllte Aqueduct opened, the two men whom Telford might have hoped would be there to celebrate with him – the Eskdale men who shaped and made him – were dead.

There was no fuss, no show of mourning. In his surviving papers there is no mention of the death of Andrew Little on 4 May 1803, aged forty-eight: parish records show a funeral in Langholm three days later and he was buried in Westerkirk, where his grave, remembering him as a 'schoolmaster', still stands.[1] Nor does Telford seem to have made much of the death of Sir William Pulteney in the same month two years later, aged eighty-five. Like his later biographers, he allowed the pair to slip from his story. It is as if their friendship and support was substituted for others with the calculated precision of an engineer.

In truth it cannot have been anything like that easy. Little – the boyhood friend who understood as no one else did the private energy that lay inside the full-grown man and who, because of this, knew what he meant when he said his life was driven by the 'old Way of stirring the Lake' – cannot ever have left Telford's mind. Nor could Pulteney – the patron who plucked out a young stonemason and gave him a platform from which he could launch himself as an engineer, the man of whom Telford once wrote 'the world says I am young Pulteney, I wish I was' – have disappeared fully from his thoughts.

Of the pair, it was Little whose absence he will have felt the most. With Pulteney the puppyish adoration of his early years in Shrewsbury had given way to a wary recognition of his patron's peculiarities. When Pulteney died, newspapers reported with delight the gossip that his house had been searched, unsuccessfully, for a will; that his frugality extended to refusing fires for heating; and that at the onset of illness a month before he had predicted his death almost to the hour. When he was buried in Westminster Abbey (the first man to be honoured from their part of Eskdale, with Telford the second) there was little ceremony and no monument was put up.

There is no record of ceremony for Little either and if Telford went to the funeral he did not mention it. He had known his friend was unwell for years. The first sign that something was wrong came in December 1798. Little had written complaining about rheumatism and Telford fretted in reply. 'You perhaps eat too little animal food,' he advised. 'I believe a glass or two of pure water taken during the night while in Bed is a good thing – it can easily be placed by the bedside and taken when you wake.' A year later Little was in agony and Telford from faraway Shropshire was desperate to help. 'If you will state your case, I will get the opinion of our best Physicians in London and the country, this would be satisfactory – and might be the means of procuring your relief from pain, which is a dreadful thing. It interrupts the operations of the mind.'[2] If the suggestion was taken up, it failed; so did his offer of help from Dr Robert Darwin from Shrewsbury (father of Charles, born a decade later): 'I believe he would afford his best advice.'

By late 1800, Telford's words imply he knew that Little was dying. 'I in fact consider you as so much a part of myself that I can never help thinking as if you were rising or sinking along with myself,' he told him.[3] He visited a few weeks later, perhaps for the last time, though he intended to return. In 1801, travelling south from Scotland, he explained that 'it was with reluctance that I passed you on my return to England'.[4] In 1802 he promised 'if I can stop in Eskdale I will write you, but I suspect that pleasure may be deferred'.[5] His last surviving letter to Little was sent on 18 February 1803, in response to one that Telford explained, poignantly, he had lost.[6] Weeks later, his best friend was dead.

And perhaps more than his best friend. Anyone who reads Telford's wonderful letters to Andrew Little may eventually come to ask themselves about the nature of their friendship, just as anyone who considers the emotional tenor of his life may wonder what lies behind it: unmarried; restless; drawn to passionate friendships; close to many young men; poetic; radical; nonconformist. There is no evidence of a physical relationship with anyone, ever – man or woman – or indeed much interest in such things. It is striking, for instance, that Telford's copious poetry often mentions friendship but rarely love and that even as a footloose young man he seemed more moved by the landscape in which he lived than by the people who shared it with him.

It is reasonable at least to consider whether this part of his character may have shaped his professional behaviour; whether it could have been one of the things which fired him to work so hard, the drive, the searching for respect. Do they suggest a man who felt himself in some way to be on the outside of society? It is notable that an unusual proportion of his surviving letters seem to have been written at Christmas or New Year – when Telford was free from work but had no domestic celebrations to occupy him. Of Telford's earlier biographers, Charles Hadfield suggests that his subject's ambition was 'helped by being a bachelor, very much so, for he never seems to have looked seriously at a woman'. Later, he adds that 'unmarried. . . half a normal man's life is closed to him'.

Perhaps Telford's deep affection for Andrew Little was just a very good friendship, rooted in childhood and literature. There are moments, though, in the letters when something close to love might seem a better word for what they appeared to share – even if only the love of two intimates, who knew everything about each of their characters and missed each other's company deeply.

'I wish I were near you I would tickle your sides,' Telford wrote to Little in 1787.[7] Perhaps this is just the teasing of boyhood friends. But it suggests a certain proximity. And what of that mournful line sent to his ailing friend thirteen years later: 'I in fact consider you as so much a part of myself – that I can never help feeling as if you were rising or sinking along with myself'? Again, perhaps just friendship of the very closest kind. But the words he wrote in 1789, at the end of the long passage praising Jonathan Swift, whose *Life* 'by Sheridan' he was reading, are intriguing. 'Great Swift! I venerate thy memory. I pray to be possessed of thy virtues – to thy talents I dare not aspire!' Telford declared, in full poetic passion.[8] And then he added something else to conclude the letter; something telling and short: 'I am so far happy as to be certain of a Stella, in my Andrew Little – to whom I hope always to remain his Tho[s] Telford.'

It is a remarkable sign-off. Stella was the pen name of Esther Johnson, the woman Jonathan Swift was widely and famously believed to have married in secret. Telford will certainly have known the story. Sheridan, after all, tells it in the book Telford had just read and describes in great detail Swift's love for her, hidden at his request to the outside world. So for Telford to describe Andrew Little as his 'Stella' is a particularly striking way in which to end a letter, even to a close friend and even in an age when correspondence was at times more personal than it is now. It is possible to interpret this as a suggestion – or a wish – that Little could be Telford's secret partner, though whether that means Telford was describing a covert proximity or whether, equally possibly, he was referring to the fact that they were dear friends whose exchanges were carried by post (just as sixty-five of Swift's surviving letters to Stella were) is unknowable. One can only speculate. Beyond these glimpses in his letters to Little this side of Telford's life will always remain closed.

———

Despite Little's failing health, the depression which had at times swept over Telford and many others in the grim 1790s was lessening. Britain was making things and building things and conquering things and anything seemed possible. 'I do not believe that any spot of the Earth, of equal dimensions, ever exceeded what Great Britain is now, I mean with regard to wealth and useful Arts,' he declared as the century came to an end.[9] It was a bold, patriotic spirit which was shaping national politics under William Pitt, but it shaped Telford's world too: the drive to build in iron, to dig canals, fix roads, create power from steam was a conscious expression of national glory, every bit as mighty as the Royal Navy.

The European war remained a backdrop; but now there was hope. The threat of mutiny, famine and invasion had passed. 'Bonaparte. . . raises his impudent voice again, and boasts of the deeds of Arms,' Telford noted in 1799, but by now his tone was mocking and French victory no longer inevitable. 'As to the Drama – the Germans seem in the very zenith. . . the sighs of The Great Nation itself, bears witness of this truth.' In that letter, characteristically, Telford praised Goethe, as well as German troops.

In his correspondence with Little, despite the undercurrent of worries about illness, the chatter of settled friends continued to the end. Both men continued to read hungrily. 'Having no leisure for reading I have found the *Encyclo: Britannica* a perfect treasure, it contains everything and is always ready,' he wrote, but in reality their reading was far from constrained.[10] 'You have an inexhaustible fund in the lives of the Poets. . . since the time of Chaucer,' he told Little. 'What a Troop when they all get together but they will not be long together they will all fly off like Skye Rockets.' He kept writing verse, too, including a poem written at Stratford-upon-Avon, 'where I was detained all night as I went up to London'.[11]

'Let me have your opinion. . . as to the merits and demerits of my Stratford verses,' asked Telford, '– they Employed me very pleasantly in composing, as they have also this evening in copying one for you.' Little seems to have been unimpressed (rightly so, the surviving draft suggests). 'I think your observations respecting the verses are correct, it is seldom I have time to rectify,' Telford accepted politely in reply.[12]

There was time to study politics and economics, too. 'I have just read twice over *The State of Europe* by Gentz a Professor, translated by Nanicus – Except perhaps Adam Smith's *The Wealth of Nations* and one or two more I have never received so much delight and information from any book.'

By April 1802 the European war was over – or at least it seemed to be, for a year anyway: 'we have got the peace, the Income Tax repealed, a good loan and plenty of food which put all the good subjects of this Realm into as good spirits as if there had never been War or Scarcity or Taxes', Telford wrote.[13] With peace came expansion. 'Everything wears the appearance of plenty and my canals are full of water and full of business.'[14] Soon, the first general election of the United Kingdom (following the abolition of the Irish parliament and inclusion of its members at Westminster) was under way. 'This warm weather will sweat the candidates,' Telford noted, and for the first time he took a detailed interest in the contest, not only because he happened to be in London when Parliament was dissolved but because his career was coming to depend on the actions of parliamentary committees and the granting of government commissions and funds.

It was a significant advance for him. Even during his time on the Ellesmere Canal, he had climbed from being the servant of a single scheme to an engineer of growing national professional regard, if not yet fame, consulted and employed in London to promote projects across the country and coming into proximity with politicians, even ministers, and their leading officials. It was a standing of which he was proud. 'It is certainly gratifying,' he wrote in September 1800, 'to be admitted as a leading member in the profession in which a Man has been bred, and more especially in objects of magnitude and difficulty – and where the improvement of a great Kingdom is materially concerned.'[15] It was also a position he intended to use and to enhance; no longer a supplicant, he dealt directly with men at the head of the nation's political, intellectual and economic life. He had arrived.

He was busy before, but now he became busier still. The dominant theme of the last years of his correspondence with Little is constant movement and a lack of time. 'I have been kept in continual motion from the Country to the Town, and from the Town to the Country,' he wrote in 1802, a passage typical of many others for the next thirty years. From now and for the rest of his life there are no neat dividers between one project and another; no series of promotions or formal contracts to which he dedicated all his time; no distinct passages. His story tumbles forward in a great array of overlapping schemes, some of which came to nothing, many of which flourished, and all of which took years to complete.

In each project he played multiple roles – the common theme being his flexibility and drive. He was a great one for gathering his forces and making things happen. He could still turn his mind to small problems – to the challenge of designing a particular form of bridge in a particular place, or to the nature of a specific type of rock, or to the way drains should be built to ensure a road being built across a slope was not washed away – but this was only part of his talent. The other – the skill that lifted him more and more above his talented contemporaries – was to paint on a broad canvas the great scope of a project and its national purpose. There had been good bridge builders before, of course, but their interest often lay more in the practical construction of the structure than the lives and businesses that could be joined together by it. By contrast Telford was able to judge the sort of country that his schemes were helping create; to weigh up the factors that led one town to succeed and another to fail; and to do so without roots in one place, which might have distorted his judgement. He had that gift that politicians still seek today: of vision, the ability to make a series of actions add up to a greater whole. This command became his principal attraction to others – and in turn, attracted him to his work. Money was never the lure. He was well paid but not overpaid and never became rich.

From now on Telford's role became political; he was a man who could build things to fulfil national administrative goals. In Westminster he served as an expert, a skilled, trusted adviser whose presence in committee hearings gave good advice to MPs and investors and who did not let people down. They must have enjoyed his company too: bluff, amusing, ready with a supply of stories; a strong, confidence-giving personality, who had been to all the places that were being talked about and who could give substance to ideas. With him involved, you would think something might actually be done. And with this came allies and associations of a new kind, not with rich county patrons like Pulteney but with leading members of a new national breed of bureaucrat-politician, brought on by the growth of the state during the war. They included in particular Nicholas Vansittart, an ally of William Pitt's successor as prime minister, Henry Addington, undistinguished in character but powerful in the Treasury and from 1812 to the end of 1822 the Chancellor of the Exchequer. He remained a well-placed ally for years and the key to the next stage of Telford's career.

Even as Pontcysyllte Aqueduct was being finished, Telford spent more and more time in London, staying at the Salopian Coffee House. 'Telford is a most useful cicerone in London,' wrote Thomas Campbell, a young Scottish poet who stayed with him at the Salopian in 1802 and remained in contact long after.[16] 'He is so universally acquainted, and so popular in his manners, that he can introduce one to all kinds of novelty, and all descriptions of interesting society.' Campbell, introduced by their mutual friend Archibald Alison, was so taken with Telford that he named his first son, born in 1804, Thomas Telford. The engineer, in return, predicted great and unfulfilled success for his friend. He told Alison he would surpass 'your Pindars, your Drydens and your Grays'.[17] 'I am quite delighted with him,' Telford wrote to Little at the same time, 'he is the very Spirit of Poetry. On Monday I introduced him to the King's Librarian, and I augur some good to him there.'[18] That little boast about the royal librarian does more than promote Campbell. It shows, too, that the Eskdale stonemason was moving in elevated circles. 'I have got into mighty favour with the Royal folks. I have received notes written by order of the King, The Prince of Wales, Duke of York and Duke of Kent,' he added.

For all of his politicking and socialising with the men who grew rich in Pitt's Britain, however, he did not become a pompous grandee. A different character might, in his mid-forties, with the Ellesmere Canal nearly done, have settled in London and secured a well-paid government post. He might have bought a house; even a small country estate. He might also, of course, have married well and sought a title. Telford, though, never did any of this and it is this difference that makes him unusual. His activities in London came on top of rather than in place of his old occupation as a practical engineer, and his old interests in books, history, landscape and commerce, and of course his old habit of relentless journeying. He never stayed in London long. He still got wet and dirty on construction sites when he left. Each year, from late in the eighteenth century until his health began to fail three decades later, he travelled and surveyed hundreds of miles, mapping out roads and canals, designing bridges, securing harbours and even, much later on – in Scotland – building more churches. He knew first hand the places and schemes he over-saw because he had visited them and because he also sustained the habit, begun in Shropshire, of corresponding in intricate detail with his trusted advisers and contractors and his clerks. These included men such as Matthew Davidson, John Simpson and William Hazledine, whom he knew well, but as his projects developed new names were added to the roster. Many of these men, too, became immersed in his life and work.

Their importance, and the continuing role of senior figures such as William Jessop, has given rise to a fair question: what proportion of the great works of civil engineering that Telford built in the opening decades of the nineteenth century can be honestly said to be his own and how much was done second hand by others? In his autobiography, encouraged by John Rickman, a parliamentary official to whom he grew close through their shared work on his schemes and who revised the draft after Telford's death, he claimed it all for himself and underscored this point in the lavishly illustrated *Atlas* that accompanies the volume: page after page of beautiful engravings of almost everything he did. There is no doubt these are meant to be seen as Telford's personal achievements. A more persuasive source can be found in places such as the National Archives of Scotland, which retains hundreds of letters from Telford micromanaging progress of the Scottish roads and canals he began building from 1800 onwards. The grasp of detail they show is astounding. No one, after reading them, can doubt that these were Telford's projects, being done in Telford's name, to Telford's instructions. He was the conductor who held a great orchestra together. He chose the players. He set the rhythm. He did not play all the notes, but without him there would have been cacophony.

———

The most haunting illustration in any book about Telford, the one which most compels, is of a bridge that never existed. It would stand today, if it could have been built and kept in place for two centuries, a little upstream from the present site of London Bridge. It would have spanned the Thames from Southwark Cathedral on the south bank to Angel Lane on the north, between what are today the offices of two international banks. It would have been beautiful and impressive beyond belief; a fine, arched, metal latticework, 600 feet across and 65 feet high at the centre; designed with a sort of perfected, pinched purity, so that despite its bulk it would have seemed light, with none of the lumpy pretend-baronial of Tower Bridge a little to the east. Its principal author was Telford and his role in the proposal was a sign of his growing significance in London. For a short time, it occupied much of his energies; and it nearly came off. It would have been the Eiffel Tower of London, only built a century before.

The scheme's origins were practical. If the cramped medieval London Bridge could be replaced then ships would be able to pass beneath and unload further up the Thames and pressure on the hugely crowded waterway lower down would be eased. A parliamentary committee had got nowhere in the late 1790s, but in 1800 the idea was revived. By May 'a Mr Telford of Shropshire' had sent in a proposal for an iron bridge, with new embankments and wharves to support it.[19] His involvement grew quickly, as he told Little later the same month. 'I have twice attended the Select Committee on the Port of London. . . The subject has now been agitated for four Years, and might have been for as many more if Mr. Pitt had not taken the business out of the hands of the General Committee and committed it to a Select Committee.'[20] There were several competing proposals to help shipping, he wrote. One, to build docks on the Isle of Dogs, was already under way (under Jessop's direction, becoming the West India Docks and now the site of Canary Wharf). A second was to build docks at Wapping and a third was 'to take down London Bridge, rebuild it of such dimensions as to admit Ships of 200 tons to pass under it'. This, he said, 'has been taken up a great measure from statements I made while in London last Year – and I have been called upon to explain'.

Was the idea of replacing the bridge really his? The obstacle caused by old London Bridge had been obvious to everyone for years, as he told Little. 'It is a great national object to render the Port of London as perfect as possible.' And others, including the social reformer Sir Frederick Morton Eden (an ancestor of the prime minister), had already proposed an iron bridge. Telford was, as so often, getting above himself in claiming all the credit. But he was central to the next steps in the story.

'The business is as yet in a very unformed State, and it is very uncertain how far the improvements will be carried,' he told Little, the main reason being that while what was needed – a higher bridge – was clear, no one had yet proved how one could be built. One challenge was the width of the tidal river. Another was the low land to either side: there were no embankments to support the structure or carry a road without very steep approaches, which would be beyond the capability of horse-drawn vehicles.

Telford did not put forward his solution on his own. As almost always, he joined forces with a younger colleague: in this case James Douglass, 'The Eskdale Archimedes', who in April 1799 had put forward his own plan for an asymmetric five-span iron bridge.[21] Telford had great hopes for Douglass, his fellow countryman; he first heard of his talents in 1797 in a letter from Eskdale and he had asked Little to sound out his background among their friends – cautioning him: 'Do not say anything to any body else untill [*sic*] we have heard further, in case there should be some mistake.' He had shown 'great proofs of his Mechanical Genius' in America and been sent back to Britain to prove himself. In 1799, Telford told Little 'I. . . employ him to do some business for me in London. . . he is a very clever young Man and will, I hope, do well.' He was certainly keen: before his bridge, Douglass had promoted a machine for making bricks, a machine for shearing cloth and a cannon ball he had invented for 'destroying the Rigging of Ships'.

Crossing the Thames by the greatest iron bridge ever built on the planet, however, was something else. Even for Telford – keen on Eskdale, keen on novelty, keen to promote and work with talented people, darting about, always unabashed by the scale and variety of the things he was proposing – it was a challenge. In the summer of 1800 the parliamentary committee published a report containing several alternatives, including a plan for a three-arched iron and masonry crossing drawn up by Thomas Wilson, who had built a pioneering iron bridge at Wearmouth in Sunderland in 1796 (using some of the parts from Thomas Paine's experimental structure). Telford and Douglass submitted plans for both a three- and a five-arched bridge, with long approach ramps on either side. At the same time Jessop, consulted by the committee, suggested embankments which could deepen and narrow the river, improving its flow. This was Telford's opportunity. A narrower crossing, he decided, could be managed in one single arch, and late in 1800 he and Douglass put forward a plan for their revolutionary bridge. They had a model made which was, with Nicholas Vansittart's support, put on show in the Royal Academy. The committee was impressed. 'The obvious Advantages which would be obtained if the Communication could be effected by means of a Single Arch,' it concluded, 'as well as the Magnificence of the proposed Structure, appeared to give the Design a particular Claim to the notice of your Committee; yet the Attempt was of so novel a Nature, that they thought it absolutely necessary. . . to request the opinions of some of the Persons most eminent in Great Britain for their Theoretic as well as Practical Knowledge.'[22]

Telford was happy to be sending sparks flying, as he told Matthew Davidson in a letter early in 1801. 'The Plan and the Model have been universally admired, only the unprecedented Extent startles people at first.'[23] He set about convincing the doubters through systematic inquiry, drawing up a formal list of questions 'To Scientific Men', which he put to a group of the great and the good, including the Astronomer Royal, professors from Oxford, Cambridge and Edinburgh, James Watt, William Jessop and two Shropshire ironmasters.[24] This is, in itself, a tribute to Telford's standing: he was being taken seriously, his idea more than a crackpot dream even if the diverse group's conclusion was contradictory. 'I think your plan for the Iron Bridge excellent and do not forsee the least difficulty in its execution,' William Reynolds told him in one of the first replies. By April he had been flooded with advice. Some thought casting the massive iron pieces would be too hard; others questioned the strength of the arch, though they did not rule it out completely. Reporting from Edinburgh, Archibald Alison reported that the city's professors 'all agree in conceiving the plan free from any scientific objection'.[25] Their private reaction, he reported, was more enthusiastic still. One was 'lost in astonishment at both the grandeur and the Beauty of the Conception'; a second suggested that there was '<u>nothing like it in the whole solar system except the Rings of Saturn</u>' – 'is this not <u>good</u> from an Astronomer?'

By the summer of 1801 Telford had refined his bridge further and published an article championing the scheme and himself. This offers much more than a detailed description of the design. It roams widely to cover what might be called his philosophy of infrastructure: an intellectual underpinning which can be applied to all that he did.

The article falls, broadly, into three parts. First, he offers a Whiggish sense of historical progress which places his projects at the apex of national ascent. 'It would be an amusing task to trace the progress of this useful art from the rude efforts of the savage in his unassisted state,' he writes, 'to the magnificent works of civilized nations, when science, wealth, and increasing population have united to overcome difficulties considered before as unsurmountable.'[26] Second, he is resolute in his belief that good transport supports a strong economy and national unification, in what he saw as the emerging nation of the United Kingdom. 'Nothing tends so much to promote the improvements of a state, as the establishing of an easy and uninterrupted communication through all its districts,' he argues. And third here, as always, he advocates innovation. Iron, he writes, is 'the most abundant, cheap, and generally useful of all metals' – and lack of experience in its use should not hold work back. 'In great works we should proceed with caution; but the very principle of improvement must be wholly abandoned, if the demonstrations of science and the evidence of practical knowledge are to be disregarded.' As the architectural historian Pedro Guedes points out, 'to Telford the future was an extension of the past and improvement through the application of accumulated knowledge was the way forward.'

This argument stands as a defence of everything Telford did: a respect for dramatic projects, using new technologies to achieve big things. But in London, for once, the recipe failed. Late in 1801 he issued a beautiful aquatint engraving of the planned bridge, which stands in itself as a piece of fine art, printed by Rudolph Ackerman, a London lithographer. In 1802 he reissued it (following his notes 'from the Royal folk') with a dedication to George III, and its popularity helped secure his growing reputation.

But the bridge itself came to nothing. One reason, as Telford bemoaned to Davidson, slipping into colloquial English, was the economic effect of war. 'The French are like Paddy's Racer – Driving all before them,' he wrote.[27] 'How far we shall partake of the blessings or woes of the impending negotiations is beyond the Reach of my telescope – I am afraid they do not argue favourably to my projects, and to tell you the truth I shall not be surprised if the whole scheme is abandoned until some accident happens at London Bridge.' That was not the only problem. The City of London had doubts about the enormous scale of the ramps needed to reach the crossing, and the investment in the new docks on the Isle of Dogs and at Wapping also counted against his scheme: ships might not need to come so far up the Thames, after all. So did the sudden disappearance of his ally James Douglass in June 1802. As Telford complained to Little, in a phrase of the time, he 'played us a Pilskey, however by good management I think there will not be any great loss – <u>no thanks to him</u> – where he is gone I know not'.[28] It turned out he had joined the fashionable dash to France during the short summer peace. 'I am not surprised at it, he was peculiarly fitted for that country,' Telford mused when he found out the truth many years later. 'I wish Douglas may deserve his success, altho' they might have been exercised in a more respectable manner, but he was impatient for distinction and Wealth, for which in this country he found too many able Competitors.'[29]

Perhaps this was, for Telford, a piece of fortune. A. W. Skempton, in the first engineering assessment of his plan for almost two centuries, questions whether the abutments could have taken the strain. Even John Rennie's conventional stone replacement, built in the 1820s, began to sink. Telford's London Bridge, just like the old one in the song, might have fallen down and with it the prospects for its originator's reputation.

He was always scribbling. Wherever he was, he carried a little notebook, jotting down reminders, questions and tasks. He also filled up a separate, much longer, formal 'Memorandum Book' with the rules of his profession: the useful sums, ready-reckoners and measurements which assisted his projects. These books, which were not written to be published, are like the over-heard mutterings of a clever man talking problems through in his sleep. The unfiltered outpourings of a burbling brain, they expose the things that mattered to him most.

He began the habit early. 'You must know that I have a book for the Pocket, which I always carry with me,' he had told Little in 1787; 'I have cramm'd into it. . . Hydrostatics, Pneumatics, and all manner of stuff and to which I keep continually adding: now it will be a charity to contribute your Mite.'[30] He begged him to share his thoughts on science – 'for you know I am chemistry mad' – promising to tease him 'til all your philosophy vanished in smoke, and after you had laughed yourself into the common man, I would then make you promise that you would communicate every article that you thought would be of service to your friend'.

Three of these little books, at least, survive: in the library of the Institution of Civil Engineers and the contents of his engineering Memorandum Book were also published as an appendix to his *Life*.[31] Pick up one of these informal, day-to-day notebooks today and you can feel an immediate connection with the life of the man who wrote in them; the red-leather covers, battered by travel; the confident inscription in ink on the opening page in case it was lost, 'Thomas Telford Civil Engineer' – even the smart shop Telford bought it from, 'Made and Sold by Williams Stationer to the Duke and Duchess of York No 44 Strand' – all stand out. The contents are varied, an engaging mix of engineering wisdom, science, maths and historical learning. There are sketches of canal locks; a description of 'a composition for making paper transparent and which will bear colours'; and a note that 'Boulton and Watt calculate one Horse Power equal to raising 3300lbs ten feet high per minute'. There is also a record that 'At Gordon's Wharf Deptford' an iron 'Chain put up in the year 1745, and [has] been in daily use (Sundays excepted) ever since'. It was, he observed, 'still in excellent order'.

These were more than the magpie-like pickings of habit: the entries show serious application and skill. There is, for instance, a demanding description of a mathematical exercise to find the square root: 'Find the cube of the first period, subtract and bring down next period, call this Resolved. Draw a line under it. Under the Resolved write the triple sq, of the Root, so that units in the latter under the place of hundreds in the former, under the triple square of the Root write the triple root expand one place to the right and the sum of those two lines call a Divider under which draw a line,' it begins – and there is more before the sum is resolved. And that by a shepherd's son.

The brain which could work through sums such as this was clearly supple and wide-ranging; and it was as drawn to the beauty of objects as it was to the beauty of numbers. At the back of the book is Telford's Architecture Memorandum, added to over many years. It shows a man – and a society – with fast-expanding horizons. 'Endeavour to get specimens of Egyptian tombs,' he urges himself. 'Consult Wood as to the different styles. . . at Balbec and Palmyra'; 'see Arthur Young's travels through France'; 'See Geo. Ramsden's China'. Babylon, he observes, was 'Built of Stones fastened by clamps of iron'; meanwhile the Aztec 'Aqueduct of Champoralan [Chapultepec] [was] executed from the 16th century under the direction of a Franciscan missionary. . . its length was 30 miles. There were three great bridges over ravines, one of 47 another of 13, a third of 67 arches – the largest was 100ft high and 61 broad.' He sets himself the task of finding writers who had visited 'Persepolis about one day's journey from Shiraz in Persia' and discovering the means by which the Pillars at Alexandria had been constructed, 'it is said that it is formed upon a fragment of an ancient Egyptian obelisk'.

His principal architectural fascination, though, was, as with so many of his contemporaries, with the classical world. 'The Amphitheatre in Rome was begun by Vespasian and finished by his son Titus AD 65 it is above 600 feet in length 500 in breadth and 200 in height cost ten millions of crowns and held with ease 87000 spectators,' he notes. He must investigate 'the origin of the Doric name'; read 'Addison travels in Italy'; and consult 'Swinburne travels in Spain for the Aqueduct of Segovia'. He had combed through Gibbon's *Decline and Fall* for lessons on Roman structures: 'Herod son of Atticus, constructed an aqueduct,' he quotes.

And yet all this is simply entertainment when set against the still more serious contents of his Memorandum Book. This is less a ready-reckoner than a distillation of the scientific and engineering learning of his age, running to several thousand words, with tables and erudite mathematical formulae. It shows beyond question that Telford was much more than a builder who operated on the basis of practical experience or the detailed knowledge of his contractors. Over time, he developed a deep and wide theoretical understanding. He was comfortable with calculations and with abstract reasoning and knew how to deploy the answers. There are instructions on how to calculate the curvature of the earth and the strength of machines ('the power and resistance are in the ratio of the spaces passed through by each in a given time') and rules on measuring the velocity of water and of air. His experience in Shropshire, and with bridges, shows in sections on the strength and malleability of iron and the strength of masonry arches. Among much else are sections on matters as varied as the performance of carriage wheels ('Coachmakers' rule, never to allow the fore wheels to have but two spokes less than the hind ones'); the angle of roads against the horizon; the efficiency of railways compared to canals ('Horse works three times as much on a canal as on a railway') and the speed with which tides swept around the coast. It is a sort of commonplace book of engineering, the notes in the margin of a bigger story.

And such curiosity, such a retentive memory, made him a man who could turn his hand to almost anything. Over time, he became less not more of a specialist: to his early skill with masonry, he added architecture; to bridges, he added aqueducts; to canals, he added roads; to waterworks, he added harbours. On top of this, as the notebooks show, he was able to observe and record accurately an immense amount of information quickly and draw upon it later. He was never tied down by the details of any one particular issue or place; he could juggle topics and summarise them and then return later to organise a response. He had a fine memory. He kept his notes in good order. And soon all these skills of observation and implementation, this flexibility, this diversity, this enthusiasm for new places, were to be put to use in a task of daunting scale.

My Scotch Surveys

As night fell on Wednesday 20 January 1790 a group of men gathered by candlelight in a new and rough little settlement called James Town. It had been built just below Telford's birthplace at Glendinning in a lonely valley above the River Esk. The men were miners come to cut into the hills where he had once watched his flock. Their task was to extract antimony from the Louisa mine, a shiny, grey metalloid used among other things as an alloy with lead in forming metal letters for printing books; and this use was appropriate since on that winter evening in 1790 the miners met not to drink or gamble but to start a library. They kept a careful record of their meeting, noting that money 'Received from the Westerkirk Mining Co' would be used to buy 'the following books for our mutual improvement': a collection of seventeen volumes which was as wide-ranging as it was dutiful.[1] Robertson's *History of Scotland* sat on the shelves alongside Seneca's *Morals*, Henkel on *Pyrites*, Fourcroy's *Chemistry* and Ferguson's *Lectures*. 'Hearing of the Company's good intentions,' they added, various gentlemen had 'thought it fit to present the miners' with additional books. Among the donors was William Little, minister at Westerkirk church and Andrew's brother.

There is something endearing about this drive for self-improvement; and perhaps something particularly Scottish too, for it is questionable whether English miners, perhaps less literate, were doing the same in, say, Shropshire's new collieries. This upper Eskdale library and the employment of the men who created it were representative of the boiling intelligence and energy of Scottish culture.

Three years later they expanded their collection. 'We the miners in this place, finding the Books sent by the Company and others will tend greatly to our Improvement, have thought proper to advance Five Shillings to each man for purchasing more books,' they wrote in their account of the minutes, ordering a copy of Burns's newly published *Poems* and Ridpath's *Border History*, among other volumes, and agreeing to impose on themselves an annual charge which amounted to at least half of one week's wages. 'The Miners this night having met and exchanged books thought it fit to form themselves into a society,' they decided, before voting at one of their monthly meetings a little later to 'buy some candles paper and pens'. By 1800, members who did not attend were being fined 8d each for their lack of application in supporting an expanding collection which stretched from two copies of *The Wealth of Nations* to *The True Nature of Imposture Fully Display'd in the Life of Mahomet*.

The Louisa mine, for which its owner Sir James Johnstone (Pulteney's older brother) had great hopes, did not last. It shut in 1798 and despite brief attempts to restart work in the late nineteenth and early twentieth centuries it now exists only in a few hillside ruins and waterlogged tunnels. Of the miners' cottages in James Town, only a few footings remain. But against all chance their library has survived and still contains the leather-bound book which recorded its establishment and was used to catalogue additions for over a century.

Today the library sits in a trim little building that looks something like a chapel, right by the River Esk and not far from a cumbersome roadside memorial to Telford, cut in rough granite and put up in 1928. It is still acquiring books, its two-centuries-old collection tracing contemporary British enthusiasms from the Boer War to Margaret Thatcher; and its connection with Telford is much stronger than the mere chance that he came from the same valley. The bound volumes have 'Telford Legacy' embossed on their spines, for when he died in 1834 he left the little library a bequest of £1,000; and more came later from the residue of his estate. It was invested, the minute book shows, in '3 per cent Government stocks' and according to the present librarian the fund still makes the occasional disbursement to support the institution today. Now restored and cherished, it opens for an hour on the first Monday of each month and among the books on its shelves are survivors from those acquired by the Westerkirk miners in 1790. To pull them off the shelves is to join those few miners back in their first meeting more than two centuries ago.

The creation of this library and the creation of the mine which was its origin show how the nation that Telford left for London in 1782 had advanced. Scotland was richer and more ambitious; better connected; industrious. Its economy was growing at a faster rate than England's. It was also, through empire, reaching out to the world. 'It was pleasant to observe the thriving appearance of the country,' Telford noted when he returned to his home county of Dumfries in 1801, 'roused. . . from the Lethargy in which it slumbered for some centuries'.[2] Visiting a year later, in search of Burns's family, William Wordsworth saw the same growth in a less benign light. 'We were glad to leave Dumfries,' he wrote in his journal, 'which is no cheerful place to them that do not love the bustle of a town that seems to be rising up in wealth.'[3] The forces of enterprise and industry were spreading north through every part of Scotland, through the Borders, along the Clyde, and up the east coast towards Aberdeen.

It was only a matter of time before they reached the Highlands.

When James Boswell and Samuel Johnson made their famous journey through the north of Scotland in 1773 they left 'fertility and culture behind'.[4] They passed through a land with many bad roads, or none at all; without safe bridges to cross its wild rivers; and in places without law. But even as they travelled, the Highlands were changing and their journey, a romantic curiosity, was itself a sign of that. They came as tourists to see a culture which was dying. The estates of clan chieftains who had supported the Young Pretender, Charles Stuart, had been confiscated after 1745 and although they were returned in 1787 the gap had been long enough to rot old feudal ties. The tribal system of clanship was giving way to a new hunger for wealth, just as Gaelic was being replaced by English. Landowners began removing people in favour of more profitable sheep; and Highlanders – who only knew how to run cattle – found themselves pushed out of their homes to make way for shepherds brought up from the Lowlands. This was the start of the clearances, seen by Boswell in Skye where he watched 'a ship, the *Margaret* of Clyde pass by with a number of emigrants on board'. It was, he thought, 'a melancholy sight'. At their height, homes were burnt, families forced out and forests and farms gave way to vast acres of rough grazing.

It was often brutal, engendering hostilities from both sides. In 1792, in Ross-shire, Highlanders fought back in what became known as *Bliadhna nan Caorach*, or the Year of the Sheep: but by then their cause was lost. Economic necessity, official paternalism and military alarm in London and Edinburgh about the risks from this only half-governed space had made what politicians in our age would probably call the development of the Highlands an urgent political imperative. The difficult harvests of the late 1790s and the blockades that had restricted European trade drove a new interest in enhancing and extending the national economy. The war encouraged the government in London to throw a protective arm around a part of the country which had never been fully under anyone's control. Good harbours would be useful to the Royal Navy as well as to fishermen and good roads were needed to move troops as well as the herring catch.

Over the next three decades Telford was to become one of the principal instruments of this change. There is hardly a spot in the Highlands and Islands today in which you are not near one of the roads or harbours or canals or churches whose construction Telford oversaw. He stood near the apex of a system that combined an intense bureaucratic drive with heroic human effort, planting in the wild landscape the backbone of a modern economy. It was 'the greatest achievement of his career', writes one biographer, 'judged in terms of the sheer magnitude of the work involved and its historical importance'.[5] It was also a duty of military severity. One of the men he raised up to oversee his work on roads travelled on average 9,600 miles a year, on foot or by pony; another, working in just one county, Argyllshire, estimated that he had walked 100,000 miles in eighteen years by the time he attempted to retire in 1828 (he was told his duty was to stay on, not least because Telford, by then in his seventies, showed no sign of letting up). It is no wonder that of the eight resident superintendents or general inspectors, effectively deputies, who served Telford on the ground in the Highlands over three decades on his road and canal schemes, five died in service (two of them were replaced at Telford's request by their sons). Across mountains, moors and wide rivers, these men worked in miserable conditions. As one of them, who followed his father into the job, recalled: 'Many a snow-storm and bitter blast and wet jacket I had to endure; still I had generally good roads and tolerable inns with the advantage of youth and health. How different was my poor father's life! Traversing the country where there were no roads of any kind, crossing dangerous rivers and streams, travelling in wet clothes and for shelter living in small and wretched huts where oatcakes with whisky were the chief and only refreshments.'[6]

Telford, at least, did not impose such arduous service only on others. He suffered too. He was in Scotland at least once or more often twice in every year for thirty years from 1800, each time for many weeks, always moving on foot, by pony or by chartered boat from the southernmost parts of the Lowlands to the Highlands, Islands and as far north as John O'Groats. His travels were relentless. 'Do you think nobody works but you?' he once exploded at a superintendent who had complained about his burden.[7] 'I have examined and settled the Glasgow and Carlisle and Lanarkshire roads, a road near Edinburgh and the waterworks. Examined the river Clyde at Glasgow... perambulated and set a-going a survey of the present and new line of road between Glasgow and Port Patrick. Travelled through the Galloways... Is that idleness?' Of a journey deep into the Highlands in 1828, by which time he was almost seventy-one, he reported: 'it was with difficulty and not without danger that I could scramble along a rugged, broken, sandy shore or by narrow tracks on the edge of precipices frequently interrupted by rude and inconvenient ferries; and having for lodgings only miserable huts, scarcely protected from the inclemency of the weather'.[8] In the winter it snowed, in the summer it rained; and if the problem was not ice, then it was being bitten half to death by the midges.

——————

The scale of the task to come would have been impossible to guess when his connections with the Highlands began in the

1790s. At this time Telford seems almost never to have said no to the offer of extra work. His letters record activity in counties such as Leicestershire and Worcestershire; in 1800 he worked to improve navigation on the Severn; and his fruitless efforts to build a replacement for London Bridge were matched by schemes in other cities. 'I am at present engaged in conducting a Plan for supplying the Town of Liverpool with water, by means of pipes. . . this business is of some magnitude and we have some opposition to contend with,' he told Little in 1799. The plan, which involved pumping stations powered by Bolton and Watt steam engines, to serve the booming Georgian port, fizzled out, though he was left with both a strengthened interest in steam power and an admiration for the 'young, vigorous and well-situated' city which, he said, 'will become of the first commercial importance'. Such dead ends must have been frustrating but even as his Liverpool and London projects were coming to nothing, another was bursting into life. He took on an unpaid position assisting the Highlands, which at the time must have seemed to promise only modest additional duties, as engineer to the British Fisheries Society.

Eskdale connections provided the opening, of course, as so often in his career. The position came from his patron Sir William Pulteney, who was, among much else, the society's governor. In April 1790 Pulteney reported that he had asked 'Thomas Telford, a practical Surveyor lately employed at Portsmouth and now at Shrewsbury, to know whether he would be willing to visit the Society's Settlements'.[9] The do-gooding joint stock company had been founded in 1786 with the aim of improving the productivity of the industry by securing government support for better harbours to bring in the catch and better roads to take it south to hungry cities. In the 1780s it had worked to develop plans for improving the Highlands and a series of engineers and inspectors had been sent north, whose ideas Telford drew on when talk finally turned to action. 'A very considerable part of this island was lying in a state of nature,' reported one of them in 1786, 'the riches of its shores, tho' more important to great national purposes than the Mines of Mexico and Peru were scarcely sought after.' The Highlanders 'were rendered torpid by idleness: they were frequently exposed to famine; and many of them forced, through necessity, to abandon their barren but beloved wilds'.[10]

This report, by the Edinburgh geographer and bookseller John Knox, described the British Fisheries Society as a kind of salvation. 'To you the naked, the hungry and the helpless. . . will look up in transports of gratitude.' Telford seems to have taken a more practical view. There is no comment in his surviving letters to Little of his work, but the Society's records contain letters and reports of his inspection of potential sites for improved settlements including Lochbay, on the Isle of Skye, and Ullapool, on the mainland to the north. They are striking examples of his role as an architect and planner, overlooked later (including by himself) as his career as a civil engineer developed.

Telford travelled to both Skye and Ullapool in June 1790, filing his report, from Langholm, six weeks later. In it he criticised work being carried out by the Fisheries Society's contractor – and he was listened to: 'Telford's integrity [has been]. . . established. . . by your experience of him,' the society's secretary told Pulteney. He drew a plan for the site at Lochbay and corresponded regularly over the next two decades with the society's officers over plans for a new school, a better harbour and pier, a church and an inn – the last of which, the Pultneytown Inn, he designed and which was built in 1808 (only for it to be adapted as a home instead by the society's contractor). Telford declared the building to be 'an example of neatness'. It is, the historian Daniel Maudlin writes in a revelatory paper on Telford's early work with the society, evidence of his eagerness to experiment with contemporary trends in architecture, with its 'hipped roof with over-sailing eaves, central roof-ridge chimneystacks and advanced, bowed window bays'.[11] Most of his plan for Lochbay was left unbuilt, but even on paper it was striking, with a curved street in the form of a crescent, influenced by John Wood's Royal Crescent in Bath (which at this stage Telford had yet to see) and Palladian terraces derived both from Bath and from Robert Adam's unfinished pioneering planned settlement at Lowther in Cumbria (which he had visited in 1789).

Through his involvement with the British Fisheries Society Telford had put himself in the right place at the right time to present himself as an engineer with unique expertise in improving a region of growing interest to the government. Among the officials and politicians he cultivated was 'my friend Sir John Sinclair', an ardent agricultural improver and prolific writer (and, in passing, the first man to use the word 'statistics'). Dubbed 'the Great Scot Rat' by *The Times* for his opposition to Pitt's plans for extra taxation (the paper has always tended to hunt with the governing pack) he was a dominant authority in Caithness in the far north of Scotland, much of which, until his eventual bankruptcy following a scandal over stolen Indian diamonds, he owned. He was also the President of the Board of Agriculture, a body similar to the British Fisheries Society, which also advocated government and private action to increase food production. This was the organisation for which Telford had written his paper on mills while in Shropshire in the 1790s (which never seems to have been published in his lifetime) and it was either for this or his work in Scotland that he received a letter of commendation in May 1797 from Sinclair 'for promoting the Objects of the Board of Agriculture', which, he said, 'merits not only the thanks of the Board and of the Public, but also that you should have it in your power to wear some Public

Mark of our approbation – I request, therefore your Acceptance of the Set of Buttons herewith sent'.

Whether Telford ever wore these baubles is not known. What counted was the support he was now getting from very power-ful men. Sinclair and Pulteney were both well placed to help his next move. Sinclair was a leading member of the Highland Society, which promoted the cultural and economic advance of the region and included among its members many of its richest men. His political battles with Pitt – who removed him from the Board of Agriculture in 1798 – did not seem to harm his friend Telford's chances of advancement at a time when political atten-tion was turning to the securing of British influence in places such as Ireland and the north of Scotland; and the new income tax was beginning to give the state new funds with which to do it.

Everyone agreed the Highlands needed better communica-tions. That much had long been clear: in the early eighteenth century, George Wade, commander of the army in the north, had been the most famous (but far from the only) contributor to what became a network of several hundred miles of so-called 'Wade's military roads' through the Highlands, including a link between Inverness and Fort William and more than one impressive bridge. In peacetime these roads also carried the first tourists to the Highlands, among them Boswell and Johnson; but they were, as Telford later observed, 'generally in such Directions and so inconveniently steep, as to be nearly unfit for the Purposes of Civil Life'.[12] Wade's roads were built with a soldier-like disregard for human preference, marching up the sides of steep valleys and climbing over bleak passes. The network was random, incomplete and uncared for. In 1799 an official report commissioned by the government confirmed the dreadful state of roads in the region and, worse still, the absence of bridges, which meant that even the journey from Edinburgh to Inverness required perilous crossings by ferry, and the sole track into Sutherland involved crossing a rocky beach passable only at low water.

And all at once Telford was in place to do something about it.

On 27 July 1801, Nicholas Vansittart, not yet the Chancellor but already a powerful force at the Treasury and an established ally of Telford on his London Bridge plans, wrote asking him (in his capacity as engineer to the British Fisheries Society) to carry out a general report on Highland communications.

The task he set was sweeping. Only a Telford could have made something of a mission that required him to begin his travels by finding 'the most commodious and productive stations for the Fishery both of Herrings and White Fish', as well as 'the means of establishing a safe and convenient intercourse between the Mainland of Scotland and the Islands' – before going on to 'ascertain the practicability of forming a complete inland navigation from the Eastern to the Western Coast'. Vansittart's letter also asked him to study 'the making of the Harbours on the North Eastern Coast' and as if that were not enough he was told: 'You will also examine the most convenient situation for the Erection of Towns or Villages, and for the building and repairing of Ships, as far as your Judgement goes, consider how far any of these Harbours are capable of defence against an Enemy.'

Fishing. Roads. A great canal from coast to coast. New ports. New towns. Defending the nation. His orders read as if the Lords Commissioners of the Treasury were despatching a colonial administrator on a mission to plan a model economy in some romantic, rebellious and savage land. That is not coincidence. It is exactly what was happening.

The speed with which Telford reacted suggests that the general order cannot have come as a surprise. More precise orders were to follow, Vansittart instructed, but Telford, presumably already fully aware of what they would ask, was already in the far north before these arrived in September and he had completed much of his first Highland survey before he came to read them. He disappeared northwards for three months, taking with him reports from a previous generation of engineers employed by the British Fisheries Society and others to advance change in the Highlands. Among them was work on a proposed Caledonian Canal, by James Watt, the steam engineer, in 1774, and by John Rennie, whose reputation then much exceeded Telford's as a civil engineer, in 1793.

It was a tough trip, in the late summer of 1801, even for a Scot bred in the hills, trekking along Wade's military roads, crossing rivers and mountains, always observing and noting things down and taking care to flatter the powerful: seeking advice from the Highlands Society, for instance, in a manner designed to encourage the Highland lairds who led it to support his plans. He was lucky with the weather. 'Never was there a Season more favourable for making Surveys,' he told Andrew Little when the journey was over, 'I passed along the Western and Central Highlands, from thence to the extremity of the Island, and returned along the Eastern Coast to Edinburgh and scarcely saw a Cloud upon the Mountain's top.'[13]

The trip was a triumph. 'It would require a Volume to specify anything like the particulars of this journey,' he reported, 'I shall therefore only say that every part of my Survey exceeded my expectations and I did not leave anything unaccomplished which came within the compass of my Mission.' But, as he told a Liverpool friend, Dr James Currie (the publisher of Burns's verse), in a detailed letter sent from Peterhead on his way back towards Edinburgh, it had also been relentless. 'I have carried out regular Surveys along the Rainy West through the middle of the tempestuous wilds of Lochaber, on each side of the habitation of the far famed Johnny Groats, around the shores of Cromarty, Inverness and Fort George, and likewise the Coast of Murray. The apprehension of the weather changing for the worse has prompted me to incessant and hard labour so that I am now almost lame and blind.'[14]

He recovered in Aberdeen and Edinburgh, where he 'enjoyed one week of all the heart of Man could wish', spending time with the luminaries of the city's university including four professors, led by the venerable John Robinson, a physicist and inventor (of, among other things, the siren), and by this time a dotty conspiracy theorist who challenged what he saw as the secret works of the Freemasons and Illuminati. Telford had first met Robinson in 1796, describing him as 'a douce old man and cantie – religious and an aristocrat but candid and moderate'. He stood out, one contemporary remembered, with 'a pig-tail so long and so thin that it curled far down his back' and 'a pair of huge blue worsted-hose without soles, and covering the limbs from the heel to the top of the thigh, in which he both walked and lectured'.[15]

Weatherbeaten and tanned, Telford looked indestructible. It was at this time in Edinburgh – if not after a similar expedition a year later – that he sat for Sir Henry Raeburn, Scotland's leading portrait artist. The vigorous picture which resulted captures him better than any other. Strikingly youthful (he still looked exceptionally young in a second portrait a decade later), he has the bulk of a man whose body had been hardened by travel and work, with broad cheeks, unkempt hair, a plain white cravat and dark cloak: a practical look which commanded respect for his actions, though close enough to the emerging, austere, un-dandified style of the age to suggest he was someone of means and significance. The picture was probably commissioned by Archibald Alison and remained in his family for a century, before ending up in the Lady Lever Gallery in the Wirral, where it now hangs, stranded among Lord Lever's famous Pre-Raphaelites.

After a week in Edinburgh he returned to England in mid-November full of cheer. In Shropshire he slaved at preparing his report as well as managing his business in the county and on the Ellesmere Canal, 'which had accumulated during my long absence'.[16] He worked 'almost day and night' to clear the decks so he could concentrate on making 'Charts Plans Estimates and Reports of my Scotch Survey'. From now on the project occupied his mind more and more. In December he wrote to tell a new acquaintance, Professor Patrick Copland, whom he had met at Marischal College in Aberdeen, that 'I am now about laying down in charts the results of my northern surveys'.[17]

Copland, a mathematician and physicist who had advised on providing his city with a clean water supply, was soon added to the list of people Telford cultivated as useful. 'I have not the scientific knowledge which your position has afforded to you,' he told him in a letter from Bath sent on New Year's Eve 1801.[18] 'Early in life, I perceived that science and practice looked shy at each other and I have with unremitting exertion been endeavouring to bring them better acquainted.' Telford pestered him with questions about the nature of magnetic north and boasted of a new design for bridge building that he had developed, 'which will probably have the effect of introducing a considerable change in the forming of Stone Bridges'. He also encouraged his educated friend to submit a piece for the *Philosophical Magazine*, 'rising in reputation', Telford assured him, 'and an excellent vehicle for conveying useful information'.

The draft report was finished by December 1801, but Telford continued to work on it over the winter. In April he turned to Little, apologising for his silence over the winter. 'It cannot be hard to conjecture where my leading object has been and where my chief anxiety centred,' he explained after the break.[19] 'Never when awake and perhaps not always when asleep – have my Scotch surveys been absent.'

'My plans are completed, the Draft of my Report has been made out and has been perused by the persons I wish,' he wrote when it was almost done; 'It is now in the hands of the Copier and I hope to deliver the whole in the course of a few days. I shall be fully committed, and then, the L- -D have mercy upon me – or rather, may they be productive of good and that good benefit Scotland.'

'There is,' he added, 'at present no reason to doubt': a confidence which was encouraged by the close care he had taken to ensure Vansittart was in lock step with him from the start. In a flattering letter sent from Scotland towards the end of his tour in 1801, he had told his political master that 'it is a most singular and fortunate coincidence that the result of my Surveys, goes to prove that the whole of the objects which their Lordships have in view are not only practicable, but are capable of being formed into one intimately connected system, which would very evidently have a striking effect'.[20]

To Little, he was, as always, more honest and bullish. Politicians had their place, he wrote mockingly, 'if they will only grant me one Million to improve Scotland or rather promote the general prosperity and welfare of the Empire, all will be quite well and I will condescend to approve of their measures'.

Telford's first survey of the Highlands set the pattern for everything which followed: a practical plan for improvement across tens of thousands of square miles, a blueprint for reconstruction which could be followed – and was – for the next two decades.

It was based on a blend of observation, conversation and research. Telford did not travel without taking care to find out the views of the landowners on which the success of his schemes depended. He drew heavily, too, on previous detailed proposals, in particular Rennie and Watt's work on a Caledonian Canal cutting through the Great Glen, a route which he followed north-east to Inverness after travelling through Lochaber to Fort William. In his letter to Vansittart he summarised his proposals: to build 'a complete Inland Navigation' from Inverness on the east coast to Fort William on the west; as well as new harbours at Oban; at Wick in the far north; and along the east coast at Peterhead. It was a plan, he said, which would improve the Highlands 'by the introduction of habits of Industry, and Capital into the Country'.

This was just what the powers in London wanted to hear. Telford's first report was submitted to the Lords Commissioners of the Treasury, who were, he was told in the middle of 1802, 'perfectly satisfied with the manner in which you have executed the survey'.[21] That summer they sent him back to the Highlands to expand his plans. Exploring the remotest parts of the country on a government commission like this would, for most people, have been a full-time job; but Telford was always able to compartmentalise with hardly a ripple of anxiety. Nor did his engagement in one project seem to concern his backers and allies in others; so even as Hazledine was opening his foundry at Plas Kynaston in Wales and preparing to cast the iron ribs for Pontcysyllte, Telford was heading north for a second season of surveying.

This time he was asked to focus on the need for new roads to link British ports serving Ireland, as well as to re-examine plans for the conjectured inter-coastal canal and to ask about the causes of emigration, 'which is said to prevail from the Highlands and Western Islands' – although, as his biographer Rolt points out, 'one might think that these would be obvious enough to even the most boneheaded government official'.[22]

Telford also attempted to add a twist to his Scottish work, away from the Highlands. As he told Little, his aim was 'if possible, to get some aid towards the Esk Bridge' – a plan to build a new crossing over the river a few miles below Langholm which, a few months before, he had decided 'cannot be done unless the government grants aid'.[23] In public he promoted this extra work as a plan to help travel to Ireland by opening a new route to the coast, 'to facilitate the communication between the two Kingdoms – or rather the different provinces of the United Kingdom', since, as he pointed out, the sea passage to Ireland here 'is only a sort of wide ferry' and travellers would put up with a longer road journey to avoid the perils of the crossing. But as he explained to Little, in a sort of stage whisper, that argument might 'possibly have a greater weight with you', as it included the new bridge across 'our beloved Esk' – a comment, he wrote, 'which must not be published. . . our general arguments bear us out and that is sufficient'. Pork-barrel politics is not new and nor is it always successful: Telford's enthusiasm for this route was never shared in London. The road he wanted was left incomplete. The Esk did not get its extra crossing, only an upgrade of the old route over Skipper's Bridge, which still carries the modern A7 just south of the town.

This, of course, was a diversion from his main purpose: opening up Highland communications. By February 1803 he was writing his second report (having in the meantime somehow found time for 'a very extensive, tho' rapid survey into the counties of Leicester and Northamptonshire').[24] 'I am now putting that last hand to the Report of my last Survey in Scotland,' he told Little in what also turned out to be his last surviving letter to his dying friend.

Of all Telford's reports and journeys in the Highlands this was the one which mattered most. 'For the past two years this business has seldom been out of my mind,' he wrote as he compiled it. He presented himself not only as an engineer advocating specific schemes but also as the champion of the Highlands and their integration into a bustling new country, the United Kingdom. 'I have endeavoured to make the northern proprietors sensible of their own Interests, and to convince Government and the public that the nation at large is deeply interested in the proposed improvements,' he told Little. Those clan chiefs were crucial. 'I wish the Clans do not burst out unto open rebellion,' he explained, 'they are mad bodies but matters are changed.' Whereas not long ago they would have seen his plans for new roads as a threat to their traditions, now they saw them as a source of potential personal enrichment. 'The chieftains may fight each other, the de'el a highland man will stir for them – [but] the Lairds have transferred their affections from the people to flocks of Sheep – and the people have lost their veneration for the Lairds.' He could see that change was unavoidable even as he mourned it. 'It is the natural progress of Society, but it is not a pleasant change. There was great happiness in the patriarchal state of Clanship – they are now hastening into the opposite extreme. It is quite wrong.'

From this stands out not only Telford's emerging affection for the Highlanders but his sophisticated understanding of the forces that were altering their homeland, and of the fact that clan chiefs as much as outsiders were powering them. In his dislike of emigration he differed from James Loch, a lawyer and estates commissioner who served as a commissioner on both Telford's roads and the Caledonian Canal, and became hated for driving tenants from the Sutherland estates which he ran. Both men, however, hoped that the changes they advocated would replace feudal servitude with a secure economic existence. The story of the Highlands over the following century, as the clearances and emigration continued, suggests that Telford was to be disappointed.

In early 1803 he completed his second report on the Highlands. 'I shall have to go again to London in a few days to give my Scotch Report &c – great is the anxiety about it,' he wrote in February. He comforted himself with the thought that 'everything is at present promising but if from some unforeseen occurrence, we should be totally disappointed – what has been done will smooth the way for future exertions'. But things looked well set. 'All parties, at present, view my personal exertions in a favourable light' – and more than that 'I have just been elected a Member of the Royal Society of Edinburgh. . . I have had the thanks of the Highland Society – The Countys of Inverness and Ross-shire and many of the Highland Chiefs.' With backing like that in Scotland and from Vansittart in London, there was every chance that this second report would do more than languish on a shelf in the Treasury.

The document, printed and bound in a large volume, with its author's new Fellowship of the Royal Society of Edinburgh part of his proud title on the opening page – 'Civil engineer, Edin. FRS' – stands as one of the most significant and perhaps under-appreciated interventions in recent Scottish history.[25] It was not the first to identify the challenge of northern roads, as Telford knew. 'The Obstacles which at present obstruct the Communications in the North of Scotland, are numerous and well-known,' he wrote, 'previous to the Year 1742, the Roads were merely the Tracks of Black Cattle and Horses.' But it was the first to propose a plan of action in circumstances when this could be carried through and was, almost to the letter.

His main theme was the improvement of the roads between the Lowlands and the north; this required new bridges and – in the far north beyond the Great Glen – entirely new routes, up the east coast to Wick and west towards Ullapool and Skye. Most of these links were eventually built, to his plan. So was the Caledonian Canal, one of the great struggles (and eventually, when its trade failed, regrets) of his life. In this second report he confirmed that the best route for the canal was the one that had first been proposed by Watt in 1774 and endorsed by the eminent Rennie nine years later.

Yet on the causes of emigration, which he had also been asked to consider, he had nothing but anxiety to offer: 'Three Thousand went away in the course of last Year, and if I am rightly informed three times that Number are preparing to leave the Country in the present year,' he wrote. The cause was clear: the policy of 'converting large Districts of the Country into extensive Sheepwalks' (looked after by experienced shepherds sent from the south of Scotland, including Eskdale; in a different life, Telford might have been one of them). This was an issue, he wrote, 'which I enter with no small Degree of Hesitation'. Was it right for the government to intrude into the activity of the legitimate owners of private property? At best he thought his proposed 'works should be undertaken at the present Time. This would furnish Employment for the Industrious and valuable Part of the People.'

And now the pace picked up. The unreformed British Parliament could move fast if the right men were pulling the strings and in this case they were. The select committee which examined Telford's survey and proposed the legislation that unlocked funds for the government's response reported in June 1803. The roads and the great Caledonian Canal were to be built at the same time, under the guidance of two interlinked commissions. By 4 July an Act appointing the commissioners for Highland Roads and Bridges had been passed and on 7 July they met for the first time. They had one easy decision to make: the choice of engineer. 'Having been informed that Mr Telford... was now about to be engaged in directing the execution of the Caledonian Canal,' they declared, 'he was judged to be the fittest person for this trust.'[26] Given that the commission was filled with his patrons, the decision can have surprised no one. The select committee also endorsed proposals for the canal as 'of great importance to the general interest of the whole United Kingdom' and the Act establishing a second commission to build it was passed on 27 July.[27] There was no pausing for the local consultations or environmental impact assessments which hold up infrastructure projects today. By the time anyone in the Highlands can have known properly what was being proposed, the whole scheme was law.

Some turning points in history may be the consequences of irresistible forces and have a kind of inevitability about them; others come about by chance. Telford's Highland schemes were a combination of both. There can be no real doubt that if he had not built good communications into the region then someone else would have done, at some point. The forces of economics and technology were too strong to hold off indefinitely, but might have arrived later in a more disjointed or piecemeal fashion. Telford's great project was a heroic example of what today might be called joined-up government, and few students of politics would see any historical inevitability about that. Luck was on his side in the timing of his plans, and fate was on Scotland's side in providing an individual with the force and capability to make them. Telford was equally fortunate in the men who gathered now to work with him. They can have had no idea at all what was about to hit them. Had they suspected the scale of the task they might never have started it; but once they began they stuck to it admirably. The work aged them; for some it was a burden they carried to their graves. A different set of men in different circumstances might well have faltered long before, but this group proved the very definition of determination.

The most elevated were the commissioners. They were the tough men of government in the militant Britain Pitt had created. Many but not all sat on both commissions, for roads as well as the canal, among them Vansittart, who stayed in his posts even after he rose to be Chancellor of the Exchequer, and Charles Abbot, the Speaker of the House of Commons, who continued to serve after being elevated to the peerage. They were joined by Sir William Pulteney, until his death soon afterwards. It was an unwavering and well-connected group which applied itself seriously, and its work is a tribute to the strengths of the late Georgian system.

If the commissioners provided money and authority, then it was the trio sitting beneath them that actually carried out the administrative work. The first of these was of course Telford, the engineer. His restless, never-say-die energy was the fuel which kept the Highland schemes burning. But he was given both support and leadership when he needed it by two unusual men.

One was James Hope, a successful Edinburgh lawyer who became the commission's man of business in the city. He found himself in the middle of a tangle of land rights and construction contracts, paperwork, accounts, bank bills and potential legal conflicts which began in 1803 and did not let up until the 1830s. Privately he despaired at the volume of work, which came on top of his own legal practice; he was 'sensitive and, for all his ability, extremely diffident', one historian writes.[28] In his *Life*, Telford pays tribute to Hope's 'skill and integrity in managing the perplexities', and that is the least he deserves.[29] His handwriting is everywhere in the archives. 'A rainy day in Argyllshire,' he wrote to Telford in 1811, pestering him to read one of his endless bundles of paper, 'may afford you time for considering them.'[30] Many more letters and packets followed.

Hope had inky fingers. The third member of this essential and active trio had numbers running through his brain. John

Rickman was in title only secretary to both commissions but in reality, as secretaries so often are, he was the guiding spirit to all their work. He was a quick, curious, completely honest civil servant and statistician. An original man, he had put the case for a national census in 1796, saw it into law with Vansittart's help and then oversaw the census every decade between 1801 and 1831, dying in 1840 just before the most elaborate of his surveys got under way. He began his work as secretary to the two commissions as a young man of promise; he became a life-long bureaucrat and elderly grandee, serving as a secretary to the Speaker of the Commons and later as an assistant clerk, all the while keeping the commissions running. His dedication, working deep into the night in his rooms in the old Palace of Westminster, is visible not only in the hundreds of his letters to Hope and Telford, which survive in the dense, practical, demanding scrawl of a man with far more work than time in which to do it, but also in the fine printed annual reports he wrote and published.

He was also obsessive, awkward, outwardly cold and privately social. 'The only subject in which Rickman truly took an interest was what is now called economics, though he would have hated the word,' wrote his Edwardian biographer.[31] His character was in some ways similar to Pulteney's and he formed a bond with Telford that was just as strong. He influenced the latter heavily, leading him more and more into crusty Tory circles far removed from the radical friends of his Shropshire youth. He was more respected than liked; bigoted even, in his many letters to his friend the Poet Laureate Robert Southey, full of condemnation for political reformers as 'devils', 'blackguards' and 'vermin'. 'I have lately imported a wife by means of experiment; I think it will answer; we shall see,' he wrote soon after his marriage.[32] Yet such bewigged, claret-stained, beef-eating harrumphing was perhaps as much for show as substance and it did not stop him promoting radical change through infra-structure. 'Manners like the husk of a cocoa nut,' said Southey, whose friendship persisted despite Rickman's detestation of his and everyone else's poetry, 'but his inner nature like the milk within.' He had strange habits, such as spending his time driving his coach around England – his daughter estimated that she had seen every cathedral in the country by the time she was eight. But he was not without friends. Charles Lamb, the writer, who lived opposite him in London, remembered a man who was 'the finest fellow to drop in a-nights, about 9 or 10 o'clock – cold bread and cheese time – just in the wishing time of night when you wish for somebody to come in'.[33] Most exceptional of all, though, was his intelligence. 'Oppressively full of information,' recalled Lamb, 'you need never to speak twice to him.'

This trio – Telford, Hope and Rickman – intermeshed so tightly that the loss of any of them might have brought down everything that they set about doing in the Highlands. As Rickman observed in the 1820s, 'it is lucky that Mr Hope, Mr Telford and I have all lived 17 years as the death of any one of us would have produced a terrible derangement' – and they were to serve another decade together after that.[34] When Telford was in London, Rickman was always on hand to discuss decisions, costs and controversies over an evening glass of wine; when he passed through Edinburgh, Hope hurried to catch him to do the same, and when Telford hastened north to see the men who were doing the work on the ground, they too had their questions and concerns. And when he was away from all of them they missed him. 'I long for his return,' Hope wrote to Rickman when Telford was making the second of his two visits abroad in 1813.[35]

Over the three remaining decades of Telford's life, this relationship led to construction the scale of which is hard to compute: the building of over a thousand bridges; 1,200 miles of good road across tough terrain; and, since the commission also took on responsibility for harbours and fishing ports, forty-three of these, not counting his private work on other Scots ports, his work in Shropshire, on canals in England and – eventually – the new road across Wales to the Menai Bridge and Holyhead. In the Highlands, he eventually added to that great tally a network of new state-funded churches and manses. He was, punned Southey, the 'Pontifex Maximus'. When it was done, the road that carried fishermen to the village and the fish to the cities, the church in which they prayed, the port which landed the herring, and the harbours from which some of them emigrated to new lives in North America: all of them were his.

9

Telford's Tatar

In the summer of 1819 a boy on the brink of manhood watched as the Highlanders gathered. Alexander Ranaldson Macdonell, the 15th chief of Clan Macdonell of Glengarry, stood before his men by the shores of Loch Oich, in the Great Glen which slices from sea to sea in the upper part of Scotland. It was as if the defeat of the Jacobites had never happened; as if roads and steamships were not intruding into his lands; as if he was not – himself – already a curiosity for travellers captured in caricature in the novels of Sir Walter Scott and a martial portrait by Sir Henry Raeburn, which has ended up since then on countless tins of shortbread and bottles of whisky.

That hot day by Loch Oich there was wildness. Highland games were played, not just of the sort which continue today – dancing, piping, lifting a heavy stone, throwing the hammer – but those of a violent past, too. A cow was slaughtered before the crowd and men challenged to twist the legs from the carcass. One succeeded after an hour, to be rewarded with a fatted sheep and a speech in Gaelic from Glengarry. There was a race from the ruins of the old family castle, burnt by British troops after Culloden. As the boy who watched it recalled in later life, 'the young men who ran came in exhausted, and almost in a state of nudity, for they had thrown off their kilts on the way, and arrived in their shirts only. A blanket was cast over them and a glass of whisky administered.'[1] There was more drinking, 'riotous orgies were kept up in the house the whole night', basins of

8

broth, boiled and roast mutton, beef, fowls, salmon, potatoes, and oaten bannocks for which 'there were no knives and forks but the men's dirks and skean dhus were called into requisition'.

All this demented celebration had a purpose. In part it was a piece of vainglory by Glengarry, 'a man of excitable disposition, desirous to be considered the type of an old chief, absolute in his commands, litigious and sometimes hurried by his ungovernable temper into acts of the most serious nature'. He was rumoured to have killed at least one man on his estate. In part it was proof that Highland culture still lived. When Glengarry's personal bard read Gaelic verse that day to the crowd, the words had power. They seemed 'to excite the feelings of the hearers, and to make them vibrate at will from mirth to sadness', wrote the young witness. But most of all, the display was a piece of political artifice. Glengarry was dressed in Highland costume, but underneath he was a man from a modernising time who had attended University College, Oxford and served, rather unsuccessfully, in the British Army; a man whose brother had won the Duke of Wellington's praise for service at Waterloo; a man who was felling his forests for cash and encouraging his people to emigrate abroad so he could clear his land for sheep; and a man who, even as the Highlanders partied, was writing furious letters to Telford demanding absurd sums in compensation for the new roads and great canal being built through his estate.

Everything was changing in the Highlands that summer and Telford's men were at the heart of it. One of them was the boy who saw Glengarry's games and wrote about them in his old age, Joseph Mitchell, the son of the general superintendent of Telford's Highland roads, John Mitchell (and following his father's death, his successor, too). In Mitchell's view, 'nothing could be more fortunate for the Highlands than the appointment of Telford'.[2] He remembered the region as it was in his youth with limited affection. Inverness was 'a very poor and primitive place' with a few good buildings in the centre, but otherwise 'wretched thatched hovels', a town where a popular entertainment was to see poor Highlanders hanged for sheep-stealing. Born when there was no through road from the south – it was broken by risky ferry crossings – he died after the first steam locomotives had linked the town to Edinburgh and London in just a few hours. These changes began before Telford, of course. The military road above Loch Ness had been built fifty years earlier. But it was the work of the two commissions set up in 1803, in both of which Telford served as engineer – the first to oversee the building of the Caledonian Canal and the second a new network of Highland roads and bridges – which made the biggest difference.

These projects ran in parallel. When young Joseph Mitchell went to watch Glengarry's summer games it was as an escape from his time training as a mason on the Caledonian Canal nearby; but his father's loyalty and later his own was to the creation of a network of roads. Each project had its own set of accounts, its own chain of command reaching to Parliament in London (and to Telford by mail wherever he was to be found) and its own elaborate, printed annual reports, but their stories cannot be disentangled. Both took up the first two decades of the nineteenth century and when Telford arrived each spring or autumn to see progress on one he moved on to inspect the other. All the way up to Rickman, Hope and most of the commissioners in London, the decision-makers were the same. The core task on each was to shape Highland rock and water to speed travel and trade in the service of the newly created United Kingdom.

Telford was working for a different sort of master now: the state. He knew how to manage the private investors, aristocratic patrons and county authorities who had funded his work in Shropshire and the Ellesmere Canal. But in Scotland he became subject to orders from bureaucrats, the very type of men he had despised as 'official insects' in the first draft of his memorial poem to Burns. There were advantages: it brought him greater prestige, secure funding, wide powers, the loyal support of clever men such as Rickman (in Parliament) and Hope (their lawyer in Edinburgh) as well as a roll call of impressive commissioners which stopped only just short of the prime minister. But it also tied him into an unfamiliar, constraining system which sat uneasily with his free-wheeling style.

He almost tripped up at the start. In October 1803 Telford found himself charged by the fastidious Rickman with failing to account properly for funds on the Caledonian Canal. For a man so little interested in money and never corrupt, the charge must have stung. He sent a desperate letter to his accuser apologising for 'my not having strictly adhered to the letter of instruction'.[3] 'I hope the Commissioners will admit to it as a mistaken zeal,' he wrote, pointing to 'the difficulty of getting an entirely new business of this nature brought into the form which it is both proper and necessary it should be carried on in'. He promised to prove himself through 'incessant exertion'.

And from now on, he did. His official correspondence from this point – of which, though incomplete, an immense quantity survives – took on an increasingly formal structure. Telford often ordered the points he wanted to make in tidy, numbered lists and he submitted detailed monthly accounts to Rickman who, in return, came to trust him utterly. 'I do not think that the Commissioners acting for the public. . . can deviate from the plain rule of seeing the public money well spent,' Rickman told Hope in 1809, 'and that by the only means in their power, Mr Telford's eyes.'[4]

'Mr Telford's eyes' is a neat way of putting it. With work on this scale, it is easy to imagine that his involvement in Scotland was no greater than that of an old master painter whose output owed more to the 'school of' than the artist himself and whose personal hand is only suspected in some fine bit of brushwork on the eyes, or detail on a lace cuff. And there is truth to this. Telford did not stand by every river crossing and culvert as it was built. He was active elsewhere, on the Ellesmere Canal and later the great road to Holyhead. But he was immersed in the Highlands, too, and frequently present. Over two decades he selected, watched, walked with, checked, cautioned and encouraged a network of men who knew him personally. He was familiar not only with the rivers that his bridges crossed but with the best source of stone; whether the land nearby could support the farming of black cattle; and probably the names of the farmers affected. He was welcomed in every inn and was a friendly and familiar face not only to the innkeepers but to the servants too. His letters are full of intricate details about things such as the right sort of nails to use and the price and availability of oatmeal to feed workers.

They are also packed with details about budgets and costs. The problem was not so much a lack of capital as establishing a mechanism to spend it well. Each of the commissions took a different approach. From the start, the commissioners for the Caledonian Canal were clear that their project had to be paid for in full by the government – and as costs rose, so did the call on the taxpayer. No investor, after all, would take on the risk of a possible strategic folly whose economic potential had not been proved (and in the end, turned out hardly to exist). The Commissioners for Highland Roads and Bridges did things differently. Roads in remote places increased the value of the land through which they passed. They could not be built 'without the aid of public liberality and encouragement', the select committee had agreed when the commission was set up, but 'it seems just that the landowners should assist the public to the

extent of their abilities' – an early and successful form of public–
private partnership which remains the goal of infrastructure
planners two hundred years later. What these abilities were and
how payments could be squeezed out of men in the far north
of Scotland was left to Telford, Rickman and Hope to decide.
The commission was given an initial grant of £20,000 and Hope
designed a system of indemnities to raise the rest: a system
whose administration gave him headaches for years to come.

They got moving fast. In the summer of 1803 the commissioners for roads met frequently in London and advertised for
applications for their funds in two Edinburgh and two Glasgow
newspapers. This set the pattern for what became a settled
process. A landowner would first make the case for a road and
ask for government support to have it built. Telford – and his
assistants – would survey the proposed route. If it was agreed it
made sense, then Telford's team would go on to oversee detailed
designs. Finally tenders would be let to private contractors who
carried out the work under the scrutiny of a network of official
superintendents, answering to Telford.

Roadbuilding was exacting work and he wanted it to be done
well. He drew up detailed standards: contractors were to make
sure their roads were at least twelve foot wide and mostly more
than that, built 'in a substantial and workmanlike manner, to
the satisfaction of the Engineer employed by the Board', and
bridges set on firm rock foundations where possible.[5] He was
also careful to specify a road surface which could protect the feet
of the Highland cattle 'whose accommodation and good condition when brought to market is a primary object of all Highland
road-making'.[6] But local conditions could defeat even these
precautions. Gullies which seemed dry in the summer could
turn into torrents in winter, destroying the new bridges. Snow
was a challenge; so was a lack of tools, one of Telford's first
tasks being to assemble a collection of useful equipment in Fort

Augustus at the western end of Loch Ness. In remote places the very idea of a road was foreign. 'Is it that the surveyors will not listen to the desires of the inhabitants,' Rickman asked Telford in 1806, 'or that the inhabitants do not know what they want?'[7]

One surveyor, sent to Caithness in the late autumn of 1808 to oversee work, was not heard from again until the following spring: Hope speculating that whisky, not hard work, was the culprit. 'Your hospitable Country ruins him,' he wrote.[8] The man's daughter had to be sent north to drag him home.

Though the budget awarded by Parliament was large the work turned out to be more difficult than anyone had thought. Costs rose. Construction was slow. Among those proved to have been over-optimistic was Telford himself, who had predicted in his second survey of the Highlands in 1802 that his proposals, which involved four major bridges and over a thousand miles of road, could, if 'undertaken by proper persons. . . be executed in three years'.[9]

In the event, it took almost two decades: something that Telford must have suspected from the start. Like engineers on so many projects over so many centuries (including our own) he was more interested in persuading politicians to begin building than in the financial consequences of doing so, and he was quick with his excuses as projects began to slide. Progress was 'less than expected', he admitted after inspecting one of the earliest schemes in 1805, seemingly unconcerned.[10] 'In works of this kind,' he told Rickman a few years later, 'widespread, executed by contractors indiscriminately employed and amongst a people just emerging from barbarism, misunderstandings and interruptions must be expected'.[11]

Part of the problem was that no one knew what it really cost to build roads in the Highlands. Rickman, Hope and Telford soon found that the most optimistic, cheapest bid from an inexperienced contractor was not always the most realistic. They began to award work on performance not price, but even so, many of their eager new contractors soon struggled. 'His agitation and misery, poor creature, excite my compassion,' Hope wrote to Telford in 1804 about one who had got his sums in a twist and whose winning bid was £1,000 lower than he had meant.[12] He, at least, was allowed to increase his charges to cover his costs but many like him lost money, caught by rigorous obligations to complete roads that turned out to be far more expensive to build than they had ever guessed. This was not always their fault. Labour prices went up fast as contractors competed to find men who could do the work. 'Convince the people of the Highlands that if necessary the work may be carried on without them,' Telford ordered one; 'This will soon dissolve any weak combination that may at present exist, and I have no doubt that in a very short time plenty of men will offer for employment.' Telford had a traditional view of loyalty and no time for combinations – nascent trade unions.

Slowly, things began to turn around. By 1812 almost forty road schemes had been completed or were under way, at a cost to the government of £120,000. They ran not just in the central Highlands but across islands such as Arran, Jura, Islay and Skye. Each year Rickman drew up a parliamentary report containing an elegant hand-tinted map showing routes that were being surveyed, constructed or built, with the southern boundary of the Highlands marked in a sweeping blue line that ran north of Stirling and just south of Perth. The length of the completed routes on the map grew with each publication, joining on to Wade's old military roads, marked in red, to provide the northern part of Scotland with something close to the main road network it has today.

Even making this map proved a challenge. When Telford and his team set off they found that the most commonly used plan of the Highlands put some of the Western Isles in the wrong place, misread the coasts of Skye and Islay and only guessed at county boundaries. The commission searched the archives and soon turned up records from a pioneering but overlooked survey carried out decades before by William Roy, the founder of the Ordnance Survey. Using this it was able to draw a more accurate – but less charming – map showing the boundaries of the land that it intended to tame.

These borders mattered. As the work grew, Telford's road-building team expanded with it. He organised it into six regional divisions, each led by a superintendent, with a general superintendent or chief inspector as his resident deputy at their head, eventually living in Telford Street in Inverness, which runs out from the town towards the entrance to his canal (and which is today home to one of the lesser memorials to the engineer, the Telford Retail Park). This was a lonely and exacting position. James Donaldson, the first general superintendent, died in the job in 1806, soon after starting it, and the second, John Duncombe, was destroyed by the role. Telford had imported him from Shropshire where he had thrived on the Ellesmere Canal. The north of Scotland was another matter. 'Duncombe seems to be getting into his dotage,' Telford complained less than three years after his arrival, 'there is no getting him to finish things in time. I have for ten weeks stopped his salary and shall pay him only by the mile for what he really does.'[13] Perhaps drink was the problem. Duncombe 'was ill of a liver complaint', Matthew Davidson – by now working on the Caledonian Canal – told one of his sons in March 1809.[14] Soon after, he was dead in the worst of circumstances. 'I am vexed about the old fool – his dying will be no matter of regret; but in a jail at Inverness is shocking,' Telford wrote to Rickman – an exceptional note of unkindness from a man who was almost always generous to his juniors.[15]

Telford made sure that Duncombe's successor was made of the best granite. John Mitchell, the father of Joseph, who watched Glengarry's Highland games, was a man of superhuman strength and good character: Telford called him his 'Tatar'. To him more than anyone else went the honour of getting the work on the roads done. His loyalty to Telford, who had picked him out from the ranks and promoted him, was total. 'He could scarcely read or write,' when Telford discovered him as a mason, Robert Southey, who met him in 1819, wrote, 'but his good sense, his good conduct, steadiness and perseverance has been such that he has been gradually raised to be Inspector of all these Highland Roads.'

Mitchell must indeed have been a man of terrifying solidity, 'of inflexible integrity, a fearless temper and an indefatigable frame', wrote Southey. 'No fear or favour in the course of fifteen years have ever made him swerve from the fair performance of his duty, tho' the lairds with whom he has to deal. . . have attempted to cajole and to intimidate him equally in vain. . . they have not unfrequently threatened him with personal violence. Even his life has been menaced; but Mitchell holds right on.' This was no exaggeration. 'In the execution of his office he travelled last year not less than 8,800 miles,' the Poet Laureate wrote admiringly, 'and every year he travels as much, nor had this life and exposure to all winds and weathers and the temptations either of company or of solitude at the houses at which he puts up led him into any irregularities or intemperance. . . he is still the temperate, industrious, modest, unassuming man as when his good qualities first attracted Mr Telford's notice.'

Yet even for Mitchell, the work was health-destroying, jolting over rough tracks in a cart where it was possible, otherwise travelling by pony or on foot. He died in service, still young, in September 1824, 'a sacrifice to over-exertion', Telford admitted, 'with such unremitting zeal as destroyed his vigorous constitution'.[16]

He knew this first hand. Each summer Telford joined his inspectors for hundreds of miles of these difficult journeys in conditions which eased only slowly as their road network grew.

One of John Mitchell's last contributions before his death was to design a large caravan, big enough to sleep sixteen to eighteen men, with a fireplace, that could shelter workers and be moved from site to site: an improvement on the military canvas tents used previously, so hot and cramped men 'could not endure fires to cook their victuals, and dry themselves'.[17]

If the roads were hard, an even bigger risk lay in the wide crossings of the fast-flowing rivers that cut deep into the Highlands, particularly in winter when they were alternately flooded or frozen. Even in summer tidal estuaries posed a risk. In August 1809, Mitchell had a lucky escape when one ferry 'with between 100 and 120 passengers went down about the middle of the river, whom only 5 persons of the whole number were rescued'.[18] He had arrived at the shore only an hour after the boat had set off.

The bridges which Telford's team threw across such rivers made the biggest difference of all to Highland travel. One of the first of them still stands at Dunkeld, a town on the broad River Tay north of Perth. Telford had begun work here just downstream from the ruined riverside cathedral even before the formal commission process was in place. The Duke of Atholl had offered to pay half the cost and give up his right to the old ferry dues so long as he could charge tolls, and secured his own Act of Parliament to do so. A price was agreed by the commission 'as estimated upon Oath by Mr Telford'. By 1805 he could report that 'considerable progress has been made' and its completion in 1809 made safe a crucial crossing on the route north to Inverness, 'Dunkeld being as it were the portal of the Central Highlands, and more remotely the access to all the Northern Roads' as the commission put it later.[19]

The bridge at Dunkeld is an elegant seven-arched structure in cut sandstone, grand enough to grace a city centre, with a little prison house under its northern end whose rusted iron bars are still there, locked shut. It is the finest and most ornamental of Telford's masonry structures in the Highlands, but in design and appearance it is traditional, as are the thousand or so less sophisticated crossings, which his contractors threw across streams and smaller rivers to a general specification he had drawn up.

Dunkeld was one of four strategic crossings identified in his 1803 report as places where 'accidents frequently happen' on ferry crossings. At the other three, all on the route completed through the spine of Scotland from the Lowlands to the far north – over the Spay at Fochabers, over the Beauly and over the Conon near Dingwall – stone bridges were eventually built. But they did not prove ideal. 'In highland rivers, where in general the stream has not before been noted with any correctness,' Telford reported, 'we have already found that the water-way which at the time of building was supposed to be unnecessarily ample has experimentally proved too small'.[20] The question was how to build bridges that were higher and lifted clear of the torrent. The answer was to turn to iron.

The best of the wonderful bridges which resulted is in place today, though no longer used for vehicles, at Craigellachie, on the River Spey. It leaps across the foam-flecked, pebbled river like a deer springing from crag to crag, the elastic energy of its iron struts held firm at either side by the medieval pretence of fortified towers, with arrow slits and castellation. On the northern shore the road cuts hard and sharply to the right, climbing out of the valley in a deliberate model of a picturesque landscape which still impresses, even though the romance of the scene has been spoilt by the graceless concrete bridge which carries the fast modern road just downstream.

'As I went along the road by the side of the water [I] could see no bridge,' wrote Southey when he visited Craigellachie in 1819, 'at last I came in sight of something like a spider's web in the air – if this be it, thought I, it will never do! But presently I came upon it, and oh, it is the finest thing that ever was made by God or man!'

Craigellachie was significant for another reason too: it was a product of that network of alliances which Telford had built up over decades. The ironwork was cast by William Hazledine and the masonry overseen by John Simpson. The pair had built the iron aqueduct at Pontcysyllte and now barges crossed this great structure carrying iron parts for a second, from Hazledine's forge nearby. The logistics of this operation were testing, but the speed with which it was done shows how well Telford's teams of contractors were now working. The parts arrived at the site in April 1814. By November the bridge was open.

Craigellachie is the most beautiful of Telford's surviving iron bridges, a precursor to many similar designs. But it was not quite the first to be built in Scotland. That was put up north of Inverness, at Bonar, between 1811 and 1812, as a replacement for the ferry crossing in which Mitchell had almost been drowned. This bridge, too, was a Telford team effort: the ironwork was cast by Hazledine and the bridge erected by what became an established pair of contractors, Simpson and John Cargill, who first joined them on the Caledonian Canal and went on to work with Telford until the 1830s. Bonar was a pioneering scheme. Pre-erected and tested in Wales before its portable pieces were shipped north, its success restored national confidence in iron bridge building after a series of earlier structures by other engineers – at Staines in Middlesex and Yarm in Yorkshire – had collapsed. It was damaged by a flood in 1892 and replaced, but the white stone plaque which stood next to it remains. 'Traveller, Stop and Read with gratitude the Names of the Parliamentary Commissioners appointed in the Year 1803 to direct the Making of above Five Hundred Miles of Roads through the Highlands of Scotland and of numerous Bridges' it begins. The gratitude was real and lasting.

And it can be felt, too, along the north-east coast of Scotland, where Telford supervised the reconstruction of harbours, from Peterhead to Aberdeen and Dundee. The most striking of these waterside schemes was built even further north, in Wick, with the support of the British Fisheries Society, the organisation that had first brought him to the Highlands in 1790. In that year he also surveyed ports along the north-east coast for the society, travelling to the northern tip of the mainland. He recommended that a big new harbour be built by the older settlement of Wick. It was not until 1807, however, that he proposed detailed plans for the site, including a bridge over the Wick River and a new fishing settlement on its southern shore. Pulteneytown, as this site was called in memory of Sir William Pulteney (who had died in 1805), is a gently astonishing place, one of the world's earliest planned industrial towns, little known only because of its isolation. It is one of Telford's most significant works and evidence of his continued interest in architecture as well as engineering.

His design breaks free of the industrial grid of streets which the society favoured elsewhere to draw on what the historian Denis Maudlin calls 'a fashionable and aesthetically considered plan', including a central square whose angular crescents appear to echo Sydney Gardens, in Bath. By the time he worked on plans for Pulteneytown, Telford knew Bath well: he had worked as a surveyor on the architect Thomas Baldwin's Bathwick estate in the town. But he took ingenious care to present his plan to the commercially minded British Fisheries Society as practical, not ornamental. 'One great objective in forming the new plan was to exclude the north wind. At present the wind blows right across the flat evenly but it will hit the wall of the crescent and be forced down a side street.'[21] Visitors to blustery Wick may doubt this works, even as they admire its elegance. In 1810 and 1811 Telford expanded his plans to include a large double harbour, which, by 1818, supported 822 herring boats and remained until fish stocks collapsed in the 1930s one of the largest herring ports in the world.

Today there is a sense of dignity and purpose to the surviving buildings of Pulteneytown, with its innovative separation of industrial and living places; its grey stone storehouses and curing yards and cottages speak of a generous confidence that fishing families should live in something better than shacks. Out of stone and soil and the sea, Telford was helping to carve a new world in the north of Scotland. His surveys were turning into solid structures.

It was enough to drive a man's ambition to overreach itself, even his. And that is what happened next.

From Sea to Sea

A ny schoolchild fresh from learning how to read a map might survey the charted world and identify without hesitation a handful of places which seem to call for a canal cutting from sea to sea. Panama would be one. So would the Isthmus of Corinth, where a canal was first attempted two thousand years ago. Suez was already dreamed about when Christopher Marlowe's Tamburlaine says (on his deathbed):

And here, not far from Alexandria,
Whereas the Terrene and the Red Sea meet,
Being distant less than full a hundred leagues,
I meant to cut a channel to them both,
That men might quickly sail to India.

And should Tamburlaine have reined in his ambitions and surveyed a map of Great Britain, two possible plans for a Scottish canal would have leapt from the picture. By Telford's time the Forth and Clyde Canal, completed in 1790, was already open. But a wider, deeper sea-going canal further north linking the Irish Sea to the North Sea, from Fort William to Inverness, though often contemplated, had not been attempted.

The watery cleft through the upper part of Britain lies on a fault so sharp that if a model of Scotland could be picked up and twisted this is where it would snap. It is all but filled by three lochs (the longest of them Loch Ness) and at its highest point rises only ninety-four feet above sea level. Engineers long before Telford realised that if a series of canals could be built between the lochs – adding roughly twenty-two miles of new water-way in total – ships might be spared the risky and unpredictable passage around the top of Scotland through the Pentland Firth. It was a simple idea. In reality it turned out to be an overblown by-product of seductive geography.

Telford seems to have sensed something of this risk at the start. Before coming down in favour of what became the Caledonian Canal, in his second report on Highlands communications in 1803, he took advice. He sought out both James Watt, who had advocated and mapped a canal through the Great Glen in 1773, and the greatest civil engineer of the age, John Rennie, who had put forward a similar plan in 1793 and who gave evidence to the select committee in June 1803, together with Telford's old chief on the Ellesmere Canal, William Jessop. He managed to bring the pair together to endorse the route. In 1801 and 1802 he trav-elled the line Watt and Rennie had proposed (the second time most probably in Jessop's company) and all agreed none better could be found.

Laying out the route was one thing, however. Confirming there would be a use for it was another. Telford asked the opin-ions of merchants at ports along the east and west coasts. 'This sanction of experienced people,' he wrote, 'who are all deeply interested in commercial concerns, will, I trust, satisfy your Lordships that it has not been upon insubstantial grounds that I have ventured to recommend this great national object.'¹ He – or the committee – might have done well to ask whether any businessman would have said no to the offer of a large new canal at no cost to themselves; but anyway, the government was not interested only in the commercial potential. The other attraction in what Telford later described as 'political circumstances' was that a canal might allow Royal Navy ships to cross from coast to coast quickly and out of reach of the enemy: the Napoleonic war was at its height, much of it fought at sea, where a sustained blockade or successful engagement could threaten national disaster.[2]

In his second, more detailed report on Highland communications, in the summer of 1803, Telford proposed a massive route for sea-going vessels, going up and down hills, on a scale never tried before: twenty foot deep, fifty foot wide at the bottom and a hundred foot wide at the surface, climbing to and from the highest point on Loch Oich through twenty-nine huge locks. It was, writes the principal historian of Telford's efforts in Scotland, 'a pioneer work without parallel or precedent which must surely have staggered the imagination'.[3] Telford miscalculated badly, both in the time he thought it would take to build the link – which turned out to be eighteen years – and its cost. Looking back when it was finally done, Joseph Mitchell did not mean to exaggerate when he said that compared with the conjectured Suez Canal, the Caledonian posed 'construction difficulties of a more formidable kind'.[4] The project took on something of the character of Fitzcarraldo's demented efforts to carry the parts of a ship through the Peruvian jungle, in Werner Herzog's film *Fitzcarraldo*. They had started so they had to finish. 'From this I cannot escape without the art of brain transfusion could be discovered and all my memory of the subject placed on another man's shoulders,' Rickman lamented in 1816, after a dozen years of work on the canal with six more to come before it opened, 'but this cannot be and. . . I must drudge on.'[5]

The start, at least, was brisk. By September 1803, Telford was in Inverness and securing terms from landowners along the route as well as permission to use timber and stone from their estates. Everyone seemed keen. It was, one laird told him, a 'publick national work, which should have been attempted a century ago as becoming a great commercial Empire'.[6] Even the awkward Glengarry wrote to agree the canal was 'fit for the advantage of the rising generation'.[7] Both men of course had their eyes on the compensation payments they could expect from the government for giving up land and allowing vessels to travel through their lochs. Glengarry, who controlled Loch Oich, proved particularly vexatious over the next two decades (on one mad day he rode his horse into the loch to chase after a boat carrying Telford, forcing the animal to swim a mile while the engineer reportedly hoped it would sink and drown the troublesome man with it). Hardly less difficult was Major Duff, first mentioned in the canal correspondence in 1803 and still writing boring, pestering letters about the value of his fields in 1829. Duff, 'an able man but somewhat eccentric', had once been held hostage for six months in a church in revolutionary France and declared that this time counted as a down payment which excused him from ever having to attend religious services again.[8] He tormented everyone involved in running the canal. 'Mr Duff being confined to his bedroom,' Telford wrote resignedly in 1820, after calling at his house only to be told the major was too ill to talk to him, 'my communications with him have been through Mr Kinloch' – his agent. On that occasion Duff retreated, soon writing to praise negotiations 'conducted in so kind and amicable a manner by Mr Telford', but within a month his campaign for further compensation was back in full swing.[9] This sort of thing was soul-destroying and seemingly endless.

For Telford, however, the first frustrations came not from Duff but from Jessop. In the Ellesmere scheme a decade earlier he had found himself under the older man's moderate and experienced eye. The same now happened on the Caledonian. On 4 August 1803 the commissioners – clearly not wholly trusting Telford – asked for Jessop's 'opinion and assistance'.[10] Luring this diffident man to meetings, however, proved as hard for the commissioners as it had for the proprietors of the Ellesmere

Canal back in 1793, when Jessop's absence from a crucial vote had allowed Telford to secure his position. First his son replied to the commissioners saying his father was away and could not be contacted. Then he got in touch but declined to attend, citing business in Hereford. On 29 September he wrote again to explain that he had 'been peculiarly unlucky in being two or three times disappointed of the honour of attending at the meeting of the Commissioners', giving his busy duties as the new Mayor of Newark as an excuse.[11] By now Telford, in Inverness, was stamping his feet and waiting for the senior man to arrive. 'As the idea of Mr Jessop being consulted was suggested before I left London,' he wrote impatiently to Rickman while the absent engineer was in Newark trying on his mayoral chains, 'I was in hopes, if his coming down had been considered necessary, that he would have been here by this time.'[12] With or without him, Telford said, he would set off to survey the route from Inverness the next morning, before the weather closed in. It is plain from this letter that he thought the latter's involvement superfluous. 'It is possible that Mr Jessop may not judge it absolutely necessary to view the line,' he wrote in something like cold fury:

I can now show him a correct Map of the Line of Canal, with a Section of the Ground; I can describe the nature of the soil in each district, with the situation and quantity of the materials for Building. He can examine Mr Simpson, an eminent and respectable Builder, who has now examined the matter very carefully, and with whom Mr Jessop has been acquainted for many years – he can also examine Mr Howell who took the Levels and made the surveys for me, and Mr Jessop is likewise acquainted with him, if after all this Mr Jessop should think it necessary to view the ground then doing this early in the spring would be advisable, after we have opened some quarries and proved the ground near the entrances to the Sea.

Telford was laying down a marker to Rickman and the commissioners. He was the boss, or wanted to be. 'I have been constantly engaged in making arrangements as to the forms of, and the execution of the works,' he pointed out. Such eagerness brought a rebuke from Rickman for spending money without properly accounting for it and it also, eventually, lured Jessop north, in October, by which time Telford had left. The two engineers met at last in November, in Shropshire, Telford writing tersely to Rickman that 'I have received yours containing an order to furnish Mr Jessop with the situation of the Caledonian Canal. . . this order has been complied with.'[13] Jessop – not Telford – signed off the estimate on which the canal was built, in January 1804. And the following March there was surely an element of mischief-making in Telford's comment in a letter containing Jessop's charges for his work on the canal: 'Mr Jessop informs me that he has just purchased a commission in the army for one of his sons, and will be glad to receive the amount of this Bill to enable him to pay for this.'[14] This wasn't a friendly act.

Nonetheless, the pair settled into a working relationship which lasted a decade, until Jessop retired in 1813. If only, as Telford put it, 'to avoid blame', each summer he accompanied Jessop on an inspection of the work and together they then drew up a report for the commissioners.[15] 'It would give satisfaction to my mind as well as additional evidence to the Board, to have the assistance and advice of Mr Jessop, while in Scotland,' he wrote, sounding as if he didn't entirely mean it.

Was Jessop's contribution significant? Accounts differ. Charles Hadfield argues that Telford disguised Jessop's significant role in order to take full credit for the canal himself: not mentioning him, for instance, in his autobiography (which plays both ways since by then he knew the canal was a commercial flop and he might well have shunted some of the blame onto the older man). The surviving archive suggests that while Jessop was certainly active and influential, almost always it was Telford or his men on the spot who made day-to-day decisions. The more experienced canal builder shaped the design of locks and played a significant role early on in fixing the rates of pay, among other things. But as another of the project's historians, Alistair Penfold, argues: 'Jessop's role in the project was essentially that of a monitor, rather than a decision-maker.'[16]

So did Telford build the canal? A fair assessment might give principal credit neither to Telford nor to Jessop but to the four men who served in turn as resident superintendents at each end of the project, and the contractors who worked with them. Working from the coast inwards, they met some of the greatest challenges right at the start: the series of locks which lifted the canal from the sea at Corpach, just outside Fort William, and the near-impossible piece of engineering that was needed to form a navigable route from the shallow muddy shore of the Beauly Firth, where the canal began at Clachnaharry, near Inverness. They put up with terrible conditions and had to adapt plans which took many years longer to complete than anyone had guessed. At Corpach, the superintendents were John Telford, from 1804 to 1807, and then Alexander Easton. At Clachnaharry they were Telford's old friend Matthew Davidson, from 1804 to 1819, and then his son John Davidson. These were the true heroes of the Caledonian Canal.

The first pair – Telford and Davidson – were appointed by the commissioners on Telford's recommendation in June 1804, following the passage of the second Caledonian Canal Act, needed to secure funds for construction. As he explained to Jessop: 'The works being upon a scale of uncommon magnitude, and in a district of the country unaccustomed to operations of this nature, I propose. . . persons. . . as have to my own and your knowledge, for ten years past, been employed upon works of a similar magnitude whose abilities may be relied on and who are likely to enter with zeal into the spirit of the undertaking.'[17] Both men were imported from the Ellesmere Canal. Davidson, of course, was one of Telford's childhood allies and a man whose morbid intelligence rotted in the isolation of Inverness. John Telford, who led operations at the other end of the canal at Corpach, may have been an Eskdale relation: he was described as 'your kinsman' in a letter to Telford following the former's death in 1807.[18]

They faced their monstrous task with a shared cynical humour. John Telford probably had the harder time of it, his early cheerfulness soon ground down by the work. He assured Davidson that Inverness was a paradise compared to Fort William, a town so remote it only held a market twice a year. 'The colony in the snowy regions. . . desire to be remembered to their friends in the milder climate of Inverness,' he wrote to Davidson, who held – or feigned – a deep dislike of everything to do with the place in which he found himself living.[19] 'If he hated anybody he hated the Highlanders for their inert ways – or said he hated them, for I am sure he did not,' Joseph Mitchell recalled.[20]

That assessment was perhaps too generous. 'You may assure Mr Davidson that there was not one Highlandman smothered,' John Telford wrote to him after a storm had killed several men in Fort William in 1806, 'they understood self-preservation too well.'[21] Davidson's depressive misery about the Highlands appears real and deeply felt. 'My soul is tired of this accursed country – joy has been a stranger since I first saw its accursed face – if there be a Scotch highlands in heaven, I would rather be damned than go there', he wrote to Telford, 'I am so vexed I could not resist telling Mr Rickman this – excuse me – good night.'[22] Poor Davidson. 'Having been of late exposed to much wet and cold I have been unwell for some days,' he ends his letter, 'this *amor patria* will certainly be the Death of me.' It was: he died in Inverness, still building the canal, in 1819.

For such men, as they began their task in the summer of 1804, there were few comforts. 'A pipe and a glass of ale was their usual solace in the evening,' Mitchell remembered.[23] Even John Telford's good spirits seem to have suffered; in Fort William, books and newspapers were all but unavailable and when – in the summer of 1805 – he planned a visit to Davidson in Inverness there was no time for pleasure. 'I must return the next day, as my expenses the last time I visited you were deducted.'[24] Rickman's bureaucracy drove men hard.

Their first challenge had been not just to start construction, but to find the workers to do it and enough food to feed them. 'There is not a peck of oatmeal in Lochaber,' John Telford wrote to Davidson in November 1804.[25] When the harvests were poor, as in 1807, the project's managers feared famine: the commissioners considered buying enough oatmeal to feed a thousand men for six months. Conversely, when food was plentiful the workers tended to drift away. 'The herring season has been most abundant, and the return of the fine weather will enable the indolent Highland creatures to get their plentiful crops and have a glorious spell at the whisky making,' Telford wrote from Inverness in 1806.[26] Finding enough men to make progress was hard. In the summer of 1807, Telford wrote to Rickman celebrating a rare moment of success, 'the season being favourable, the days long, and the workmen disengaged from potato fields and fishing'.[27]

The seasons continued to affect work on the canal: most of those employed on it were Highlanders, despite repeated criticism that the project was too dependent on outside migrant labour. But most difficult of all was comprehending just what the project actually involved. At the Inverness end, the entrance at Clachnaharry became the scene of a drawn-out muddy struggle, which took Davidson nine brutal years to overcome. The shallow waters meant that the entrance to the canal had to be taken far into the sea, 400 yards beyond the usual high-water mark. Dredging the soft mud was impossible: when an iron bar was plunged into it, it proved to be fifty-five feet deep. In the end Telford and Davidson devised a solution which took a huge effort. To create solid foundations and sides to the entrance, they extended the land into the sea. Massive quantities of stones and clay were piled onto the site and left to sink in, squeezing out the mud beneath until a solid mound remained into which the trench for the canal and its lock gate could be dug.

There was nothing glamorous about this task and the end result was undramatic, if effective, but it vexed Davidson and Telford as much as any other project in their careers. 'I congratulate you on having conquered the grand difficulty of the Clach sea lock,' Rickman wrote to Davidson in late 1811, when the thing was nearly done and Telford was on his way north to inspect it.[28] Today, as yachts rock quietly in the marina which sits just inland of the sea locks guarding the canal, there is no trace of how difficult a piece of engineering this was.

At the other end, to the south-west in Corpach, the task was no easier. Here, John Telford and later Alexander Easton had to oversee what Telford called a 'series of the largest locks ever yet constructed': the staircase of water which lifted the canal from sea level through two locks and then, after a mile of cutting, up eight more.[29] The great chambers for these were hacked into solid rock, with a Boulton and Watt beam engine kept running almost non-stop to prevent the works flooding. What was once a silent corner of coastal Loch Linnie became a place of industry and noise, with a railway to bring stone from a quay and iron pieces for lock gates and bridges sent up from Hazledine's Welsh foundry, and other parts shipped from every corner of industrial Britain. With this work came a small town of labourers, among them at one stage the young Joseph Mitchell, who remembered how workmen slept 'in temporary beds, one above the other, constructed like the berths of a ship', some of whom 'drank continuously from Saturday til Monday or Tuesday night, without food and without sleep'.[30]

The pace of construction was relentless, too. 'Mr Jessop is arrived and I have with him examined the works at this place, they are going on well,' Telford reported from Clachnaharry in September 1805, adding that 'the Lock is completed in its Inverted Arch' – the elegant curved masonry base which prevented the structure collapsing in the troublesome soft mud.[31] A dozen days later the two men were at the other end of the canal at Corpach, where work on the staircase of locks had begun. In between these two places there was little activity yet, but from this year on, operations on the canal settled into a routine, shaped by the availability of labour, the weather and Telford's usual biannual inspection visits, one with and one without Jessop. Much relied on the good sense of the men on the spot. 'After the principal engineer has decided upon the most advisable outline of the operations,' he wrote in his *Life*, 'very much depends on judicious workmanship, and attention to practical suggestions.'[32]

The other thing that mattered was money: there was never enough of it. Operations often ran ahead of what could be paid for. 'Last Saturday was pay day,' John Telford wrote in 1804 after his men demanded increased rates, 'and a very disagreeable one it was. . . Mr Wilson and I were in eminent danger of our lives. . . they threatened much and were on the point of using violence several times.'[33] To men at the sharp end like this it cannot have been much comfort to receive letters from Telford, who found himself acting as an accountant, with vast sums passing through his account at Hoare's Bank, urging them 'to keep the monthly payments under £1500 per month'.[34] 'I have, for a long time past, kept a strict rein upon them,' he told Rickman in 1808: even a life-long friendship did not prevent him paying Davidson less than he could have hoped for.[35] 'Wishing to have kept the expense of management as low as practicable, I certainly fixed the salaries below the average of what such persons usually are paid.' A few years later, the unhappy Davidson sent a distraught letter to Rickman after being accused of mishandling funds. 'I am sorry to have occasioned so much trouble to the Parliamentary Commissioners and to Mr Rickman and have no excuse to offer in extenuation other than that the whole originated in a wish to simplify the keeping of the accounts,' he wrote, offering to repay the money from his savings.[36] Two months later – after receiving Rickman's reply – he wrote again: 'I can see no good or useful purpose to be attained by troubling Mr Telford with my presuming and indolent blundering. In the course of the last 25 years Mr Telford and I have had quarrelling enough about the conduct and execution of works, but never once a disagreeable word about money matters.'[37]

The misery which often seems to have overtaken Davidson can only have deepened after John Telford's death in the spring of 1807. 'I shall have some difficulty in replacing him with a successor in whom I have so perfect a confidence,' Telford wrote, on the day Davidson attended the funeral.[38] He managed by appointing Alexander Easton, who had begun work as a roads surveyor for Telford aged eighteen and found himself in charge of overseeing all the works on the western end of the canal aged just twenty. It was a typically bold choice by Telford, who built lifelong loyalty by spotting talent among his young workmen and promoting them fast; Easton was still working with Telford on the latter's death, in 1834. Under the new partnership of Easton and Davidson, 'every matter upon the canal is going with perfect regularity', Telford could report a year later.[39] The truth was that only the entrances at each end were well advanced and in the middle section almost nothing had been done at all. A dredger, built to deepen Loch Oich, had sunk.

The government was at least pouring money into this quagmire. 'I am much relieved by the mode of settlement now adopted by the Treasury,' Telford wrote from Clachnaharry in the summer of 1809, 'matters will go on with promptitude and regularity, it is a great satisfaction.'[40] But as he wrote soon after in a private letter to Rickman, even this was not enough: 'the time in which the Canal will be completed. . . depends upon the annual sum of money allowed'.[41] Meanwhile Davidson was driving himself into the ground. 'By working hard and carrying my own instruments I have nearly lost all that consequence attached here to office,' he told Telford in one of their few surviving pieces of informal correspondence, 'any importance remaining (which is very little) arises from the liberty taken of addressing my letters to R.H. The Speaker – you laugh – but this is not all joke.'[42] 'Mr Rickman can have no conception of the consequence it gives a man here, to correspond with an MP,' he added, 'besides, we are aware, that St Paul exhorts us to Mortify our Members.'

The canal was nowhere near finished but a lot of money had been spent. Meanwhile, though Telford continued his visits to Scotland, his attention was roaming elsewhere. Travel, for him, was nothing new: his surviving letters to Rickman between 1803 and 1811 come from almost thirty different addresses. This stood out as extraordinary behaviour even to those who knew him well. 'I hope you are safely arrived at home if a person who moves about so much can be considered as having one,' James Hope wrote to him in the autumn of 1805.[43] But now – and unbelievably, with Telford in his sixth decade – the pace picked up. 'I have had a complete perambulation of North Wales,' he told Rickman in the summer of 1810: the first sign of interest in what became his great road to Holyhead.[44] But that was not all. In 1808 he had written to Rickman about a more adventurous project still. 'I have now to mention a private matter,' he told him.[45] 'The King of Sweden [has] been convinced that an inland Navigation, on the scale of the Caledonian Canal, if carried through the lakes and valleys of that country, would be of great consequence to the prosperity of his people.' The king had 'issued an order. . . entreating my assistance in so finally adjusting the line of the canal and forming a plan for the execution of it.'

Building one vast, difficult canal from coast to coast in remote and wild country was not enough of a challenge. Telford was about to start on a second: in a country he did not know, whose language he did not speak, which was on the front line of assaults from Napoleonic and Russian forces in Europe's wars and whose winters were so wild that by contrast the extremes of Highland weather seemed like summer.

The Enjoyment of
Splendid Orders

'Sir,' the letter began, 'You will I hop excuse me for introducing myself quite a stranger to You in this manner and
I fear in Englisch not of the best sort.'[1] It was a hesitant start to
what became a deep friendship between two men separated by
class and country. Count Baltzar Bogislaus von Platen, the son
of a Swedish field marshal, was a senior naval officer and later
governor of Norway, one of the elite in a nation still famous
for its recently lost empire; he was writing to Telford, who was,
when the letter reached him in May 1808, simply an engineer
squelching through the mud at Clachnaharry.

Like much of his post it had chased him around the country
(sent to Edinburgh, the letter was forwarded first to Ellesmere,
then Shrewsbury, before finding him in Inverness), but unlike his
familiar heavy mailbag full of accounts and orders from Rickman
and Hope, this was accompanied by a formal decree from the
King of Sweden asking him to work on what became the Göta
Canal. 'His Majesty Graciously wishes that any of the most
experienced Canal Engineers in England,' it informed him, 'and
particularly amongst them Mr. Telford to whom the execution of
the Caledonian Canal is said to be trusted may be prevailed upon
proper terms to take a vieu of this undertaking and give his opinion and council thereupon.'

Von Platen's accompanying letter explained this lustrous command, which must have shone all the more when read by candlelight in Inverness, so far from kings and their courtiers. 'The reason for this You will learn of the inclosed extract of His Majesty the king of Swedens instruction to me about marking out &c. of the new canal wich is to pass through Sweden,' he wrote in tangled English, 'His Majesty wishing the performance of this national work to be made in the most perfect manner wich the national circumstances alow has naturaly looked for advices from a Country were such performances are common and particularly has fixed his Gracious attention upon the first of all these performances the Caledonian Canal and his Chief; known by several publications about his work not unknown in Sweden.'

Among the things that stands out from this passage – perhaps best read aloud in a cod-Swedish accent – is that it was Telford, not William Jessop, who was being invited to help: in reputation, if not yet quite in title, he was the 'Chief' of the canal.

Telford enjoyed the attention of the ruling classes almost as much as he skipped over their ostentation, and was flattered by the royal request that he travel to Sweden at once to survey the route. The project, von Platen assured him, was 'a work belonging not only to Sweden but in a great deal almost to the whole world'. 'As the time for acting is fast coming on in this country,' he added, 'I hope You will favour me with a speedy answer and I request You once more for the common sakes that it be a good one.'

It was. Telford replied on 2 June from 'Upon the Caledonian Canal' (an address surely calculated to impress von Platen, since he never seems to have used it on any other letter).[2] He would, he said, let 'no trifling obstacles or inconveniences' get in his way and could set off the following month. He set one condition: 'these Northern Seas are at present much infested with Danish and French privateers, this makes a passage in a common trading Vessel to be attended with a risk from these armed Vessels – however trifling. I must therefore stipulate that I shall be taken up at Aberdeen and also landed at the same place by either a Swedish or English ship of war.' Quite a demand.

In Sweden, von Platen charged into action with these orders, visiting the English admiral on Baltic service, Sir James Saumarez, to arrange an armed convoy. 'I come just now ashore from the *Victory*' – the ship that had played a famous role at Trafalgar – the count wrote.[3] In Scotland, Telford prepared for the expedition. As he told von Platen he was already 'fully aware of the general importance of the proposed undertaking' through his contacts with the Swedish-born but British-trained engineer who was working on it: 'my friend' Samuel Bagge.[4] Telford was a supporter of Bagge's (a man who ended one later letter to Telford as 'your devoted disciple') and suggested from the start that he should 'stand first in the execution of this great national work'.[5] It may well have been Bagge who first gave von Platen Telford's name.

The broad outlines of the scheme had long been familiar: to link the Baltic and North Sea coasts of Sweden with a canal that, as with the Caledonian, joined inland lakes together to make them passable to shipping. For Sweden the attraction was to avoid the narrow sea passage through the Oresund, on which Denmark levied heavy tolls; for Telford an added draw was that the Göta Canal might form a sister to his Caledonian route to provide an easy passage from the Atlantic to the Baltic. Survey work had been done intermittently since the 1770s and the first part of the route across Sweden had been completed in 1800, a canal between Gothenburg on Sweden's west coast and the huge Lake Vänern, in the centre of the country. Now von Platen was directing energies towards completing the link onwards to the Baltic in the east, the missing link in a route from sea to sea of some 380 miles. 'Its realisation became the sole aim and purpose of his life, a cause to which he dedicated himself utterly,' says one historian.[6] The count wrote to his chosen engineer as a fan might to a rock star; and the gushing tone, once begun, rarely faltered.

For a Scot who had never travelled abroad and had never made a sea voyage (despite having been commissioned by the Treasury to design harbours), the offer must have been thrilling; his worry – other than his safety at sea – being the reaction of his employers on his Scottish schemes to the disappearance to the Baltic of their indispensable engineer. This must be why he waited until the last moment to send his letter alerting Rickman. Writing from Leith, where he had expected 'the Convoying Ship of War' to leave the next day, he said: 'As no warship is arrived, that day is now not likely; and it is possible that the same cause which brought back General Moore, may change our relations with Sweden' – referring to the likelihood (soon realised) that Sweden would fall under French influence and so out of reach. 'In that case my journey will be prevented, which I shall not much regret,' he explained (disingenuously, for he was surely eager to travel), 'but in case I do go, it is right to state that every matter relating to the Roads and Canal is put into a regular train.'[7] Rickman, anyway, was left with no chance to object. By the time the letter reached him, his engineer was at sea.

The crossing turned out to be swift; but in case it were not, Telford had packed enough supplies to carry him across the Atlantic. As well as buying sets of blankets, sheets and pillows he placed an order with Bell Rannie & Co. (a famous Leith firm which had supplied Bonnie Prince Charlie with the claret that accompanied his invasion in 1745) that included two dozen bottles of Madeira and the same of port; three dozen bottles of cider; six dozen of porter; half a dozen each of gin and brandy; a range of teas; three bacon hams; three pickled tongues; and a vast quantity of sugar and biscuits. Unless (and it is possible) he anticipated a good deal of entertaining, the man who had once boasted of his abstemiousness while working for Pulteney must now have drunk with the best of them, but even Telford and the two men who travelled with him – Robert Bald, later a renowned Scottish mining engineer, and John Hughes, who had worked on plans for the Caledonian Canal with Jessop and then served as one of Telford's assistants – could not have made much of an impression on this pile before they reached Gothenburg six days after setting off.

Hopefully sober and no longer seasick, Telford met von Platen on 8 August and formed, as he sometimes did with powerful, driven men in search of a practical companion to realise their vision, an instant, intimate bond which was to last until death. The scheme the Swede set before his new friend was even more ambitious than the immense canal being dug back in Scotland. Of the route's trajectory, fifty-four miles were to be man-made and, as finally built by 58,000 conscripted soldiers, fifty-eight locks were required to raise it almost a hundred metres above sea level on a route which then as now was frozen for many months of the year. The contrast in numbers between the unfortunate, untrained horde who worked in Sweden and the far fewer but better-drilled men needed to build the Caledonian was a tribute to the management skills of Telford's Scottish superintendents: 'I foresee you will laugh at me for the great number of men we employ but it is hardly ever to be done without in this Country,' von Platen admitted: later he defended this as a way to train engineers, but the wage bill ended up all but bankrupting the project.[8]

The pair set out to reconnoitre the route immediately, before the cold weather came, and for the next six weeks Telford worked almost without a pause; only breaking off once, in Stockholm, to dine with the Secretary of State. He settled on a line that was close to the one first laid out late the preceding century, but he adjusted it and increased the number of locks, and tackled the particular difficulty of the shorter, western, section of the canal which ran between Lake Vänern and Lake Vättern, in parts steep, rocky and remote. He then retreated to von Platen's comfortable country estate to write up his report, still working on it on 1 October as he prepared to board the *Diana* home to Harwich.

Landing in England his first act was to write reassuringly to Rickman: 'after a tedious and very stormy passage. . . from Gottenburg. . . I have just arrived at this port in one of the Packets, not being able to obtain a passage to any part of Scotland'.[9] Telford wanted to show he was not neglecting his other business. 'I shall immediately proceed to Edinburgh, to Inverness and to Fort William,' he promised, where soon after he was joined by Jessop for his annual inspection of work on the Caledonian Canal. Meanwhile his Swedish survey was redrawn by von Platen's clerks – who carefully copied Telford's signature onto their new prints – and it was presented to the country's parliament in 1809. It took four years to draw up his detailed charges for this visit: sending a bill in 1813 for £416 10s 6d, including twenty-two days of consulting in Sweden at his set fee of £5 5 shillings a day (about £365 on a basic measure of inflation).[10]

From the start, von Platen showed a puppyish desire to keep in touch, his first letter reaching Telford on board the *Diana* at Gothenburg even as she waited for a change of winds to take her to England. 'A few hours after your departure I found how well it would have been to see You once more,' it began; 'I wish You would, if the wind is settled strong westerly, come once more ashore.'[11] He contrived a way to lure him back. 'My reasons are 1st that fearing Your provisions of wine and liquors will soon fall short and not knowing if the Capten has provided for of having any other,' he explained, the start of an endearing outpouring.

Telford had brought a theodolite to Sweden which his friend did not know how to use. 'After looking at the levelling instrument I found I had better take advices of you about it than standing talking nonsens last evening up stairs, but as this was not the case Self interest should bring me even for this reason to wish you back for a little while,' he said. 'You see mankind are always so foolish either to run away from or run after each other. Now, as in the beginning running away would not do for our business, we soon found since that it would not do neither for our affections; and thats the reason why I now find myself in the habitude of running after or looking after You.' It may be thought that this goes quirkily beyond what might be expected in a note inviting a friend to come ashore.

Others, too, took pleasure in Telford's cheery company; but it must have been an unusual and perhaps unsettling experience to be pursued in this way by a forty-two-year-old Swedish count only six weeks after meeting him on his first-ever trip abroad. In the years that followed, the letters hardly let up, though Telford, already swamped by the travel and correspondence demanded by his multiple schemes in Britain, could not keep up in reply (and he was not alone: Bagge once complained to Telford about the barrage of letters). This did not bother von Platen much; he was, as perhaps the most prolific letter-writers must be, used to the asymmetry. 'It would be a very tedious business if in our correspondence I always should wait upon your answer to a foregoing lettre before I made a new one,' the Swede assured him, asking for tools and equipment to be sent from England, among them railway wagons, wheelbarrows, picks and shovels: all proof of Britain's technological advantage.[12] Telford sent over drawings too, to Bagge, who had written to promise that, 'Neither trouble nor reflection shall be spared in executing your orders – in short I will be your Davidson in Sweden.' That, and von Platen's immense and fond correspondence, shows that Telford was seen as more than a consultant or figurehead for the Göta project, though it was far away and he was busy already. Beside von Platen he was its guiding mind and spirit and is recognised as such in Sweden today.[13]

They kept up a happy chatter; in the first letter to reach Scotland, in late October 1808, von Platen found another excuse to get in touch. 'I remember that I forgot to let You have the berrys of a tree which in our Country we find the best storm tree of all single trees,' he wrote.[14] 'Seeing through all my books I am not quite sure if you have it in England, tho in one I find the name of White Beam, white leaf Tree; in all cases You will now make its acquaintance as a wery resisting tree, I wish only that the berrys may reach you unrotten other wise You migt take them for Mountain ashes.' For years afterwards their letters are full of exchanges about pine seeds, which Telford wanted to plant in the Highlands, or advanced varieties of English seeds, which von Platen wanted to try on his Swedish farm.

Behind the bucolic talk lay difficulties, both political and technical. 'What I fear most is that [our] governments may one day or other be upon terms that the exportation of mens and tools will not be permitted,' von Platen wrote in early 1809, a year in which Sweden fell increasingly under French control.[15] Soon after, what he called 'a revolution' – in which he played a significant role – deposed the king, 'almost everybody being convinced that the king was in a state of mind who made change unavoidable', as he explained to Telford. For Swedes, this was the time of the Finnish War and the loss of the eastern portion of their country. Von Platen told his friend in Britain that he wanted 'to be buried under Swedish ground rather than yield to an unjust Foe'. 'Farwae my happy and healthy dwelling, my chearfull company in my comfortable house,' he added; 'Yet I hope the world shall have some more proofs that the Swedish spirit is as yet undaunted as he is quiet and orderly.' For Norwegians, who – in a policy led by von Platen – soon found their nation incorporated into Sweden as a replacement for Finnish territories lost to Russia, such quiet and order had darker consequences.[16] 'God preserve my poor wife and Children but I can impossibly behold them Russian slaves – perhaps Europe will. . . have reason to lament that the barrier against the eastern stream of barbarism has been permitted to be overthrown,' von Platen wrote: for the battle against Russia was all and Norway was to be Sweden's barrier. A year later, to secure this position, he helped place a French marshal as crown prince – and later King – of Sweden (which is why today's Swedish royal family can trace its origins to Pau, in the foothills of the Pyrenees).

Pause, for a moment, and consider this: a canal and road engineer in Scotland was being sent discursive and emotional accounts by an aristocratic naval officer: a powerful insider, deeply implicated in political manoeuvrings inside the Swedish court, which saw a king overthrown, territory lost and won and a British ally slide towards France. If Telford ever felt any awkwardness here – particularly at a moment when his main employer was a parliamentary commission led by the great and the good of the British government in the midst of a war – he did not reflect on it in writing. Instead, he continued to help his friend, whose appeals for assistance were ceaseless. 'I had no letter from You since Yours of 12 Jan, and consequently am as yet at a loose for the calculations. . . Yet I hope You will not leave Your friend helpless as I would make an ugly appearance in the house of Nobles if not duely fitted out by You,' von Platen wrote in early 1809. He was battling for funding and support for the canal from the new government and, eventually, secured it.

His thanks to his British ally came in startling form. In 1809, a letter arrived for Telford from the Earl of Kellie (a Scottish grandee who had lived for many years in Gothenburg). 'I have the honour to inform you that His Majesty the King of Sweden [has] named you a Knight of his Royal Order of Vasa' it began.[17] As a reward for six weeks of service on an as-yet-unbuilt canal this verged on the embarrassing. 'I am instructed by His Majesty to Invest you with this order,' Kellie continued; 'I understand that you are now in Scotland and shall be glad to hear when it will be most convenient for you to receive this mark of honour.'

He asked Telford to come to his Scottish castle to collect the decoration, but the engineer baulked at that. In a letter sent to Sir John Sinclair, who passed it on under his cover to Kellie, he tried to refuse. 'I consider myself as highly honoured by the expression of satisfaction which the Swedish sovereign has deigned to bestow upon my services in planning an Inland Navigation through the Central parts of Sweden,' he wrote.[18] Then he swallowed hard. 'If my residence were in that Country, the being invested with this honourable distinction would be to me as acceptable as it would be proper, but as these circumstances apply not in our Country, I trust my declining this honour will be accepted as a mark of my respect for the Order and of gratitude to the Sovereign.' He added that he might be more worthy of the title (one awarded to people successful in commerce and agriculture) when the work had progressed further.

To an extent this was the real, unshowy Telford. 'I trust you are convinced that it is more by the performance of useful work, than the enjoyment of splendid orders, [that] I wish my name to be known to present or future generations,' he explained to von Platen two years later.[19] Privately, however, he took pleasure in the award: and at any rate it turned out to be too late to turn it down. 'As for the Order conferred on you I think it well merited,' von Platen (who was behind the whole thing) wrote sniffily in the middle of an otherwise long letter.[20] 'The writs are already issued you will now be under the necessity to keep it.' As a result, in much of his correspondence from Sweden – and nowhere else – he was, from now on, 'Sir Thomas Telford'.

He did not, unlike the Swedish-born architect Sir William Chambers, for whom Telford had once worked as a mason on Somerset House, arrange, as Chambers succeeded in doing, to convert his foreign honour into its British equivalent. Nor did he ever get to dress in its formal green breeches and doublet with padded shoulders and white piping. But he did not mind letting people know both that he had been honoured and that he had tried to turn the honour down, writing back to the Borders in 1815, for instance, after four sons of the Malcolm family had been knighted, 'the distinction, so deservedly bestowed upon the Burnfoot family establish a splendid era in Eskdale and almost induce your correspondent to sport his Swedish orders which that grateful country has repeatedly (in spite of refusal) transmitted'.[21]

He was proud, too, of the portrait of the Swedish king 'set in valuable diamonds', which he was sent later.[22] This sat alongside a second foreign present, which arrived in November 1808 and about which less is known: the mystery perhaps explained by the fact that it came from Sweden's enemy, Russia, and arrived at the height of the Finnish War. Von Platen would surely have been dismayed by the gift, an inscribed diamond ring presented by Count Raumantzoff, among other things the Russian Minister of Foreign Affairs, Commerce and Inland Navigation. The accompanying letter, which was printed in an annex to Telford's autobiography, shows that Raumantzoff 'made known to his Imperial Majesty the Emperor of all the Russians, the kind attention you have manifested towards our studies on Civil Engineering'.[23] What those were is uncertain: though after Telford's death John Rickman recorded that 'the Russian government consulted him frequently on various schemes of canal navigation and other improvements' and that Telford was on familiar terms with one English-born member of the Russian nobility.[24] Telford included a drawing of a road through Poland in his autobiography and is said to have advised on the building of a route from Warsaw to the Russian frontier.

Foreign diamonds were flowing his way hardly a decade after he was being honoured with nothing more than a set of buttons by the Board of Agriculture. That was encouraging. Actual progress on the Swedish canal, however, was proving sluggish. The climate was one challenge. 'It is a pitty you stopt not here for a winter for to see us all converted to snow and ice,' von Platen had told Telford in February 1809, in one of his less inviting attempts to lure his friend to Sweden again. Securing workers and supplies was another. Von Platen – unlike Telford – took a while to trust Bagge and complained he was frequently absent. 'I am in a bad scrape, Mr Bagge coming not, and I begin to fear will not come, God knows what is the matter with him,' he wrote.[25] As a result, he badgered Telford for help. 'It is with a great deal of anxiousness that I hear not a single word from you,' he complained in April 1809.[26] Most of all, though, the challenge was political. 'Tho' engaged in the revolution and very deeply to I have never loosed out of my sight my principal aim, the Canal,' he told Telford that autumn, 'certainly a period of revolution and war is not very fit for carrying on such business.'[27]

Was von Platen manipulating the Swedish court and foreign policy to promote the interests of his scheme and the services of his British engineer? 'At last we have got a tolerable peace with France,' he wrote in 1810, 'I hope we can peaceably employ our efforts on canal digging.'[28] Work on the canal began properly that summer, but Telford was circumspect about assisting a potential British enemy. 'I am very desirous to do everything in my power to promote this great national work,' he told von Platen carefully, 'but it is absolutely necessary that I have regular authority from the English government.'[29] Authority was needed, most of all, to travel to Sweden, to which Von Platen repeatedly asked his friend to return. 'You make me highly rejoice by giving me hopes to see you here,' he wrote in 1810. 'If I had a wish to utter it would be, that circumstances might permit You once more to give us a call,' he said a year later.[30] Even in an age of florid letter-writing, von Platen at times sounded obsessed with his engineer. 'One of my last feelings against you will be eternal friendship,' he added.

Telford's surviving replies are fond but more straightforward: he mostly kept his focus on the things which needed to be done to build the canal. He also seems to have tried to signal to von Platen, gently, how busy he was elsewhere. In a draft reply in his handwriting from Bonar Bridge in the summer of 1812 he explained that he had 'been for several months fumbling about amongst the Hebrides', adding that 'our Roads, Bridges, Canals and Harbours are proceeding with great alacrity and success, we have already made upwards of 600 miles of new roads in Scotland'.[31] That year soon became significant for something else: Napoleon's defeat in Russia, which, the following year, encouraged Sweden's new French-born but independent-minded crown prince to throw his weight against France. This cleared the way for British involvement in the canal and soon a flow of engineers and equipment was on its way to Sweden, picked by Telford and Davidson in Inverness.

That summer Telford paid £24 4s for three tickets from Leith to Gothenburg, aboard the smack *Newburgh Volunteers*, making the second and last visit abroad of his life. Over a month and a half, he renewed his friendship with von Platen and saw the early stages of work, encouraging him to concentrate on particular sections of the route rather than ill-directed digging along the whole line.

There were setbacks, though: 'All the cries and roarings of nonsense and illwill, wich you Prophetized,' the count wrote in 1811, 'are going pretty well on.' But the worst news was Bagge's death in a sailing accident in 1814, after insisting, against the count's advice as a naval officer, on keeping his own boat. It was 'the very worst canal news that possibly could happen', von Platen told Telford.[32] 'I wish such calamities may not be to heavy for my advancing age and slender health!' His friend wrote back poetically. 'I can readily conceive your distress at the loss of Major Bagge,' he commiserated, 'in the language of our Shakespeare "You could have better spared a better man", but you must make men for your purposes, as I have done.'[33]

The loss of Bagge only deepened Sweden's direct dependence on British expertise. Telford arranged for a growing gang of British engineers to spend part of the year working on what he called the count's 'favourite project'.[34] But, as he warned, this was not cheap. British workmen 'would require to be paid what in Sweden would appear enormous', he wrote; 'Especially as at present there is all over this Kingdom a very great demand for their services – Canal projects appear to be again coming forward in great Numbers.' Among those sent out was James Simpson, a nephew of Telford's masonry contractor, who had travelled out to Sweden with Telford in 1813. He found the military atmosphere of the project hard to take; in return von Platen complained that Simpson 'has a deal of pride, thinks himself superior to his countrymen and endeavours by no means to sustain this superiority by a strict conduct but drinks heartily'.[35] At times such British engineers proved wayward; their purchases ran through Telford's accounts, including an order to his bookseller, Joseph Taylor in Holborn, for a library including *Clarke on Pregnancy* and *Burns on Abortions* which, as Rolt notes laconically, may have led their master to speculate 'uneasily upon the activities of his young countrymen abroad' – although Telford's papers suggest the intended recipient was a Swede. Whoever was responsible, these were not books Telford was ever likely to need for himself, though the moral tone was improved by accompanying and perhaps less-thumbed volumes including Dr Johnson's *Dictionary* and his friend Alison's *Sermons*.

In 1814 von Platen wrote with news of a second honour from the King of Sweden: the Commander of the Order of Vasa – which, he explained later, he was upset to learn Telford could not display because in Britain the Prince Regent had banned the wearing of foreign awards by anyone other than military officers. As well as that came 'a box from the Crown Prince who ordered me to add that convinced as he was that the value was of little consequence to a man like You, he wished only to give You a proof of how much he wished to be considered by You as a man who fully feeled and appreciated the great services You had rendered to a Country'.[36] The canal company, he added, had voted to award Telford a golden medal. 'As to me I need not I hope to say how much I find myself every day indebted to you.' If Telford found all this a little too much, he was polite enough not to say so. 'As tokens of the appreciation of the Government and Canal Co. by whom I have had the honour of being employed in Sweden,' he replied, 'what you propose transmitting will be particularly gratifying.'[37] He would have preferred, no doubt, to see more digging and less glorification.

By the following year, work was at last going well. 'I have a mind whilst there is. . . a fresh gale to carry all possible sail next year, beginning now to have a tolerable exercised crew,' von Platen wrote.[38] He mocked the French for their final defeat by the British – 'I pity them very little' – but that jingoistic triumph seems to have infected his imported engineers. Returning to Britain for the winter season, they ended up in a brawl on a dockside in Gothenburg, demanding to travel home on a comfortable packet, not a common vessel: the reason, von Platen suspected, being that a Russian princess was the only other passenger on the packet. The British won the day, 'to the great comfort, I suppose, of the Russian Mylady', but the result was that among his other tasks Telford found himself having to manage the fears of his team of engineers, who disliked both

Swedes and Sweden.[39] 'I know no other nations who are more like each other,' an enthusiastic von Platen suggested to him, but behind the scenes the best workman, John Wilson, was writing to complain that 'I never could think of taking my family there and it is not very convenient to live separate.' Telford strung him along for several years, refusing his pleas to find other work in England and sending him back to Sweden each summer. 'I leave it entirely to yourself knowing you will do what you think most proper for my welfare,' Wilson wrote mournfully, as he prepared to sail for Gothenburg yet again the following spring. Telford was loyal to his men and they to him, but he did not always do what they wanted.

He had particular trouble with the young James Simpson, whom von Platen blamed, perhaps unjustly, for the punch-up in Gothenburg. He 'has a decided bend to intrigue, desimulation, laziness and pride', von Platen wrote to England.[40] The man himself asked for mercy: he 'Trusts Mr Telford will have the Goodness to forgive me for what is past – and for the future will be more attentive,' but it was no good.[41] 'We have been disappointed in Simpson. I have, of course, done with him – he is unfortunately undeserving,' Telford wrote.[42] Tested, he could be hard; but as he advised von Platen: 'You are well aware that much trouble attends governing mankind. I have my share of it (in managing Overseers) and although when matters are unfortunately brought forward I never give way, yet I am often glad to seem not to know lesser improprieties unless they are forced upon me.' In managing men, Telford's best advice was to know when to look the other way.

The middle years of the second decade of the new century brought slow toil and little reward on both Telford's mighty canal projects in Scotland and Sweden. They were overambitious, over budget and under attack. Telford was, by now, hardened to it. 'You are made to such things,' the adoring count wrote, 'and I need only tell you that I have had all the necessary time to remind the different prophecys you at different times and so often have made to me.' For von Platen it was infuriating. 'What a damned noise: what an outcry,' he wrote in 1815 as he sought more funds; 'All the sence and nonsence of the Country at once in motion.' Still, he battled on. 'Till beginning of August I have mostly alone been standing in Butt in the House of Nobles. A fine and pleasing occupation when we think of ourself actuatet only by zeale for common welfar. . . it's schoking. it's odd. but it's unavoidable.'[43] He won his money, in the end, though he might not have done had anyone suspected that the project was less than a third of the way to completion. Two years later, amid yet another storm over funding, Telford wrote to reassure him once more: 'the opposition and grumbling of those who, unable and unwilling to conduct any great and useful work, are the first to bestow unjust blame upon those who manage them'.[44]

By now he was on the verge of becoming a wise old man. 'After being thoroughly convinced of the solid merit of a public work I remain equally indifferent to the praise or blame of those unqualified to judge.'

Besides comfort, Telford continued to send practical help across the North Sea. In May 1817 he chose one Scots engineer to build an iron foundry to produce lock gates. This man failed – 'more modesty in his reserve and manners would more promote his interest and respectability', Telford concluded – but a second succeeded, first receiving lock gates cast at William Hazledine's foundry and shipped from Bristol, and then copying the design at a new foundry in Sweden which proved the start of a planned industrial town at Motala. Economic woes in Sweden in 1817 and 1818, however, saw state funding for the canal almost grind to a halt; von Platen wrote a bleak letter to England predicting ruin for his family, with his only comfort that, 'You have more than once observed in general the Children of the rich are more to be pitied than those of the poor. In such a position I think I shall steadily face the events.'[45]

For Telford, this was a frantic time, as the Caledonian Canal neared completion; he perhaps felt there was little he could do to solve von Platen's woes in Sweden. That only spurred his friend to greater agitation. 'I am now determined to write you a letter every week til I get an answer. . . I am in great perplexity by your silence,' he wrote in early 1818. Telford, he said in another letter at the time, was the 'Friend without whom all my exertions would have been vain and failing!'[46] He admired British ways as well as British engineers. 'I think it is the first time since the beginning of the world that a whole country was explored and made useful after a true and general plan,' he told Telford in 1819, 'our canal finances are not so propricious as yours of the Ellesmere!. . . I hope our children will be wiser!!'[47]

In all of his letters, the count passed on his best wishes to Telford from his wife and children, and the affection was returned. When, in 1821, he asked his friend to find a place for his youngest son as a midshipman in the British navy, Telford replied cheerfully that 'I shall be very glad to shake hands with this Swedish Nelson' – underlining deliberately the point that the boy was 'his friend'.[48]

This loyal pair planned to join forces one last time, in England, and the tone in which they discussed this suggests something of the way Telford saw himself. 'If you do not come this year or next,' he wrote in 1822, 'there is a risk of the Game being over for us both. We play with so much eager- ness – it cannot last long.' Von Platen sailed for England the same summer, arriving in London to find that Telford was off on one of his inspection tours. News that the count was on his way threw the assistants in Telford's London home into panic. 'We raw lads were in great perturbation,' recalled one of them, Joseph Mitchell.[49] 'We wondered whether a count was of equal rank with a lord, and being much-puzzled took the advice of Mrs. Spinks the housekeeper' – who set them straight. He was. 'Count Platen, seeing we were the only representatives of the great engineer. . . gave us credit for more knowledge and experience than we possessed,' Mitchell remembered. To the count, anyone close to Telford was worthy of reverence and he took the youths through plans for his canal, before joining Telford on a tour of some of his greatest projects, among them Pontcysyllte.

Late that summer, von Platen said goodbye in a letter written as his ship waited to sail from Harwich: ' "Write me!!" were your last words,' he told Telford, 'and it is in perfect sympathies with my feelings that I fulfil this your demand at the first opportunity.'[50] There was never any doubt the verbose von Platen would do otherwise. 'Yes dear friend I shall leave this Shore perhaps forever but certainly these last 5 weeks stay in England will be remembered for ever by me as an immense augmentation of the considerable debt I already stood in with you,' he told his friend, 'you . . . to whom I own all the Success of latter Years, nay more, the strength of fighting the obstacles thrown in my way; but you are no Friend of words and so no more of the subject.'

He wrote a second time from Harwich, as the ship waited for favourable winds and then yet again on arrival in Gothenburg. 'I feel me happy in setting down for to converse an hour with him who along with my wife. . . is my best friend and firmest support in this world. . . to whom I own wholy one of the most interesting and the most instructive and usefull periods of my life; who only make my regret to be perhaps so near the term of activitie.'[51] There is an intense emotion to the concluding part of this letter: 'I have been a real incumbrence to you under my long stay. . . . the month passed with you will stand highest with me between all the most common periods of my life; with a warm and unbroken feeling of gratitude, tho I shall leave of this time endeavouring to express it by words, as at our parting they absolutely failed me.'

Poor von Platen: he must have taken his leave from Telford in tears. And the parting was final. Work on the canal continued for the rest of the decade: in September 1822, the tricky western section opened, at last bringing good news, as the King of Sweden sailed the route in a specially converted vessel (and yet more gold medals were struck, with one presented to Telford).[52] 'The only thing missed not only by me but by many other was the presence of you dear friend; but I know you are not fond of Ceremonys,' the count wrote.

And it was true. Among all his great schemes Telford attended only two such ceremonies: the opening of Pontcysyllte and that of the Menai Bridge (the progress of which was followed in Sweden: 'I am happy to hear the success of the Menai Bridge and long to hear it completed as unparalleled proof of British art and knowledge,' von Platen wrote).

Did Telford encourage this immoderate enthusiasm; lead his friend on, even? Or was it kindness that inclined him to hide his own troubles in Britain? By now, for instance, the Caledonian Canal was open but underperforming commercially – a source of grief to Telford. Writing to Sweden, however, he never mentioned this, telling his friend instead that 'the Caledonian Canal is effectually answering all the purposes it was intended for, the number of Vessels passing through exceeding expectations'.[53] Perhaps he knew that von Platen's woes were greater. It was another decade before the final section of the Göta route could be completed. Not least among the difficulties was clearing a way through the rocky shallows of Lake Viken, done by a regiment of men under von Platen's direct command, working in water half-frozen into limb-numbing slush. Amazingly, nobody died.

By 1827 von Platen found his energies diverted into a role as governor-general and commander-in-chief of Norway. He was unwell, too: his shaky handwriting tells of an enfeeblement. Writing to Telford from Christiana (now Oslo) in May 1829, he said that 'being so near the compleating of the Canal, I have been more than ever fighting against a violent opposition'.[54] The letter was dated a day after the (to Norwegians) infamous moment when he had, on advice, agreed to send soldiers to confront a public celebration of Norway's Constitution Day, 17 May, 'happily without bloodshed', he reported.

They were his last words to Telford. His canal, he knew, was almost complete. But it was another three years before the final link to the Baltic was opened. Meanwhile von Platen was recalled to Sweden where he succumbed to cancer in December 1829. He was sixty-three.

And hidden among Telford's papers at the Institution of Civil Engineers – the ones on the Holyhead Road – is moving proof of the passion this obsessive and noble Swedish friend felt for him. It is a small visiting card, printed in the name of 'Le Comte Wedel Jarlsberg', with 'from Norway – Jannay's Hotel, Leicester Sq' scrawled beneath. Jarlsberg himself was significant, respected today as one of the founders of modern Norway. But it is the phrase written in ink on the back which stands out: 'with the kindest rembrance from Count Platen when dying'. The count's final thoughts were of Thomas Telford. And he sent a great man across the North Sea to make sure he knew.

A Happy Life

Sweden was only ever a distraction; the task that occupied Telford most in the second decade of the nineteenth century was building the waterways, roads and bridges which the government had ordered to link Britain together. Proud, well connected and busy, his works were vast. But they were also, to a degree, unseen ones. The mountains of the Highlands and North Wales were far from the path of most visitors. To impress, he needed to bring people to them, which led, in the late summer of 1819, to one of the more whimsical events of his life: a tour of the Highlands in the company of the Poet Laureate, Robert Southey, and his close friend John Rickman. The idea was reminiscent of Boswell and Johnson's famous Highland expedition forty-six years earlier and was written up by Southey in a *Journal*, whose style owes much to their famous accounts.

The episode suggests Telford wanted to secure his place in history. It may also whisper something about an ageing man for whom children, and dinner parties, the London house, the balls, the shooting, the country estate and the whirl of high society, had never happened – probably by choice. At work he was utterly absorbed. At leisure his instinct would have been to absorb himself in something else. There was no social diary, no family obligations to clear, no wife to square, before he decided to travel. It was just another way of staying busy.

The *Journal* which resulted is a hack piece of writing and Southey was, in retrospect, a hack poet compared with his friend William Wordsworth – to whom he never seems, sadly, to have introduced the verse-obsessed Telford. It is valuable, nonetheless, for its description of a region much modernised since Boswell and Johnson travelled it; less wild, less authentically romantic, but still without the baronial shooting lodges and tourist clutter that came after. It also gives us one of the sharpest accounts of what it was like to work, travel and live with the mature, confident engineer. Published only in 1929, it is a charming book revealing Telford's humour, Southey's tolerant pomposity, Rickman's drive and most of all the immense scale of the work achieved by the commissions for the canal and Highland roads since their creation seventeen years earlier. It is also an unusual book; has any other engineer, at the top of his game, spent so long explaining himself and his works to a poet?

Or, indeed, tending to his needs: for on this tour Telford was Southey's guide and at times carer. One of the stranger scenes of his career took place each evening in shared bedrooms – which at the time often meant shared beds – in small inns up the east coast of Scotland. The poet was suffering from 'my volcano', a boil on the right-hand side of his head; Telford had to rub ointment onto the Poet Laureate's head, acting the part of what Southey called 'my kind surgeon'.[1] It was all the more notable since the pair had only just met. Rickman had written to Southey in 1816 suggesting that he should see Telford, 'a very able and very liberal man, whose plainness you will much like', but it was not until the start of their tour that they came together.[2]

For Telford, this encounter must have been significant. Southey's reputation at the time was much greater than it is today. Born into a middling commercial family, he had risen to a degree of notoriety as a radical young free-thinker. He grew close to Samuel Taylor Coleridge before sliding towards reaction and a life in the Lake District, where he cared for

Coleridge's children and earned the vitriol of Lord Byron for his cosy conservatism. There was no one, Byron once said, that he hated more than Southey apart from Lady Byron; and he satirised him brutally – as in the Dedication to *Don Juan*, written the year before the Highlands tour:

> Bob Southey! You're a poet – poet Laureat,
> And representative of all the race;
> Although 'tis true you turned out a Tory at
> Last, yours has lately been a common case;

And so it went on, for pages more. But Byron was probably right. As Poet Laureate, Southey was poor; it says something that his lasting work proved to be 'The Story of the Three Bears' (which in his version were bachelor bears and the girl was not yet called Goldilocks). As a travelling companion, however, he was cheerful and curious; not too troubled by rough conditions in the Highlands, to which Telford was accustomed, or by rough food either; for Southey, something of a glutton, found himself, like the Three Bears, faced with a diet of porridge.

For Telford their expedition was, consciously or not, something close to a valedictory, a display of achievement by a man approaching the age at which many retired. He had no intention of retiring, but by 1819 his work in the Highlands was mostly done. 'The Harbours, Canals, Roads, and Bridges which have nearly for 15 years been carrying on under the Parliamentary Boards. . . have now reached a degree of perfection which I contemplate with much satisfaction,' he had told von Platen not long before, 'a greater blessing has seldom been conferred on any country.'[3]

In the summer of 1819 the touring party began to assemble. Telford wrote to his friend Rickman, the indispensable organiser of the commissioners' business, as the latter prepared to sail from London to Edinburgh to join the group. 'I trust the Northern Ocean will smooth his rugged features and your watery way,' he pronounced in atypical purple prose (perhaps intended to impress the poet). 'Experience obliges me to confess he is somewhat wayward. He can enchant with delight or toss you about with abundant incivility.'[4]

Rickman did not record whether his journey was rough but he made it to the city where, on 17 August, he was joined for breakfast at MacGregor's Hotel by Southey. The poet had travelled overnight from the Lakes on the Carlisle Mail, a coach which followed the line of the River Esk inland past Langholm, near Telford's birthplace, through landscape that had 'a quiet, sober character, in accord with autumn, evening and declining life' – the sort of Romantic burbling which led Byron to scoff at the writing of 'the Lakers'.[5] That morning Southey also met Rickman's wife, two of his children and a young woman who must have acted as nursemaid; the latter group joined for at least some of the journey to come but make little impression on Southey's account, except for a few instances such as when the seven-year-old boy Willy picked up a full ink-well in a bedroom and 'made a very black discovery by turning it up to look at the bottom'.

In the afternoon Telford arrived from Glasgow. Southey's account of their first meeting concurs with that of similar encounters: Telford was immediately open, put the stranger at ease and made him smile. 'There is so much intelligence in his countenance, so much frankness, kindness and hilarity about him, flowing from the never-failing well-spring of a happy nature,' Southey recalled, 'that I was on cordial terms with him in five minutes.' The poet himself was curmudgeonly; he saw the sights of Edinburgh (including searching for the curiosity of 'golf-players'), but found it a smoky, dirty city hardly saved by the overblown magnificence of the New Town, 'the Scotch and the French being undoubtedly the two ugliest nations in Europe'. His mood was not helped by sleeplessness: a woman he called 'the Night Mare' kept everyone awake. 'I wish she had put up at any other hotel during my stay,' he noted gloomily.

On 20 August the party set out for the Highlands with hired horses pulling their private coach, starting a routine which continued until Telford left the group more than a month later. In his account of the next few weeks Southey mixed admiration for the engineer's works – the first of which they saw the same day, a bridge being built at Stirling – with asides about Scottish life and culture, not least the quality of the food and drink. Scottish small beer, he found at the first lunchtime, 'was as weak as Mr Locke's metaphysics. . . and when the cork has been drawn a few minutes as vapid as an old number of the *Edinburgh Review*'.

As they travelled, Telford was amused by his new friend's unfamiliarity with the Highlands: his surprise at hearing Gaelic spoken in the streets of Callander; the eagle's claw used on a bell pull on one front door; the sight of women digging fields with spades in bare feet; or 'those mists which had a right to wet R[ickman] and myself, as Englishmen, to the skin'. He played the part of the genial host and guide. Wanting to impress Southey and Rickman with their first taste of a kipper, he refused to let them try one until reaching Aberdeen, 'lest this boasted dainty. . . be disparaged by a bad specimen'. When Southey was appalled by a guidebook written by an enthusiastic local – 'the first sentence – if sentence that may be called which hath no limitations of sense or syntax' – 'Mr Telford insisted on adding this choice piece to my collection of curiosities.'

But it was roads and engineering which mattered more. Clattering through Perthshire as they headed north, Telford explained that the tracks had not been improved because Perthshire's stubborn authorities had refused to let the parliamentary commissioners work in the county. ('What's the matter?' Southey records one traveller exclaiming as his coach jolted; 'the driver cooly answered, "Perthshire – we're in Perthshire, Sir."') In Dunkeld they examined Telford's elegant stone bridge over the Tay, by now a decade old, taking in the beauty of the setting by the waterside ruins of the town's cathedral. In Dundee a few days later they saw the opposite end of the range of his works, the huge reconstruction of the harbour. From the *Journal* one gets an echo of Telford's conversation, acting the eager tour leader, explaining, for instance, to amateurs as they travelled on a new road how he had had it built: first level and drain; then lay a pavement of large stones, points upwards; then a layer of small, walnut-sized stones which lock into them; then gravel. Later on in the expedition, Telford explained his rules for laying out a route: not least that 'the road always be defined, if it be only by a line of turf on either side', so that contractors had no excuse for letting it slide into disrepair. 'Every precaution is taken to render the work permanent in all its parts,' Southey noted – in contrast to what Telford must have told him were General Wade's 'miserably constructed' bridges.

Before heading deep into the Highlands they diverted east. In Montrose they saw one of the new harbours funded by the parliamentary commissioners, with iron railways and steam pumps aiding the furious construction; and here they also met 'two of Mr Telford's aide-de-camps' – Joseph Mitchell and John Gibb – the iron-hard men who made Telford's plans real. 'The whole line of coast is in a state of improvement, private enterprize and public spirit keeping pace with national encouragement,' Southey observed, adding that 'the Government is to blame for not making its good works better known' – which only served to prove Telford's point in organising the sightseeing expedition.

As they moved north, the trio carried an air of celebrity with them. In Dundee, 'a Bibliophile asked if I was the P. L.' – and having confirmed he was the Laureate, the bookseller gathered a crowd the next morning to wave off the coach. Meanwhile Telford was dragged away to meet the Provost: 'They would fain have given him a dinner, but this would have consumed the remainder of the day, and time was precious.' Their procession must have been quite something for the Highlanders, and it made an impression on Southey, too, who was by now under his companion's spell. 'Telford's is a happy life,' he wrote in what became a renowned encomium, 'everywhere making roads, building bridges, forming canals, and creating harbours – works of sure, solid, permanent utility; everywhere employing a great number of persons, selecting the most meritorious, and putting them forward in the world, in his own way.'

In his *Journal*, Southey also kept up the chatter of a dissatisfied tourist – the bed he shared with Telford in Fochabers (where they saw another of his new bridges) exemplified 'all the faults of bedmaking', the sheets wrapped around the bolster and the mattress hard and sloping; or the vulgar guest house they stayed in, furnished 'in true Portsmouth taste' with a clutter of pictures of the royals and mahogany chairs. In a different inn, he records Telford complaining about the noise that kept him awake until 1 a.m. A chambermaid was summoned, who explained it was being made by the masons, but the noise never stopped. 'T. exclaimed that this was too bad; and as it was impossible to sleep, we began to talk about it,' Southey recalled the next day, assuming in his 'disturbed and half-dreamy state' that the masons were builders; 'but T. laughing at my mistake, told me the Freemasons were holding a lodge upstairs. . . at midnight, the aspirants were going thro' what in their Diplomas are called "the great and tremendous trials" of initiation.' He does not say whether Telford confessed that in his Shrewsbury days he had been one of them.

One gets the sense that these grumbles were really part of the pleasure of travelling somewhere different and with a man he liked – and the food was at least sometimes good. In Rothes they breakfasted on 'broiled salmon from the Spey, butter both potted and fresh, honey and preserved gooseberries'. More impressive still was Craigellachie Bridge, Telford's lovely iron structure which still floats above the river in which their breakfast salmon had been caught. Southey was drawn to picaresque details of their expedition, such as the infant 'in the full dress of a Highland Chief' who rode past on a pony, or the whiskered 'Asiatic' wearing a turban: an encounter that spoke of a region caught between new and old. Far north, in Dingwall ('a vile place' where work was being done for Telford by 'a strange poor fellow. . . who is so fond of whisky that. . . he cannot be trusted with the money he earns'), he was touched by the reaction of a woman they addressed politely as 'my good lady'. 'Nay, I'se nae gude Lady, na but a poor woman,' she answered.

Of all the places changed by the new roads, it was remote Caithness and Sutherland, previously inaccessible except on foot, which had altered most. Southey found that a regular mail-coach was now running to the tip of Britain at Thurso, crossing Telford's Bonar Bridge and a second of his structures, just as impressive, further north at the entrance to Loch Fleet. Only just completed when they visited it in 1819, the Mound is still one of the most exceptional of Telford's creations. In design it appears at first to be nothing more than a long earth causeway shutting out the sea and linking the shore at either side. The ingenuity lies in the series of iron floodgates at one end. When the tide rises, the gates are closed by the pressure of water against them; when it falls, they are pushed open by the river water coming down from above. This non-return valve keeps the tide below the barrier. Today the system still works; there is a tang of salt in the air, and there are seals in the sea (close to where Southey records seeing his first) and brown water swirling below the Mound crossing, which over time was converted to carry a light railway as well as a road and now takes the busy A9 north, flying above the sluice gates on a noisy concrete ramp. 'You perceive at once the simplicity, the beauty, and the utility of this great work,' Southey wrote, 'but you are not at first fully sensible of its grandeur.' He proposed calling it 'Telford Mound. . . in honour of our excellent companion who has left so many durable monuments to his skill in this country.'

This was the furthest north they got, around a third of the way from Inverness to the northernmost tip of Scotland. Here they turned in from the coast, crossing what is still known as the Fearn Road over the hills from Bonar Bridge. This was one of the routes improved by the commissioners, and if you drive it today the road engineer's skill becomes plain. The climb is never too steep for a horse and cart; and the summit crag 'which Mr Telford and Mr Mitchell call Davidson's Crag' because their friend, a nervous rider, was scared to pass it on his horse before the road was built, still looms above.

From here, Telford, Mitchell, Rickman and Southey continued inland without the women and children who had travelled with them thus far; and from the 'Scotch pint of whisky' which one landlord offered before breakfast, and the 'very best and purest' untaxed black-market spirit they tried at the house of a second, this expedition seems to have turned into something of a 'lads' tour'. Near Auchnault 'we tasted again right Highland whisky', and it was with reason that a few days later 'T. jested with me about the weakness which I confess for whisky' – which led Lady Mackenzie, with whom they were staying, to produce 'two bottles of the very best'. He took it happily. 'For myself, I can only say that on such occasions the smallest contributions are thankfully received.' The men were enjoying themselves as they headed along one of the commissioners' new routes, reaching the west coast in the second week of September. With the mountains of Skye in the distance, they edged their coach close to the new pier at Stromeferry. 'Ours was the first carriage which had ever reached the ferry,' Southey noted with pride.

———

On 11 September they clattered into Inverness, the epicentre of Telford's Highland exploits, where memories were still fresh of the heroic, mournful Matthew Davidson, who had died months earlier while overseeing the eastern end of the Caledonian Canal. Southey set off the next morning to see the entrance at Clachnaharry, which had caused Davidson such worry. Soon after, the party sailed on the completed section of the canal towards Loch Ness. The locks, Southey found, had 'walls of perpendicular masonry. . . the situation might have afforded a hint for a Giant's Dungeon'. The metaphor was apt: in the fifteen years since work had begun, it had all but imprisoned its creators. Though Southey does not describe it as such, the canal was becoming a nightmare.

One of the best guides to the horror is the series of loving letters that Davidson wrote to his sons, as he struggled to fulfil Telford's orders in the years that led up to his death. They make for sad reading. Life in what he called 'these dreary wilds' was awful; a matter of plodding on with obtuse workmen, obstructive landlords, bad weather and too little money, his heart lifted only by Telford's biannual visits and his fondness for a family that, increasingly, moved away. 'The house is very quiet for a new-years-night,' he wrote as 1816 became 1817, 'we are two old grey headed people, left by ourselves in a strange country' – the grey hair a family joke about Davidson's decision, in 1812, to burn his outmoded wig as fashion reached even the Highlands.[6]

Again and again in these letters he expresses his hatred of the place in which he found himself, a deepening depression which must have owed much to the struggle of building the canal. 'You know I'm not over partial to Scotch things,' he wrote to his son Thomas in 1812; while in a letter to Telford he sent the same year he admitted to 'my utter contempt for every thing Caledonian'.[7]

Both men, of course, were Scottish, even if their birthplace was only a few miles from the border. Davidson's extreme rejection of his homeland was not shared by Telford, but both saw Great Britain – not Scotland – as a route to advancement and identity and both thought of England as a place Scotland should admire and emulate. 'England is the most virtuous and best governed nation that ever existed on this Globe,' Davidson instructed his son.[8] He wanted to create a little piece of it in Inverness.

Telford's relationship with Davidson is revealing. It is clear from the letters that he was an integral part of the dispersed Davidson family, as much as he was an employer. Yet in almost all Davidson's references to Telford there is a tone of respect; he wrote as an intimate but not as an equal. Only occasionally does informality break though – calling him 'the Old Gentleman' in a letter in 1817, for instance, or noting a decade earlier that 'Mr Telford was with us two Nights ago, in good health, in great spirit, on the subject of the Iron Rail-way.'[9]

As the Caledonian Canal progressed, Telford took on something of the role of a guardian to Davidson's children. When, in 1809, the eldest, Thomas (named, perhaps, after Telford), was staying with him in London he allowed him to borrow books from his library. 'Be careful to keep them clean and return them regularly,' his father ordered. He was as bossy to the boys as he was loving. 'I am sorry to observe your letters wrote in so slovenly a manner,' he wrote to the same son.[10] 'Correct this, it is very disrespectful in a Boy, even to a beggar.' Soon after he asked him to spot and correct deliberate misspellings in his next letter and return it with appropriate words written into blanks that he had left. His children seemed to have treated such pestering with a smile.

At first, however troublesome the canal, Inverness had seemed a happy home. 'I have got two new pretty Pigeons,' the younger boy John wrote to his brother Thomas, away in London with Telford in 1809, 'they are now building a nest, the Cats have killed me three, but I have now got a pair of Pistols to shoot them'.[11] Soon after, however, John caught typhus. 'By 35 days and nights of constant watching, anxiety and grief I was reduced to a skeleton,' his father wrote. The boy recovered: he 'is growing fast, your clothes quite too small for him, he is as innocent in his manners as a sheep', Davidson wrote later the same year. But it was to Telford they looked for assistance. The engineer seems to have paid for Thomas's education as a surgeon – he may well have done the same later for John, who joined the same profession – and he later took on the third son, James, as a clerk – and eventually his father's successor. 'Mr Telford. . . is very kind to James,' Davidson wrote thankfully.[12]

For a lonely man in Inverness these letters, and his sons' proximity to his great engineering friend, were a link with a world beyond the immediate drudgery of canal building. 'I wonder much how you and Mr Telford can leave so much white paper in your letters without saying how my few friends are,' he complained to Thomas when he wrote from London; 'no mention of Mr Jessop, Simpson, Provis, Jones etc.'[13] He fretted about the health risks of his sons' medical training: 'the filthy streams of an Hospital and Dissecting room hung heavy on my mind'.[14]

But it was Matthew's – not his sons' – health which gave way first. A man who was always deep in books – who once noted that, while he amused himself with Scott's *Waverley*, 'Livy, Sallust, Caesar and Tacitus, with Thucydides and Xenophon, are my staple reading' – suddenly all but gave up.[15] 'This is the way I live,' he told his son, Thomas, the summer before Telford and Southey reached Inverness on their tour: '[I] read very little only now and then a piece of the Bible, indeed it is the only book that deserves to be read, all the rest only looking at objects through a blind man's spectacles.'[16] From his account of his symptoms, it sounds as if he had something like a stroke, affecting his left side. '[I] felt a sourness at stomach with giddiness, restlessness at night,' he wrote, 'fainting on raising the head from the pillow.' On 8 February 1819, he died, by the canal, at home at Clachnaharry.

There was perhaps little to be said and Telford was not a man to say what did not need to be said. That he mourned we need have no doubt, but nor have we any idea whether or to what extent he blamed himself and his project for what – he must surely have reflected – will have contributed to this dear and trusty friend's early death. Indirectly, Telford drove men very hard and scarred as well as motivated many lives. In his most private times, how much did he think about this, and how sanguine did he feel? We cannot know. Andrew Little, too, having died, there was nobody in whom to confide.

By the end Davidson will have known that the project to which he gave so much was at least going to be finished. Loch Ness, twenty-two miles long, was connected to the sea in the east. In the west Easton had completed the great staircase of locks from the sea at Corpach, and work was advancing on the middle section. 'There only wants two iron swing bridges to make the canal Navigable from the sea at Corpach to the Regulating Lock near the west end of Loch Lochy,' Easton told Rickman in the spring of 1819, 'which I understand are casting by Mr Hazledine, and expected here by about mid-summer.'[17]

When Telford, Rickman and Southey travelled the route a few months after Davidson's death, they could see both the scale of what had been done and how much remained. Between Inverness and Loch Ness, the river had in places been diverted to a new course and the canal run along the old river bed next to it – still an arresting sight. The loch itself impressed Southey, for its length and depth and for the views from the road on the southern side, built by General Wade, which still scrambles into the hills high above the shore, as Wade directed. Today most traffic uses the northerly lakeside road, built by the commissioners; on the old route, Southey noted 'the General's Hut, in which Wade is said to have lived' (but which is now gone) and with Telford he climbed down a steep slope to view the famous Falls of Foyers (now, since the flow has been diverted, also much diminished).

For Telford, this was familiar terrain: he had, for instance, an unkind nickname, 'Field Marshal', for the dwarf landlord of an inn in Fort Augustus, called Marshall, a man who was followed everywhere by a pet Cheviot sheep called Tom, which attacked the party at breakfast, demanding to be fed bread, and which would leap onto his master's shoulders on command.

Today the locks that lead the canal up from Loch Ness are an admired centrepiece of the town; but in 1819 they were still being built, by 'an engine of tremendous power, bringing up its chain of buckets full of stones and gravel, or whatever comes its way'. Here the party met John Cargill, the Newcastle-born masonry contractor, who had worked his way up since starting on the canal in 1804 and who continued with Telford until the latter's death; a man, wrote Southey, 'with a good-natured, intelligent face, and a genuine *burr* in his speech'. The poet returned the next day to see more of the construction. It was, he wrote, 'a most impressive and rememberable scene. Men, horses, and machines at work; digging, walling, and puddling going on, men wheeling barrows, horses drawing stones along the railways.' The steam engine poured black smoke into the clear air and 'the iron for a pair of Lockgates was lying on the ground, having just arrived from Derbyshire. . . to one like myself not practically conversant with machinery, it seemed curious to hear Mr Telford talk of the propriety of weighing these enormous pieces. . . and hear Cargill reply that it was easily done.'

Work on this scale had been under way on the canal without a break since 1804 – and it continued for three more years. 'What indeed could be more interesting than to see the greatest work of its kind that has ever been undertaken in ancient or modern times,' Southey concluded, 'directed everywhere by perfect skill, and with no want of means.'

Travelling on westwards they soon reached Loch Oich, passing more locks under construction where the steam pump was dredging water with such tremendous energy that some of it boiled. Here Telford was forced to pay one of the many diplomatic calls with difficult landowners which took up his time, in this case on Glengarry, the most flamboyant of Highland exhibitionists. This was the summer in which James Mitchell – as a boy – recalled seeing Glengarry's wild games; indeed Telford seems to have timed his visit to miss the event (to the laird's regret: he explained generously that although locals were expected to appear in kilts, 'strangers would of course have been admissible in their proper habiliments'). Glengarry contented himself with demonstrating before his wide-eyed English visitors the best way to kill a man with a dirk through an overhead stab, a show Telford had by now doubtless endured many times already.

The group was most of the way along the route and in sight of the 'precipitous, rugged, stony, uninviting' mass of Ben Nevis. Soon they descended towards the other centre of canal operations at Corpach, and reached 'the Staircase, a name given to the eight successive locks'. It was, recorded Southey, 'the greatest piece of such masonry in the world and the greatest work of its kind, beyond all comparison. A panorama painted from this place would include the highest mountain in Great Britain and its greatest work of art. . . the Pyramids would appear insignificant in such a situation.' For him every part of the canal's construction was amazing: the culverts under the waterway 'just lofty enough for a man of my stature with his hat on'; the intakes and outlets which let streams cross, 'and when the cross stream comes like a torrent, instead of mingling with the canal, it passes straight across'.

Later, when the route was done, Southey composed three poems to be placed on inscriptions along the route. They are not, to put it kindly, great works: Telford could have done no worse. 'Thou who hast reach'd this level,' begins the one written for the high point of the canal at Fort Augustus (Southey was obsessed with locks):

> Fourteen times upheaved,
> The Vessel hath ascended since she changed
> The salt sea water for the Highland lymph . . .

It was something, at least, for the Poet Laureate to be praising engineering, if badly.

From Fort William, Telford, Rickman and Southey travelled some of the new roads that were being built, along with Mitchell, who rode in front of their coaches, a man 'so case-hardened that if his horse's hide and his own were tanned, it may be doubted which would make the thickest and toughest leather'. The group was fascinated by the Parallel Roads in Glen Roy: a series of long tracks marked up the side of this beautiful valley. To Telford, whose interest in antiquities had been sustained since his days excavating the Roman bath house at Wroxeter, they seemed to be the expression of some 'display of barbarous magnificence in hunting' (on a different occasion he asked his engineers on the canal to map their extent). Today, these marks are recognised as signs of the retreat of glaciation.

There was evidence of retreat of a different kind back in Fort William, where the inn was so poor that the glass was broken in the windows of their bedroom. Southey slept on the floor and Telford took the only bed – better, they concluded, 'than either of us would sleep in a double-bedded room, where the other bed was occupied by a gentleman and his son'. Adding to their misery, the landlord substituted Cape wine for sherry and the place was out of both bread and cream. This did not stop Telford, when he paid the bill, giving 'the poor girl who had been waiter, chambermaid, and probably cook in chief also, a twenty shillings bill'. 'I shall never forget,' added Southey, 'the sudden expression of her countenance and her eyes when she understood that it was for herself.' Telford's consistent generosity was one of his most endearing traits, but it is more than that: most other guests in so inhospitable a place would have responded with a general irritation towards the establishment, but Telford might have noticed and would anyway have reflected that it was not the serving girl's fault that the place was understaffed.

That was almost the end of their inspection of engineering. They headed south on the main route, marvelling at the steep sides and crags of Glencoe and staying at the Kings House hotel at its end, where the epicure Southey regretted not trying their speciality, goat ham. As the trip neared its end he praised another culinary discovery: 'this year of my life might be designated as the great Herring year. . . I [am] a true lover and eater of this incomparable fish'.

Pausing briefly in Glasgow, they went on to pay a call on the utopian Robert Owen, whose model factory at New Lanark had become famous. He paraded the factory children before Southey and Telford, a performance which shocked the poet, who saw beneath the hygienic accommodation and decent diet a system which reminded him of slavery. 'Owen in reality deceives himself,' he wrote, 'he is part-owner and sole Director of a large establishment, differing more in accidents than in essence from a plantation: the persons under him happen to be white.' He was repelled, perhaps, by a collectivist order which later on saw Owen emerge as an early socialist. For the former radical and now Tory Southey, all this was too much. 'He jumps to the monstrous conclusion that because he can do this with 2210 persons, who are totally dependent on him – all mankind might be governed with the same facility. . . but I never regarded man as a machine.'

Did Telford share his view? Certainly he never seems to have been at home in the urban consequences of industrialisation. He was happier away from towns, as he was, in the last part of their journey, along the Glasgow to Carlisle road, one of his other significant Scottish projects. He surveyed the route between 1814 and 1815 with the help of engineer William Provis, who went on to play an important part in many later Telford projects, and the work was completed by 1825. The most unusual structure on the route was a three-span iron bridge (cast by Hazledine) over the River Esk: Telford and Southey paused on their journey to see its foundations before separating at Longtown, just over the English border. 'This parting company, after the thorough intimacy which a long journey produces between fellow travellers who like each other, is a melancholy thing,' the poet concluded. 'A man more heartily to be liked, more worthy to be esteemed and admired I have never fallen in with; and therefore it is painful to think how little likely it is that I shall ever see much of him again.' He made Telford promise to call on him at home in Keswick. If he ever did, the visit was not recorded. And if he had spent an evening over dinner, with Southey and Wordsworth, he would surely have told someone all about it.

———

The Highland roads were almost done. The Caledonian Canal was not yet finished, but by 1820 early steamboats were at least using part of it, and the rush was on to join up the ends. Telford found himself fully involved, writing several times a week and assuring Rickman, in a letter from Clachnaharry sent in November, that 'all is going well. The steam boat has been very useful and has succeeded surprisingly for the first season in the present incomplete state of the general navigation.'[18] They hoped to open it in full in 1821: but it was not until September 1822 that Rickman could write to James Davidson (now serving in his father's place as superintendent of the eastern end) that 'I have just received a letter from Mr Telford, informing me that the canal will be ready to be opened for passage from sea to sea about the end of the present month.'[19]

After eighteen years, it must have been a relief, even if the canal was over budget and shallower than intended – in parts only twelve feet deep. In Scotland plans were made for formal ceremony. But Telford, as he did so often at such moments, ran away. A man who mostly thought nothing of dashing up to Scotland – and certainly could have found the time had he wanted to – instead packed his diary deliberately with other visits. 'I leave London tomorrow (Saturday) on one block journey by Gloucester and S Wales to Holyhead, return by Chester and Staffordshire and the Holyhead Rd. – until 6 Nov. letters will find me at the Post Office Bangor N. W.,' he scrawled in a note written two days after the formal event.[20] Telford, by no means always shy of publicity, had something of a phobia about being at the centre of big ceremonies; but there may have been another reason too: he knew by now that the canal could not succeed. It was too narrow and shallow for naval vessels – and anyway the war with France was over. Its commercial appeal was always questionable and could never repay the enormous cost of construction. Technology, too, was changing. In north-east England, George Stephenson was already laying the track for the first major steam railway. Perhaps trains would have been a better answer in Scotland too.

Whatever the cause, the canal opened, without Telford, on 24 October 1822. Both Alexander Easton and James Davidson sent him written accounts, just as Provis did the day after the Menai Bridge opened four years later. The event was, of course, chaotic and drunken and Telford would have hated it. 'There were some wavering and want of decision in the arrangements as to the hours of starting,' Easton explained.[21] A steamboat had travelled west from Inverness, with the band of the local militia playing. Along the way it was joined by a second paddle steamer, the successor to famous *Comet* built by the pioneering engineer Henry Bell. They reached Fort William in the late afternoon, where a bonfire blazed, guns boomed out and fifty-seven dignitaries sat down to a dinner which continued until 4 a.m. the next day. 'The evening passed with much mirth and pleasantry,' Easton told Telford, 'excepting a slight controversy between Glengarry and one of the Gentlemen from the East, as to chieftain-ship' – his Highland honour being easily roused. Hungover, after thirty-nine toasts, the nineteenth of which was to Telford and the men who built the canal, the guests were late for the return trip to Inverness the next morning – 'it became difficult to make an early start with the steamboat', Easton told Telford delicately. The boat did not set off until 11 a.m., reaching Loch Ness at six, by which time it was dark. 'Encouraged to proceed by a fine, calm moonlit night,' Davidson told Telford, the boat made it to Inverness without disaster just after midnight and the canal was open.[22]

The aftermath was miserable. In December, Telford wrote to Rickman to surrender his formal role. 'I consider this as a proper period to have the accounts, which have hitherto been kept in my name, finally closed,' he told him.[23] But in the years that followed he fretted about the project. The canal archives show the engineers kept trying to dredge it, first to a minimum depth of fifteen feet and then to twenty, but for all their efforts the economics did not work. In 1826 Davidson sent him a clipping from a local paper showing that one steamboat company had ceased using the canal because of 'the enormous extractions in name of dues'. Parliament was no longer funding the scheme but its income could not keep up with its expenses. 'In the present state of the canal funds it is absolutely necessary to postpone any thing that can with any safety be avoided,' Telford told Rickman bleakly in the summer of 1826. 'Nothing need be expected from the Government.'[24]

For James Davidson, still living in the canal house in Inverness that had been built for his father in 1804, the consequences were hard. He wrote to Rickman in the spring of 1827 begging for permission to build a new home. 'I feel aware that it is a bad time to bring forward,' he admitted, but 'I find it more and more impractical to prevent the winter rains from penetrating through the walls. . . during these Gales there is not a Chimney vent in the house that will draw.'[25] All Rickman could do was offer sympathy, 'on account of the wretched state of the canal affairs'.[26] Davidson, in ill health, fled the scheme for a time; while Telford, by now an old man and unwell himself, made pitiable efforts to drum up business. He sought to have the tolls reduced and wrote an advertisement which he hoped would encourage people to use his great creation. 'I have consulted with Mr Telford on the subject,' Davidson told Rickman at the end of 1827, 'and under his direction drawn up a form of advertisement. . . Mr Telford recommends the insertion of the <u>advertisement only</u> in both of the Inverness Papers, leaving the respective Editors to make their own remarks in the shape of a paragraph in the column of news.'[27] It was desperate stuff and no way for Telford to end his career; yet there is something endearing about his refusal to abandon the canal. Its failure upset him deeply.

By now, the locks were leaking and the embankments needed work, the problem being not the design, which has lasted to this day, but the hurried way it was finished (shoddy workmanship which George May, one of Telford's assistants, who went on to rebuild the route, once claimed had been hidden deliberately from the engineer). Rickman wrote to the Treasury to beg for funds – it was madness, he said, for 'so great a future resource to be endangered for want of the small supply of money requisite to prevent the dilapidation of the magnificent apparatus of this national work'.[28] But none came. The last paper in the archive is a copy of the printed advertisement Telford had drawn up in late 1829. 'The Caledonian Canal is found to afford important facilities for Vessels trading from the Eastern Coasts of the Kingdom to every part of Ireland as well as Glasgow, Liverpool and all the Ports on the Western Coast of Scotland and England, as far as the Bristol Channel,' it announced, ambitiously. 'It cuts off, in the former communication between these points, a distance of at least THREE HUNDRED MILES.'

Few, at the time, were convinced. By the 1840s, after Telford's death, there was talk of abandoning the route. Only slowly was it saved, with May leading the reconstruction. In 1873 Queen Victoria, setting a new fashion for the Highlands, sailed along it (the queen noting that it was 'a wonderful piece of engineering' but 'travelling on it is very tedious'). James Davidson, Telford's old pupil, was there to see it. After recovering his health abroad, he returned to oversee the canal in 1867 and did so until his death in 1877 – and that long service by father and son to the route was later sustained, in turn, by his son John who took over as resident engineer in 1885. By now the canal was busy with steamers carrying tourists, as well as fishing boats and a handful of cargo vessels. In the twentieth century, like so many waterways, it struggled, then was revived. It was strategically useful in the First World War as it never had been in the Napoleonic, but when Rolt wrote his biography of Telford in the 1950s he lamented that it was deserted. Now, that has changed. After repairs by British Waterways between 1995 and 2005 it has become popular as a tourist route. Yachts, too, can make the journey in a minimum of two days, sheltering from the rough seas to the north just as the route's designers envisaged. 'Although the Caledonian Canal locks are large there is no need to be nervous about using them', the guide reassures skippers.[29] Telford would be mildly pleased: his canal has found a purpose and a future, but recreation and leisure were far from what he had envisaged.

13

The Colossus of Roads

'The rolling English drunkard made the rolling English road,' wrote G. K. Chesterton, 'a reeling road, a rolling road, that rambles round the shire.'[1] His genial defence of drink smoothed over the horrors of much eighteenth-century travel. 'Let me most seriously caution all travellers, who may accidentally propose to travel this terrible country, to avoid it as they would the devil,' the agricultural reformer Arthur Young warned when he reached Wigan in the late 1760s, 'for a thousand to one but they break their neck or their limbs by overthrows or breakings down.'[2]

The fact that these Lancastrian roads stood out to him as notably bad was, perversely, a sign that others in England were improving. At the start of the century even main routes had been often all but impassable, maintained, if at all, by poor parishes under an obligation imposed in 1555. In the early eighteenth century the coach from London to Edinburgh ran only once a month or so and took a fortnight. Rarely direct and poorly maintained, roads were slow to travel and so deep in mud that a man was reported to have drowned between Barnet and South Mimms in 1727, swallowed up near what is now junction 23 of the M25. One writer, looking back from 1763, remembered a time when roads had looked 'more like a retreat of wild beasts and reptiles, than the footsteps of man'.[3]

It was Thomas Telford – dubbed 'The Colossus of Roads' by his poetical punning friend Southey – who made Britain's roads modern, drawing on his experiences in the Scottish Highlands to plan and, in places, build the equivalent of the first long-distance expressways. Or, more fairly, it was Telford who helped to do so, for though his roads became the best engineered and best known of all, they were also amongst the most expensive to build and when road building gave way to railway mania his final plans were left incomplete. Others, such as the blind engineer John Metcalf and later John Loudon McAdam (who gave his name to, but did not invent, tarmac), shaped the improvements too.

The story of the road in the eighteenth and early nineteenth centuries is the story of a strengthening nation, led by increasing investment to match a people and an economy on the move. By the time of Telford's birth in 1757 the network of roads in England, if not yet Scotland, was already getting better. Then, in the eighty or so years until the railways superseded them in the 1840s, it was transformed. A chain of local turnpike trusts, set up by Acts of Parliament and run with varied degrees of competence, dragged the network into something approaching a modern condition. These were self-sustaining bodies, set up with official powers to collect tolls in return for repairing their routes. In an eruption of endeavour 389 trusts were set up over two decades from 1750. By the start of the nineteenth century these local turnpikes of greatly varied quality were being modernised again, with better surfaces to allow faster journeys. With an early start in London you could reach a town such as Brighton in time for lunch: a vast improvement on the old slog to the coast which, as late as 1749, Horace Walpole had found 'a great dampener of curiosity'.[4]

Towns and remote junctions alike found themselves littered with coaching inns, efficient machines for changing horses and feeding travellers at maximum speed. They were as practical as any service station and the best were more comfortable: buildings such as the Newhaven Inn in the Peak District, a great complex built in a remote spot by the 5th Duke of Devonshire. Now closed, it is remembered locally for opening all hours thanks to a dispensation from licensing laws granted by George IV, who once stayed the night.

Travellers had roughed it a century before but by 1800 they were being presented with bloated menus, as at the famous Sugar Loaf Inn in Luton (on a road Telford improved) where the bill of fare included 'A Boiled Round of Beef; a Roast Loin of Pork; a Roast Aitchbone of Beef; and a Boiled Ham of Pork with Peas Pudding and Parsnips; a Roast Goose; and a Boiled Leg of Mutton'.[5] No wonder that, in cartoons of the time, gentlemen always seem to be bursting out of their buttons.

This world developed its own romance and when it was gone its own nostalgia. Dickens, Southey and Thackeray all described it but Thomas De Quincey was its greatest chronicler. In 'The English Mail-Coach', an impassioned essay published in 1849, he lamented the loss. One of his chapters, 'Going Down with Victory', describes how Royal Mail coaches spread news of military success:

> Horses! Can these be horses that bound off with the action and gestures of leopards? What stir! – what sea-like ferment! – what a thundering of wheels! – what a trampling of hoofs! – what a sounding of trumpets! – what farewell cheers. . . The half-slumbering consciousness that all night long, and all next day — perhaps even for a longer period, many of these mails, like fire racing along a train of gunpowder, will be kindling at every instant new successions of burning joy, has an obscure effect of multiplying the victory itself, by multiplying the stages of its progressive diffusion. A fiery arrow seems to be let loose which from that moment is destined to travel, without intermission, westwards for three hundred miles.[6]

Each year, on the king's birthday, these wonderful vehicles paraded through the west end of London, the coaches freshly varnished, the horses decorated with blue and orange rosettes, and the coachmen in new scarlet liveries. They served a heroic national network which sent dozens of coaches from the Central Post Office in St Martin's-le-Grand in London every night at eight, their guards checked for sobriety and issued with a time-piece, a blunderbuss and a brace of pistols. The sounds of travel would have been of hurry, the royal crest shaking and rattling through the dark and De Quincey – who on another occasion recalled taking 'a small quantity of laudanum having already travelled two hundred and fifty miles' – adored it. For him, beauty lay in the Royal Mail coaches, of 'dark ground of chocolate colour' with 'the mighty shield of imperial arms. . . emblazoned in proportions as modest as a signet ring bears the seal of office. . . whispering rather than proclaiming our rela-tions to the mighty state'.

De Quincey disdained the flashier private network of passen-ger express stagecoaches, 'some "Tallyho" or "Highflyer", all flaunting green and gold', which by the end of the eighteenth century competed with – and often outraced – the stately mails. But these, too, had their own drama. One traveller recalled the scene as they prepared to leave London, 'with their lamps lit, and all their smoking and steaming, so that you could hardly see the horses. . . the coaches rattling over the stones; the horses' feet clattering along to the sound of merry-keyed bugles, upon which many guards played remarkably well, altogether made such a noise as could be heard nowhere except at the Peacock at Islington at half past six in the morning'. Today, you might find a similar scene amid the hooting and revving of engines at a Bolivian bus station. The coaches' names caught the changing times too: steady English titles such as *Diligence* and *Gee-hoes* giving way to names full of patriotic bombast – *John Bull* and

Royal Union – before finally racing into the nineteenth century with names full of pace – *Celerity, Antelope, Rocket* and *Swallow*.

Telford helped make all this possible. His roads were the centrepiece of a network of which now only the skeleton of routes and inns remains and whose impact on the nation was immeasurably deep. 'Good roads are... the greatest of all improvements,' wrote Adam Smith in *The Wealth of Nations*, a book Telford esteemed. At its zenith, around 1830, the network of coaches served every corner of the country, galloping, one contemporary calculated, an extraordinary 15,604 miles each day.

Simply breeding and feeding the animals needed to sustain this network was a world in itself. Each coach changed horses every ten miles, so that a long overnight run could demand 220 horses, ten stabled in each yard: four to go up a stage each day, four to go down and two resting. This meant that every horse had an hour's work for three days and a break on the fourth: their working life was short, but the jobs they supported and the money they made meant that coaching was part of the lives of even those who could not afford to travel. By 1835 there were about 150,000 horses pulling coaches in Britain, proof of the commercial premium placed on speed and comfort.

Better roads also meant better maps and guides. In 1674 Charles II appointed the Scottish geographer John Ogilby to the post of 'His Majesty's Cosmographer and Geographic Printer', a reward for his work drawing up the first useable road atlas of England. Ogilby eventually surveyed 2,519 miles of rough roads, measured by walking the routes with his waywiser – a giant wheel. In doing so he standardised the modern mile. His vast survey, soon reduced to a useful pocket-size for travellers, was much reprinted and updated in the century which followed. It guided travellers by landmarks, shown with symbols just as in a modern road map, though the landmarks themselves were of a different sort. 'By the gallows and the three windmills enter the suburbs of York,' one edition advised travellers heading north.[7] These atlases became fatter and more detailed as roads spread, until – complete with the names and times of express coaches – they became a precursor of Bradshaw's railway handbook.

And as the network of coaches increased, so did the fame of the men who worked on it. The coachmen of the fastest services were something like sporting celebrities. Dashing young men (among them John Rickman's son) became notorious for taking the reins and racing the vehicles, to the terror of their passengers and the passing rage of the press, which deplored this outbreak of ill-discipline among Oxford undergraduates. 'Whoever takes up a newspaper in these eventful times, it is even betting whether an accident by coach or a suicide first meets the eye,' complained a correspondent to the *Sporting Magazine* in 1822.[8] Our own age, which perhaps regards travel by coach and horses as stately and secure by comparison with the fast lane of a motorway, has forgotten the thrills, spills and appalling accident rates of horse-drawn travel. Should you choose to break the law today, you would be much, much safer at 100mph on the M1 than you would have been at 12mph on the coach road to Brighton. Disasters were common: coaches overturned, horses ran wild, passengers broke their necks. Mechanical brakes only became common in the 1830s; before then guards ran out to push blocks under the wheels to stop the vehicles over-running horses on steep descents. Familiarity with the risks must be one reason passengers were prepared to put up with the intense dangers of early rail travel when it came, with exploding boilers and frequent crashes.

The speeding up of travel between 1720 and 1820 cut journeys which once took days down to hours; proportionately a greater reduction than the saving brought by rail travel, when it arrived. By the end of the stagecoach era the fastest coaches averaged more than ten miles an hour for many hours at a time, including stops and changes of horses; not far off the pace kept up by the best of today's marathon runners over a much shorter distance.

The Wonder, running to Shrewsbury on Telford's improved road, had no rival as the fastest coach in Britain over such a long distance. It once covered the 153 miles from London in fourteen hours, forty-five minutes. Crowds gathered each day to watch it arrive, guided by the famous driver Sam Hayward; horses galloping up the Wyle Cop and under the narrow archway at the Lion Hotel in Shrewsbury, one of the most celebrated of all coaching inns. The coach was renowned for never being more than ten minutes late under Hayward's control in the sixteen years that he drove it, and it fought on to the last, in 1838 leaving London at the same time as the new train from Euston and beating it to Birmingham. A toast was raised when it reached the Lion Hotel: 'confusion to the rail-roads and a high gallows and a windy day to all enemies of the whip'.[9]

Those present must have known that this wonderful world was doomed. With the advent of railways there would be no need for stagecoaches or turnpikes – and the transition would be quick. The last stagecoach from London left at dawn on 16 January 1846 and was much missed. By 1861 even the owner of the Lion Hotel had to auction off his beloved stable of horses, 'well-known on the road for their fast pace and powers of endurance'. 'Will be sold,' this sad advertisement continued, 'without reserve in consequence of the opening of a portion of the Welsh railway.' Relics lingered: the toll house on Telford's Menai Bridge posted a charge of half a crown for stagecoaches as late as 1902. But none had passed that way for half a century.

'Them as 'ave seen coaches, afore rails came into fashion, 'ave seen something worth remembering,' recalled one retired coachman who had driven the *Highflier*, another famous service.

'Them was happy days for old England, afore reform and rails turned everything upside down, and men rode as natur' intended they should. . . the coaches is done forever and a heavy blow it is. They was the pride of the country, there wasn't anything like them, as I have heard a gentleman say from forren parts, to be found nowhere nor ever will be again.'[10]

———

How were the roads on which these coaches ran built? In Telford's view, it all had to start with stones. He did not invent a new way of building fast roads for stagecoaches to gallop along, but he made sure that it was done more precisely and thoroughly than ever before. Each pebble, each layer, had to be measured by his men. There was no place for the slapdash. His notebooks contain minute designs for the iron measuring rings through which each stone used in his roads had to pass, and the contracts he drew up for road-building became increasingly absolute in their specifications.

The Telford method produced a road which was designed to mature and to last. 'From the moment it is first opened it becomes daily harder and smoother and very soon consolidates into as hard a mass as can be obtained by the use of broken stones,' wrote Sir Henry Parnell, one of its champions and a loyal friend of Telford's.[11] His *Treatise on Roads*, intended to be an instructive handbook of good practice, was not published until 1833, close to the end of the great era of coaching. Rather than spur better road building it became an unread memorial to a skill made redundant by the dash for trains, but it still stands as the best explanation of Telford's effort to turn roadbuilding into a science.

The secret of his roads, Parnell explained, lay in getting the load-bearing pavement at the very base of the road right – a layer

Telford made as strong and durable as anything put down by the Romans. In designing it he drew on both his own experience as a stonemason and on work done in France by the pioneering road-builder Pierre-Marie-Jérôme Trésaguet, whose work he never saw but studied in books. His specification was exacting and contractors were expected to follow it to the letter. First, they had to level the bed of a new road and clear it of obstructions. Next, they had to lay a solid bottom layer of closely packed stones. This pavement fell away in a gentle crown from the centre, to ensure the surface would drain properly, guarding against the appearance of puddles – the precursors to potholes. The surface had to be raised seven inches high in the centre, sloping down to five inches at nine feet from the centre; then four inches at twelve feet and finally, at fifteen feet, to just three inches. All irregularities were broken off with a hammer.

This strong pavement was just the start. Any gaps, Telford ordered, now had to be filled with stone chips, to create a convex surface and this was, in turn, coated with smaller hard stones chipped into the form of a cube, the largest able to pass through a two-inch ring. Next he ordered that these smaller stones cover the road to a depth of six inches. The first four were put on straight away, to be crushed by the first carriages to use the road, then raked, then topped with a final layer a few weeks later. The coaches' iron wheels and horses' hooves did the final steamrollering.

'When this work is properly executed no stone can move,' Parnell wrote: which was true, but did not stop some asking if all this effort was worth it. Amongst the critics of the elaborate Telford process was John Loudon McAdam, born in Scotland a year before Telford and by the 1820s a famous proponent of turnpike improvement. He made his money by fixing up existing roads with a top dressing of stones, many of which could be reused cheaply by digging up the old surface. 'It matters not if the soil be clay, sand morass or bog,' he claimed, 'I never use large stones on the bottom of a road.'[12]

In Telford's view, this amounted to nothing more than unreliable quackery, gravel dumped on mud without the all-important solid paving beneath. He was scathing about McAdam's work when both men gave evidence to a parliamentary select committee in 1819 (for which Telford seems to have prepared well: his papers contain a torn-out passage from a book on McAdam's technique, dated that year and annotated in Telford's hand, with 'good' written by one suggestion). McAdam, in return, was cutting about Telford. Asked if he had seen the engineer's famous route to Holyhead he replied: 'No: that I understand is a new road.' McAdam – whose income came from tarting up existing roads not building new ones – had little interest in new civil engineering.

And so the squabble went on over the years – at times debated in Parliament where McAdam's detractors accused him of being little more than a charlatan who fleeced turnpike trusts and built his roads using the forced labour of paupers and young children, though he did not die a rich man. Telford seems to have taken pleasure in the failure of the expensive 'Macadamisation' of Whitehall in 1827 and of Blackfriars soon after, which needed expensive repairs five months after opening. Until his death Telford continued to insist that a solid base was essential. 'I am fully aware of the strong prejudice against paved roads,' he wrote, 'but these prejudices have been created by the total want of skill.'[13] In his book on roadbuilding, no doubt parroting Telford's views, Parnell criticised McAdam, quoting the *Westminster Review*'s verdict that 'the public naturally looks upon him as a sort of magician. . . innocent of quackery himself, he has been forcibly made the great quack of the day'.[14]

This was unfair because McAdam had a point. His method was simple and affordable and so of use to private turnpike trusts, which had to pay their way from travellers' tolls. It allowed the rapid fixing up of existing routes and had an immediate effect on the speed and reliability of travel. Telford's demanding approach was slower, more exacting and much more expensive. He rebuilt and realigned roads, rather than simply resurfacing them; and when the two techniques came into conflict it was Telford's which came out best.

He selected, for instance, a section of the dreadful, slippery Great North Road as it ran up Holloway towards Archway and Highgate Hill (a route which carried much of the traffic north out of central London then, as it does today) and rebuilt part of it with firm stone paving underneath and part without. The paved section, his allies insisted, lasted longer and was cheaper to maintain. 'It is most surprisingly improved since you adopted the plan from Mr Telford,' a grateful stagecoach company wrote soon after, 'in fact so much so that four horses can perform better their journey now than six horses could do previous to such plan being adopted.'[15] A little later McAdam's son, James, took on responsibility for maintaining a section of the route to Birmingham near St Albans, which Telford had improved, and it fell rapidly into disrepair.

Telford's allies delighted in such apparent proof of the success of his method. 'The main cause of these roads being in an imperfect state, is Mr McAdam's being the surveyor of them,' Parnell wrote after observing the failure at St Albans.[16] 'Mr Telford in all cases recommends this paving and the opinion of a man of such experience cannot be treated slightly,' added William Provis, Telford's chief collaborator on the Holyhead Road, gravely.[17] 'He has made more miles of new road than any engineer in the kingdom; and having myself studied for nearly 15 years in his school, and made a considerable extent of road under his direction, I may venture to say that his practice is not unsupported by experience.'

In the end there was and is no way of resolving such disputes. They are disputes about accountancy, not civil engineering, and depend upon the relative costs of capital versus upkeep spending, and the quantifying of inconvenience to traffic while running repairs are made. But Telford's lesson in life was that it was always worth putting the effort in at the start. 'I cannot too often repeat that a surveyor should not feel satisfied that he has done his duty until the whole breadth of ground belonging to a road between the fences is put into perfect order, as this shows skill, attention and good workmanship,' he wrote, in the hectoring tone of an ageing and increasingly inflexible man determined to see nothing but corner-cutting in the efforts of his rivals.[18]

———

When the United Kingdom came into being on 1 January 1801 the Irish House of Commons in Dublin was abolished and replaced with a united Parliament in Westminster. As a result, Irish politicians faced a problem. The journey to London, once an expedition attempted perhaps a few times in a lifetime, turned into a regular trip. Unsurprisingly, they wanted it improved. 'The frequent journeys made by the Irish members produced constant irritation and complaint,' Telford wrote, 'and gave rise to warm discussion in Parliament.'[19]

They had much to complain about. As it does today, the main sea crossing from Dublin served Holyhead, on a small island itself just off the larger island of Anglesey. The voyage could be long and violent but there was no respite on shore. 'When arrived on the English side,' wrote Telford of conditions in the early 1800s, 'the passengers were landed on rugged, unprotected rocks.' Crawling up from the boat, sick and soaked, they then faced an epic, raw journey to London.

The first challenge came on Anglesey. Today cars on the A55 Expressway roar across in minutes but the unimproved journey across this relatively flat island was, wrote Telford, 'a succession of circuitous and craggy inequalities'. Next came the short but risky ferry crossing over the Menai Strait to the mainland and then the long haul through North Wales which was, wrote Telford, 'generally speaking narrow, steep, and unprotected by parapets'. Finally, after arriving in Chester or Shrewsbury, travellers joined the main route south to London: a road that stuck close to the original Roman Watling Street and which, though turnpiked and in places fast, was anything but a pleasure to travel.

Jonathan Swift, who made the journey in the first part of the eighteenth century, remembered Holyhead as 'an unprovided and comfortless place'.[20] The journey to London, he told his friend Alexander Pope, took him 'through many nations and languages unknown to the civilised world'. He mocked the horrors of the trip in verse, starting with the early rising:

Rous'd from sound sleep thrice call'd at length I rise,
Yawning, stretch out my arm, half-closed my eyes,

and finishing with the company:

When soon, by ev'ry hillock, rut, and stone,
Into each other's face by turns we're thrown.
This grandam scolds, that coughs, the captain swears,
The fair one screams and has a thousand fears;
While our plump landlord, train'd in other lore,
Slumbers at ease, nor yet asham'd to snore;
And Master Dicky, in his mother's lap,
Squalling, at once brings up three meals of pap.

Sweet company! Next time, I do protest, Sir,
I'd walk to Dublin, ere I'd ride to Chester!

The government had long taken an interest in this strategic route to Ireland: there had been efforts to establish a reliable postal service under Queen Elizabeth I and James I, and by the late seventeenth century travellers as well as the mail were using Holyhead as a port in place of Chester, where unreliable winds could trap shipping for days. But Ogilby's guide shows that the route along the coast was still basic, part of it along sands awash at high tide, where if the horses got stuck in soft ground travellers risked being drowned.

It was not until 1765 that the first turnpike trust was established on Anglesey. It was soon joined by others in Caernarvonshire. In 1776 rough stagecoaches began running to Holyhead and soon a reliable service was provided from the Lion Hotel in Shrewsbury, run by the pioneering coachman Robert Lawrence. A General Post Office was established in Dublin in 1784 and a year later the Royal Mail began to use coaches in place of ponies and post boys. At first these travelled the coastal route, but by the end of the century they went inland through the mountains, via Capel Curig. The road, improved as a turnpike, was good enough for the Duke of Richmond to travel it in 1807 on his way to take his post as Lord Lieutenant of Ireland. A year later the express Irish Mail coach from London began making the journey, though parts of the road were still in terrible condition.

The journey was becoming easier than in Swift's day. But for political grandees forced to rattle along with the rest it remained a bleak experience. Local turnpikes could not raise enough in tolls to meet their legal obligation to keep the route in decent repair. In one week in the early 1800s, three post horses – with riders carrying the mail in saddlebags – fell and broke their legs. 'Many parts are extremely dangerous for a coach to travel on,' a parliamentary report explained in 1810, 'stage coaches have been frequently overturned and broken down from the badness of the road.'[21] Something, the politicians chuntered, had to be done.

The first person they turned to for help was not Telford but John Rennie. He was both slightly younger and somewhat more celebrated at the time than Telford, his only real rival as a civil engineer of national standing. His weakness here was that he had little interest in roads. His expertise lay in harbours and he proposed improvements to those at both ends of the Irish crossing, at Howth, near Dublin, and Holyhead. He also set out plans for a spectacular arched iron bridge across the Menai Strait, which never went further than a blueprint. And there things were left.

For Irish politicians – who, as Telford noted, still 'dreaded' the journey from London – this was intolerable. In 1810 they had a second go at improving the route, encouraged by John Foster, the Chancellor of the Irish Exchequer. His post had lingered on after the creation of the United Kingdom and he arranged for Telford to survey the whole route and propose a solution. 'Mr Telford is to proceed upon the survey as pointed out in the attached paper,' his curt orders read.[22] 'He will commence his survey at the town of Holyhead. . . his attention will necessarily be directed to the erecting of a Bridge across the Menai.'

The choice of engineer was deliberate: Telford was by this time not only well known for his bridges and roads in Scotland but, through his work for canal companies and the Highland parliamentary commissions, a practised hand at winning allies at Westminster, too. And with Foster's backing things at first seemed to move fast. In his memoirs Telford records that he surveyed a variety of possible routes, including a direct one over the Berwyn Hills, which would have involved impossibly steep climbs, before recommending, in a report published in April 1811, the reconstruction of much of the existing route from Chirk along the River Dee to Llangollen and the village of Betws-y-Coed.

From there coach traffic had traditionally diverted north along the River Conwy to the coast, a long and indirect crawl which avoided the heights of Snowdonia. In its place Telford put forward a spectacular and shorter new route straight through the mountains. Parts of this had been surveyed and built through Capel Curig, past Llyn Ogwen and over the Nant Ffrancon pass but even these sections were tortuous. 'The old road,' he concluded, 'was hilly and crooked. . . in fact quite unfit for wheeled carriages.' In Telford's plan everything was to be new: bridges, ascents, inns and tollbooths.

In 1811 he presented his survey to a parliamentary committee, which at first seemed keen to act. Then, as is so often the way with infrastructure, things stalled. The Irish chancellor who had championed the work found his post abolished and he retired, packed off with a peerage. 'The demand. . . for improvement remained equally urgent, but no one undertook the arduous task of arranging a practical scheme,' Telford wrote of this lull.[23]

For a time, he occupied himself in Scotland. But Telford did not give up and nor did the Irish politicians who wanted a more comfortable journey. In 1815 the scheme was revived and this time he found a supporter who would stay by his side until the end.

Sir Henry Parnell, an Irish landowner and MP with a liberal mind, was one of those singular, restless men with whom Telford associated easily and without whom his greatest schemes would have come to nothing. It was as if some part of his psychology needed such support. As with Pulteney, as with Rickman, as with von Platen, so it now happened with Parnell, who for the rest of his life became the administrative driving force behind the Holyhead Road. In him, Telford wrote, 'was found a degree of intelligence, zeal and perseverance which overcame every obstacle'.

Educated at Eton, he was, in the 1790s, a strong opponent of the proposed new Union with Britain and he remained after its creation, as an Irish MP, a supporter of Catholic emancipation and an intelligent authority on all sorts of subjects, including financial reform. In the 1830s he served in Lord Melbourne's cabinet and in the 1840s ended up with a peerage. But he never settled easily into the Westminster game and in June 1842 hanged himself in his dressing room at his Chelsea home, aged sixty-six. 'Attacked with fever and delirium, followed by insomnia; he lost all interest in everything,' one account of his death recalled, and it is hard not to suspect that the coming of the railways, which the year before had killed traffic on the Holyhead Road to which by then he had given so much, helped bring on his depression.[24] Perhaps it is because of this melancholic and at the time shameful end that Parnell's place in history, other than as a great-uncle of the Irish nationalist Charles Parnell, has been overlooked. For him, no flattering Victorian multi-volume biography. He was not the first of Telford's patrons to find himself bound into a cause that consumed his life.

In January 1815 Telford wrote to William Little in Eskdale. He had come to London, he said, in order to set before Parliament 'the last mode of improving <u>Pats</u> intercourse with England, which you will readily perceive will be no very easy or cheap task'.[25] He was flat out with work elsewhere: the Caledonian Canal was enough for any normal man, but he was not normal and found a curious satisfaction in overburdening himself with projects of intense political and logistical difficulty. He did not need, want or ask for much money. His reward came from a competitive awareness of status and, as in his Scottish work, from a pride in work that underpinned a strengthening United Kingdom. The 260-mile-long London to Holyhead Road was a symbolic as well as a practical project – akin to the projected

HS2 rail link in our own age – and a product of public rather than private enterprise.

Telford and Parnell had seemed to pick their moment well. Peace with France had freed up funds and government energy. There was a hunger to spend on something impressive other than the military. And Westminster had the men and systems in place. In 1815 Parnell set up a committee to champion the road and, soon after, a board of parliamentary commissioners to guide it. The model was familiar: it replicated the ones behind the Caledonian Canal and Highland roads and some members served on both. The commission's long-serving members, among them Robert Peel, later prime minister but then chief secretary to the Lord Lieutenant of Ireland, became the project's champions. They were to publish annual reports which included exact financial accounts and an intricate first-hand description of the whole route, written each year by Telford.

Now aged fifty-eight, with seemingly indestructible good health and humour, and the confidence of someone who had seen his previous projects go from paperwork to reality, Telford charged in. Armed with £20,000 from the commission he set out in the summer of 1815 with two assistants to survey the route.

The project fell easily into two parts. From London to Shrewsbury, Telford and his assistants travelled a route – large parts of which make up the modern A5 – which was already busy but maintained by sixteen different turnpike trusts. There was no need here for a wholly new road. What was needed was a guiding mind to insist on consistent maintenance and to sort out the worst sections; and serving records show that this was done with regularity, each trust being tested on the width and quality of its section of the road's surface. In places the work went further, with large-scale construction overseen by Telford on new bypasses or diversions to ease gradients which tested horsepower, such as at Archway past Highgate or between South Mimms and Barnet, where embankments and cuttings were built to his design.

These were significant undertakings in themselves, involving major construction works to strict criteria. Even the fencing was specified, together with the planting of new hedges along the roadside using strong saplings that had been grown elsewhere (Telford insisted) for at least two years. But such work was as nothing when set against the much greater challenge Telford and his team faced after Shrewsbury, as the road entered Wales, and from the start he showed greater personal interest in this rugged section.

It ran – or was intended to run – from Chirk, near Shrewsbury, through Snowdonia to Holyhead; and included a second section along the coast from Chester to Bangor, in order to link fast-growing Manchester and Liverpool to Ireland. On these sections, Telford proposed not just the improvement but the almost total replacement of the old road. This was expensive and controversial. The seven turnpike trusts that controlled the Welsh route – and which had been investing in improvements – did not cooperate. In 1819 the commissioners were forced to abolish them, effectively nationalising the road from Shrewsbury to Holyhead. From the start Telford intended the road to be a showpiece, a new marvel of infrastructure: not privately built, like the old, haphazard turnpikes, or part-funded by landowners with state assistance, like his Highland roads, but planned centrally on a magnificent scale and paid for from Treasury funds.

Nothing like it had been proposed in Britain since the Romans and there was to be nothing like it again until the building of the first motorways in the 1950s. It was, before the railways came, certainly the most impressive piece of modern transport infrastructure anywhere on the planet: 107 miles of express road between Shrewsbury and Holyhead and 153 miles between Shrewsbury and London, all redesigned and rebuilt, including a radical suspension bridge at Menai, the long Stanley embankment carrying the road across the tidal channel that divided Holyhead island from the main part of Anglesey and a second section of coastal road from Bangor to Conwy, with a great iron suspension bridge there, too.

If this was to be his monument, he made sure it was seen as such. In the 1820s one fat and much-republished atlas of roads, packed with maps and guidance for passengers, was dedicated 'To Thomas Telford Esq, a gentleman whose works as an engineer continue to panegyrise his name'. 'I am obliged alike for his extreme politeness upon all occasions, and liberal supply of materials,' wrote the author. 'The effects of his genius, carry with them the conviction of superior ability, and incontestably prove that [faced with] his perseverance no obstacle, however formidable, proves insurmountable.'[26]

Was such glorification excessive? The Holyhead route was unquestionably 'the most significant achievement of Britain's greatest civil engineer', Quartermaine, Trinder and Turner argue in their definitive account.[27] But they note too that others played a role in bringing it about and that Telford, with that undercurrent of vanity which was always present in him, 'tended to exaggerate his own achievements in building the road across North Wales to Holyhead, both in disparaging those of his predecessors and exaggerating the horrors of travel in the region before 1810'.

What is certain is that without him the road would not have been rebuilt on this scale. Regardless of cost or complication he was driven as much by the splendour of what might be created as by any study of what might be economical or necessary. At the start, he estimated that the road would cost a total of £269,000 – paid as £53,000 a year over five years – including £57,000 for the section from Holyhead to the Menai, £92,000 from Chirk to Menai and £120,000 for the bridge to link them. But along with the Caledonian Canal it was the other of his great nineteenth-century projects to go over budget.

As always, his habit was to push on before his supporters could change their minds. 'I left no time in making proper arrangements and commencing practical operations,' he reported to the new commissioners in 1816, by which time although the 'period of the year has been unfavourable [in] mountainous country and the workmen at first were unpractised' he had up to 400 men labouring away.[28] They were working even before the full survey of the route to London was complete, a task which began in September 1815 and which was not finished until March 1817.

From Shrewsbury to Holyhead, Telford drew up 123 separate schemes (or 'lots') each to be awarded as a separate contract to 'the lowest tender, if supported by character and security'.[29] He and his team drew up the overall guiding principles of budget and design and a detailed specification for each lot, listing the start and end point and a yard-by-yard description of what should be done. He numbered the lots not for their position on the route but the order in which it was intended to build them, and the order never needed to be changed. The terms of the contracts were demanding, setting out the dates by which work had to be done, the level of monthly payments and the role of referees to settle disputes, with all the risk carried by the contractor.

He also made sure each lot was part of a unified whole. Telford and his team specified not only the make-up and dimensions of the road surface but everything needed to speed the Irish mailcoaches on their way: the tollhouses with their distinctive sunburst-patterned iron gates (designed by

Provis, not Telford); the coaching inns to change horses; weighing machines; supply depots for materials to repair the road, every quarter of a mile (still visible today as indents in the walls); and the even gradient, mostly no more than 1:30 and never exceeding 1:20. The iconic feature was the series of milestones, each with a cast-iron plate painted black, specifying distances and the nearest inns, set four feet off the ground on a pillar of red limestone cut with a triangular top so that it could be read easily from inside a passing coach. 'I never saw a proper milestone that I could copy,' Telford recalled, 'I looked for three years all over England trying to find one as a pattern and after all I could not find one that looked like a decent milestone.'[30] Not until the first motorways were constructed in the 1950s was such a systematised system of signalling the distance to the next place of refreshment and fuel adopted.

By breaking the route into lots, Telford could ensure early action on some of the most troublesome sections. Lot 1, for instance, a steep line of road at Nant Ffrancon, south of Bethesda (just inland from the coast at Menai) had always proved particularly vexatious and its rapid reconstruction was soon a showpiece for travellers. The showpiece could soon be contrasted with the section of former turnpike which ran just below it.

Much of this work happened, of course, in Telford's absence. From start to finish the project was led by William Provis, who had been bred to engineering as the son of a canal surveyor and had almost certainly been Telford's first pupil, joining him in 1808 or 1809. He was Telford's creation and later his champion, working his way up first on the Caledonian Canal and then surveying the Carlisle to Glasgow road. Like Matthew Davidson before him he became more than just an employee. After Telford's death he took over his house in London and lived there in 1870. But as with Davidson there was always a reserve: they were allies and friends but never equals.

You can sense this in the politeness and speed with which he replied to Telford's incessant flow of letters; for even when away in Scotland or England the latter chose to involve himself (as surely he did not need to do) in the minutiae of managing landowners. Thomas Jones, for instance, a farmer whose grazing was affected by work on Lot 15, wrote to Telford in London to complain that 'the intended Fences are not sufficiently high to protect our lands from the sheep on the mountain above and beg leave to request you to alter them'.[31] Telford's fawn notebook (with 'Holyhead' inked in stained writing on a cover marked by damp fingerprints) also shows that he was much more than a figurehead. A day-by-day diary and working notebook of bills, improvements, to-do lists and scrawled estimates, it carries with it a sense of pace and involvement. 'From English Bridge remove Houses on the north side opposite to the Lion [hotel]. Straighten the street and fill up,' reads one entry for 1821.[32] He always enjoyed knocking things down (although in this case the direct route was not built).

Telford's Holyhead papers breathe bureaucratic order; they show a man wanting to show his employers they were getting something for their money. He was studious in preparing his plans and meticulous in carrying them out. 'The portions of Road which have been improved under the Board's directions are in a perfect state,' he assured the commission in March 1817. 'Two of those precipitous and rugged parts which formerly were most dangerous, are now rendered easy and safe. . . during the whole time these improvements were in hand, the weather was, with little interruption, usually wet and stormy.' Two years later he reported again: 'The works are let by contract, at the end of every month what has been performed is measured by the Commisnrs Inspectors and a statement thereof is transmitted to their secretary.' He was involved in the small concerns, such as how to pay workmen when, as he reported, 'Bank of England notes could not be found at Shrewsbury.'

None of Telford's surviving accounts and records carry any hint of corner-cutting, corruption or the turning of a blind eye, and no suggestion by third parties (not all of whom loved Telford) that the engineer was less than scrupulous in questions of professional probity. Opportunities will have been many but Telford was by reputation a man to whom you would not have dreamed of trying to suggest any sort of impropriety. Doubtless uncorrupt by nature, he must also have known that probity was his brand, and scandal would have wrecked it.

———

Drive the route of Telford's road today and the scale of what was achieved leaps from your path. It is a pleasure to race along what is now the A5, hardly ever having to change gear, since its engineer kept the gradients even. Passing from Chirk (almost in sight of Pontcysyllte) it dives rapidly into the still-wild heart of North Wales. Beneath a skin of modern surfacing and signage, Telford's design is still there, a ghost beneath the tarmac: the low walls along each side, the depots he had built for road materials and the inns (some of them now farm barns) which sheltered passengers and their horses from snow and rain. Of the 83 distinctive milestones on the Welsh section of the road, only five were missing when the route was surveyed in 2003; of the 18 sunburst toll gates, 11 survive and every one of Telford's bridges is still in use.

The road climbs through cuttings and along embankments west of Llangollen, with today's preserved steam railway puffing alongside. From here, it heads west and upwards along the edge of the Glyn Diffwys gorge, above which Telford built a viewing place for travellers to see the Pen-y-Bont falls. These are now lost in a thick forest below but were once celebrated as one of the principal sights of North Wales. This section of the road, bypassed in the 1990s, and now wholly disused, is an evocative place. It is quieter today, with weeds forcing apart the old tarmac surface and road markings, than it was even in the 1850s when the travel writer George Borrow described the view as 'grand, beautiful and wild. . . the most prominent objects of which were a kind of devil's bridge flung over the deep glen and its foaming water'.

Telford's route then races on to Betws-y-Coed. Here it crosses the River Conwy over the most ornamented of Hazledine's iron bridges, the Waterloo bridge, a boastful masterpiece of design which cast into solid form the explicitly political purpose of Telford's great British road. 'There was a celebratory and congratulatory feel about this great governmental building project,' its historians say, and this was never more true than of the Waterloo Bridge. Based on a standard lattice design drawn from Bonar Bridge, it carries the emblems of the four nations that formed the United Kingdom, the rose of England, the leek of Wales, the thistle of Scotland and the shamrock of Ireland, and with them the crowing reminder that 'this arch was constructed in the same year that the battle of Waterloo was fought'. In fact, that is not strictly true: the bridge was finished later. But this does not diminish the majesty of what was, when it was built, still only the seventh major iron bridge to be erected anywhere – with most of the others also the product of Telford's alliance with Hazledine.

After this, the real wild Wales begins, further west, up into the bleakest open country of all, amid rain, clouds and now wind turbines, before dropping down to Llyn Ogwen, where the route interlaces with the old pre-Telford turnpike below and the remains of a Roman road. It then passes through the grey streets of the quarrying town of Bethesda before reaching the sea just beyond Bangor. On the other side of the water lies Telford's new route across Anglesey, and the impressive embankment he built to reach the quayside at Holyhead.

But first comes the Menai Bridge. Compared to that, building the road was simple.

In 1801 Telford toured the Scottish Highlands, at official command, to plan a network of new roads, bridges and harbours. This map, from a report three years later, was one result, with the southern boundary of the Highlands marked in green, his proposed new roads in red and older military roads in yellow. At the time, the shape of islands such as Skye were still uncertain – and on this map, incorrect.

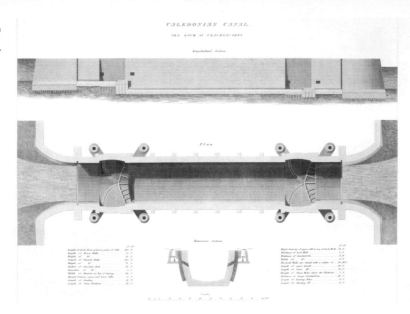

Even as Telford worked on new Highland roads he was planning and building a great waterway from coast to coast. These locks were designed to guard the eastern entrance of the Caledonian Canal near Inverness.

The canal's design made use of existing waterways, including Loch Ness, linked at its western end at Fort Augustus by this chain of locks. Traffic never lived up to Telford's hopes, though as this early twentieth-century photograph shows, steam boats kept it open until the recent rediscovery of the route by private yachts.

ne 1820s Telford returned to church building, overseeing simple designs which still dot the
hlands and Islands, as here at remote and beautiful Kinlochbervie in the far northwest.

The Menai Bridge was Telford's mightiest creation, the brilliant achievement of a talented team. In this engraving from his *Atlas* to his life's works he showed how the great chains which supported the bridge were put in place, hauled to the tops of two towers with pulleys. Without ingenuity like this the first large suspension bridge could never have been built.

...oday, the bridge is still open to traffic, its stonework and silhouette a North Wales landmark. Its ...ains were redesigned and replaced in the 1940s by Sir Alexander Gibb, whose great-grandfather was ...e of Telford's young assistants.

The 'Tally-Ho' London–Birmingham Stage Coach Passing Whittington College, Highgate, James Pollard, 1836.

Before railways, Britain was linked by a network of express coaches running on turnpike roads, the best of them built by Telford. This picture from 1836 shows the famous Tally-Ho racing through Highgate on his road to Birmingham. By then, Telford was dead and soon the coach trade would be too.

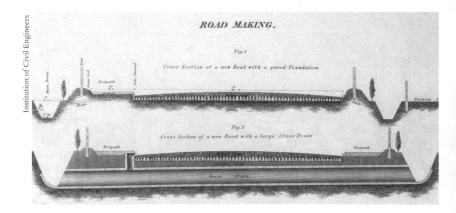

Other engineers repaired roads more cheaply and quickly but Telford insisted on perfection, making the modern road through design and good drainage, locking small stones into large so that the surface held. In this drawing from his *Atlas* he showed how to do it.

The 1830s, the last decade of Telford's life, brought failures as well as success. His attempt to fight railways by using steam carriages, such as this one built by Goldsworthy Gurney, came to nothing after one faltering journey from London to Birmingham.

Still more wounding to Telford's pride was the rejection of his design for a suspension bridge across the Clifton Gorge in Bristol. This odd, ugly Gothic proposal was swept aside in place of a winning scheme championed by the young Isambard Kingdom Brunel.

Telford began life as a stonemason but was later proud to put 'Civil Engineer' under his name in his bold, clear handwriting.

Thomas Telford, William Brockendon, 1834.

The last image of Telford, by then very ill, was draw in pencil and chalk in 1834, his hair almost gone and h body diminished. He lived only a few months longer.

This was the only home Telford ever owned, 24 Abingdon Street, bought when he was sixty-four and soon filled with his pupils. Since demolished, its site in Westminster opposite Parliament has been turned into a garden, College Green.

Pontifex Maximus

In 1815 Telford posted a letter to William Little, Andrew's surviving brother in Eskdale, full of happy chatter. 'Poor Rob. Laidlaw,' he gossiped of an old friend, 'his industry and sagacity were more than counterbalanced by childish vanity and silly avarice which rendered his friendship dangerous and his conversation tiresome. He was like a man in London whose mouth is walking by himself along the streets. . . constantly pronouncing <u>Money! Money!</u>'[1]

This is Telford's voice: jovial, worldly and judgemental. He was showing off, in the victorious Britain of 1815, with Napoleon finally beaten. 'I am projecting a Tour in France and Italy,' he continued, 'if time can be spared from my various avocations in our own unrivalled and beloved Island.' British engineers, Telford among them, had once learnt from French road design and admired the Romans. Now they could hope to teach France and Italy. 'My objects are [of a] very liberal cast,' he boasted to William Little, 'I want to be enabled from observations to describe the inferiority of the French and Italian canals, Bridges and Harbours which from their own Accounts I know to be facts but which from the Rapsodys of Frenchmen and vapouring ignorant Englishmen have long been impressed upon us as wonders of Art and Perfection.'

'If they discover my views I shall probably have the honour of being assassinated in protection of the Great Nations claim to superiority,' he concluded. He was only partly joking. Britain could now claim to be Europe's top dog and like many of his countrymen Telford was overcome by patriotic energy. 'Woe be to the memory of that race who suffer the British glory to be tarnished!!!' he boasted to the Eskdale-born military engineer Charles Pasley soon after; 'I hope that the Battle of Waterloo will not lull you into supreme security. . . the effervescence of the great Nation has not yet subsided, and if wholly extinguished, the bestial Harmony of the European Sovereigns will, in a few years, furnish you with appointments of future triumphs.'[2]

In the same year he and Hazledine designed and cast the iron parts for the Waterloo Bridge over the River Conwy, with its brash inscription recording the victory after which it was named. Work on the road, of which the bridge was a part, was moving fast. 'We have made such progress as will show that we can match the French in Road-making as well as fighting,' he boasted to Pasley, remarks suggesting a little resentment at the way glory had been heaped on soldiers and sailors rather than engineers.[3] Now, though, he could benefit from the fresh flow of government funds released by the end of the war. 'After much ignorant and mistaken vociferation about economy, the public works, in which I am concerned, have been amply provided for,' he noted in the summer of 1816, listing £75,000 available for the Caledonian Canal, £20,000 for the Crinan Canal in Kintyre, £50,000 for the road from Glasgow to Carlisle and £10,000 newly released for the Holyhead Road.

By 1819 he could report that 'many of the most dangerous portions of the route were rendered commodious and safe'.[4] The ferry route from Ireland was improving, too. John Rennie was overseeing work on the harbour at Holyhead and in 1821 the first steam-powered packet boats, the *Ivanhoe* and *Talbot*, began making the crossing.

But a third challenge – bridging the Menai Strait to join Anglesey to the mainland – remained unsolved. The need for a fixed crossing was clear. The Menai is a tight tidal channel and although at its narrowest it is only 1,300 feet wide and at the lowest tides the depth can drop to only a couple of feet (shallow enough for drovers to ford their cattle across), by ferry the trip from shore to shore was at best a burden and often dangerous. Vessels had to strain to avoid being wrecked on the 'Swellies' – exposed rocks which lurk between Telford's bridge and Robert Stephenson's later rail crossing a little to the south. In 1282 Edward I was said to have sent troops across the Menai on a temporary raft of wooden boats roped together. But it was only by the late eighteenth century that thoughts turned to a proper bridge.

Parliament had discussed the matter in 1775 and Telford's papers contain copies of a proposal from 1783 for a causeway to be built across the Swellies and fitted with lock gates to allow ships to pass. He took an interest, too, in plans assessed a year later by William Jessop for a range of crossings, including a solid embankment with a lock, a partial embankment with draw-bridges, and 'the third a bridge of wood or stone'.[5] He may have discussed the possibilities with Jessop when the pair worked together soon after on the Ellesmere Canal.

Accidents increased the pressure for action. One evening in December 1785 an overcrowded ferry, the *Tally Voyle*, was wrecked on a sandbank in the middle of the channel. The passengers scrambled out, calling for help, but strong winds and a huge sea forced rescue boats to turn back and around seventy people were drowned. Only one man, an exceptional swimmer, made it ashore alive. The occupations of those who died, among them a clergyman and his wife and many local farmers, reflect the everyday importance of the crossing to North Wales. In response *The Times* called for a bridge as 'so desirable and necessary a measure which, when accomplished, will be the saving of many lives in crossing those dangerous and diabolical ferries which are always attended with great delay and inconvenience'.[6] But nothing came of it, just as in the early 1800s John Rennie's proposal for an iron bridge, of either three arches or one great single span, was left unbuilt – and was perhaps, given the technology of the time, unbuildable.

Telford had been set the task by Parliament, in his initial Welsh survey of 1810, of finding a way to bridge the gap for good. His thoughts turned first to an iron arch along the lines drawn up by Rennie. It seemed the natural solution, given his success with light iron arches in Scotland. The challenge was to find a way of building one in so wide and exposed a spot. He came up with an answer which might have worked and which was later adopted by engineers elsewhere on structures including the Sydney Harbour Bridge: using temporary wooden supports strung from a high frame to hold the pieces in place while they were being joined together. This, however, could not overcome a different objection: that any bridge had to be raised far enough above the water to allow the masts of naval ships to pass beneath. Few large vessels ever did sail that way – Telford's colleague William Provis noted that locals could remember only one sloop ever attempting the dangerous passage – but the Royal Navy was insistent the route through be kept open in case any real-life Jack Aubrey wanted to sneak through the Menai. And that ruled out any sort of an arch.

Nor was the prospect of a new bridge popular with everyone in North Wales. 'A strong opposition to the erection of a bridge arose,' Provis recalled of Telford's first plan of 1810, 'and was obstinately maintained, by some of the commercial and trading interests of Carnarvon and its neighbourhood, who contended that the bridge would cause additional eddies, wind and water and thereby increase the difficulty and danger of passing the Swellies.'[7] Although Provis insisted that 'this design for a Bridge at Ynys-y-Moch [the 'Island of Pigs', a rock on which part of the bridge was to rest] appeared to have met every reasonable objection that could be urged by the most jealous guardians of the Strait', local sailors swore that they would never pass under it. Local shop-keepers, he suspected, enjoyed profiting, too, from the delays the ferry caused travellers. Telford was used to such troubles. 'I am in truth a slave to my business and of the Public,' he told William Little, 'for which I like other folks get finely abused.'[8]

As ever, he had plenty of work elsewhere to keep him busy, writing in early 1817 that he was 'hurried beyond measure'. Year in, year out, he swept through the country on a pattern as regular as a radar beam: in the winter of 1816–17 visiting first his Carlisle to Glasgow turnpike, then his Highland roads and bridges, then the Caledonian Canal and the smaller Crinan Canal (which had been designed by Rennie but needed extensive rebuilding, overseen by Telford and the Caledonian commis-sioners), then his North Wales roads. All this did not, however, stop him plugging away at finding a solution for the Menai.

The answer, it turned out, lay elsewhere. In 1814 he had become involved in a plan to build a bridge across the muddy River Mersey at Runcorn, a spot so broad that it demanded a central span of 1,000 feet and two side spans of 500 feet each. In March he travelled to survey the site, 'fully aware of the magni-tude of the object'.[9] A conventional structure with piers would obstruct navigation, he concluded, suggesting that a bridge held up by iron rods might work in its place. He carried out a series of experiments to test their strength. In his handwritten report, put together that summer, Telford admitted that there was no precedent for one so wide but he pointed to reports he had read of simpler chain bridges in China and India. That was a sign of the depth of his learning and curiosity. Such bridges had existed since at least the fifteenth century in China and in 1741 a modest chain footbridge had been built in England across the River Tees. Several, too, had been built in North America.

But these were simple, dangling pathways. Runcorn – and later Menai – demanded something much more spectacular. It would be, he suggested, a 'magnificent project' that would be practical to build but quite possibly not economical to fund. Telford drew on (but did not seem to acknowledge) the work of Sarah Guppy, a polymath female inventor who patented a design for a bridge 'supported by strong metal chains' in 1811. Telford sought her permission to use the design – for which she did not charge. Later she helped Isambard Kingdom Brunel with his crossing at Clifton in Bristol.

In 1817 he travelled to Liverpool to discuss the project once more and adjusted his plans. But he also admitted that, 'I have no great expectation that money will be found for it, the expense will, I fear, be too great.'[10] He was right. Nothing followed. Whether his design would have worked, had finances allowed it to be attempted, is questionable: some have suggested that its fragile chains would have snapped in strong winds – a failure which would have brought down Telford's reputation, too.

Others might have given up the idea at this point. Not Telford. In 1818, guided by what he had learnt from Runcorn, he took the most radical decision of his life. He would build a suspension bridge across the Menai Strait.

'The principle of its construction is as old as the spider's web,' his collaborator Provis wrote when the crossing was eventually built, but 'the application on a scale of such magnitude, the durability of the materials of which the Bridge is composed, and the scientific combination of its various parts, render it one of the noblest examples of British skill.'[11] This one project, this one episode, the building of this one bridge among the thousand and more that he constructed, encapsulated his character and everything he did. Nothing so tall, so long, so dependent on the unknown durability of forged iron links or the theory that a bridge of this scale could be suspended and move about with the wind and still be safe had been tried on this scale before. It is why the Menai Bridge came to define him just as a few years later the Great Western Railway came to define Brunel. In both men, the vision and the willingness to push the limits stands out. In both, there was the real risk of failure.

'I keep talking of avoiding employment,' Telford wrote to von Platen in 1818, 'but when undertakings of magnitude and novelty are forced upon me the temptation is too great to resist.'[12] He was intensely busy: 'my day has been marvellously protracted', he told Robert Peel in a letter that year, explaining that he was about to dash from Edinburgh to his work on the Glasgow to Carlisle road, before sailing on to Dublin.[13] 'Your business has not been forgot,' he reassured him: and once in Ireland he wrote again, offering diplomatic advice to the rising Tory politician, whose opposition to Catholic emancipation had antagonised many locals, suggesting that 'there should be one Irishman' on a new board of engineers he intended to establish. But for all these distractions he pushed on fast with the Menai Bridge.

Parnell was persuaded to back the radical design and the parliamentary commissioners soon voted to pay £20,000 to get it under way. In July 1818, Provis arrived in Bangor to oversee work. By the end of the month Ynys-y-Moch rock – the site of the first of two tall piers which were to hold the chains aloft – had already been levelled. All this caused a local sensation and among the crowds who came to watch was Joseph Pring, a local doctor. He was an unabashed Telford fan, later telling the story of the 'growing "form and feature"' of the bridge in a book. It 'has left an indelible reminiscence on the "mind's eye",' he wrote – adding that he hoped his book would sink 'into the shade of oblivion' if 'the erudite pen and luminous mind of Mr Telford' came forward with his own account.[14]

Not everyone nearby was so pleased. There was still opposition to building any sort of bridge and on 14 August 1818, even as Telford's men were blasting the foundations for the crossing, 'a crowd of pilots, masters of vessels, and Carnarvon tradesmen' met to complain that although the 1815 Act authorised the road it did not grant powers to build a bridge.[15] They were right. It took a second Act, passed in July 1819, to secure permission and even this bill faced objections from petitioners in Parliament. By now, however, momentum was with Telford. The local potentate, the Marquess of Anglesey, held firm in favour and the last objections were overcome when the commissioners agreed to buy out the ancient concession of the ferry company for £26,394: the huge sum a sign of the profitable monopoly which the bridge would replace.

With powers secured, Telford could press on. The first task was to design and build the masonry supports from which the chains would later be hung. He planned two tall pillars – which he called 'pyramids' on account of their shape – to support the chains. The design of these, like much about the bridge, evolved as they were being built. Early on, he had considered making them in iron. By 1818 he had settled on sandstone, ordering them to be built around a hollow core, buttressed on the inside, each block pinned together. He later wrote that he had insisted on this instead of the cruder and cheaper method of filling the void with rubble after being taught a lesson in Shrewsbury by the collapse of St Chad's church.

On 10 August 1819 the foundation stone of the first of the pillars, on Ynys-y-Moch, was 'lowered to its place and set, by myself and the masonry contractors', as Provis wrote later, 'whilst three cheers from the surrounding workmen closed the ceremonial'. By late 1819, 200 men were busy at the site and a temporary railway was carrying stone, quarried on Anglesey, out to Ynys-y-Moch. The pillar on the mainland side proved more complex: it rose out of the muddy, tidal shore and a coffer dam had to be sunk below the water level in order to dig down to solid rock. Workers immersed in the water were given ale to encourage them – and spirits in bad weather. In December 1819 the masons alone drank 52 gallons of ale.

This ought to have been the easy bit. It did not demand ingenuity with iron chains. But the challenge of building the bridge was already such that after nine months the contractors, who had bid for the job for a fixed price, gave up. It was a crisis to which Telford responded in the only way he knew. He turned to his loyal intimates, men with whom he had worked for decades. John Wilson, who helped build Pontcysyllte Aqueduct, was recalled from the Göta and Caledonian projects to oversee the stonework. In 1822, he was joined by Thomas Rhodes, who had overseen much of the ironwork on the Caledonian Canal, and at Menai he was put in charge of erecting the supporting chains. William Hazledine made the ironwork and Provis took overall charge.

Through these men, Telford put his stamp on the bridge, sending instructions by post, to which Provis replied dutifully, almost always the same day. Though by now growing old and with an increasingly settled life in London, Telford's efforts did not slacken. The endurance and authority which had marked his work in the Highlands was sustained in Wales, as John Rennie, the son of Telford's engineering contemporary, recalled. He served as a young assistant on the Holyhead Road and remembered an arduous day spent with the great man in North Wales.

The rain now came down heavier than ever, so that we had no alternative but to retrace our steps back to the dirty old 'public' at the Ferry, called Cross Keys, about 3½ miles distant. We got back, thoroughly soaked, about three in the afternoon. I immediately stripped and went to bed. Old Telford, being a strong hearty man of about 70, instead of following my example, ordered a large fire to be made in the only sitting room there was, called for a newspaper and sat himself down to dry. After two hours nap I was thoroughly refreshed and went down to the sitting room. When I entered there was such a steam that I could hardly see anything, but approaching the fire, found Telford had nearly dried himself, and he abused me thoroughly for being so effeminate as to go to bed.[16]

That was not the behaviour of a man who left road and bridge building to others. But neither is it fair to claim all the glory for Telford. He was a man, the Menai's historians admit, with 'a consummate ability to take and improve new technologies leaving the impression that he was a pioneer and sole innovator of the techniques used in his great engineering achievements'.[17] He depended on the enterprise of others, among them Samuel Brown, a retired naval officer who had developed an interest in forging metal anchor chains after the loss of Royal Navy ships whose hemp ropes had broken – his firm later made the vast, fat chains which give so much character to the background of the famous photograph of Brunel, with hat and cigar. Brown built a trial chain bridge at his works in London, which Telford spent many hours considering. In 1820 Brown, together with the elder John Rennie, designed and built the Union Bridge at Berwick-upon-Tweed. Although begun after the Menai, it was completed first and still stands as the oldest of its kind, the first suspension bridge ever to carry vehicles. It is not as high or as wide as Telford's but its success must have encouraged him. It is equal to the Menai crossing as a landmark in engineering history and its lessons were useful. Brown, for instance, persuaded Telford to use eye-bar chain links to support the deck of his bridge at Menai, in place of the bundles of square wrought-iron rods which he had first considered, a decision crucial to its success.

Others contributed to the design, too. In 1821, with the stone pillars and arches at each side of the bridge rising up fast, Telford's attention was drawn to an academic paper published by Davies Gilbert, a Cornish MP, mathematician and a member of the Holyhead Road Commission. Gilbert, a parliamentary champion of innovation, was a fascinating man, close to everyone – from the Cornish steam engineer Richard Trevithick to the naturalist Sir Joseph Banks.[18] His invaluable gift to the scheme was to persuade Telford greatly to increase the planned height of his pillars, so that each reached 153 feet above the high-water mark and 50 feet above the roadway. This, Gilbert convinced Telford, would steepen the angle of the chains, thereby decreasing their maximum tension and strengthening the bridge. Smaller towers would have been easier and cheaper to build but the chains might have snapped. Telford was bad at giving credit for such contributions but he was, at least, willing to take good advice when it came his way.

With the design coming together work could begin on the next stage of the bridge. The first task was to dig out the anchor points for the heavy chains. Though hidden, these were massive and difficult pieces of construction of a kind never tried before. On each side tunnels were dug down, cut at a gradient so that the chains would run in a straight line from the tops of the towers, meeting in chambers deep underground. Here they were secured to iron frames. The cavern on the Anglesey side, wrote Telford, was 'productive of feelings of superhuman agency'. On the mainland side the job was made harder still by a layer of soil which had to be supported by masonry before solid rock was found – which is why the tunnel on this side is longer, and chains from the bridge are not quite symmetrical.

All this preparation, however, meant nothing without the successful creation of the chains themselves and for this, in 1821, Telford turned to Hazledine, who produced them in his forge at Upton, in Shropshire. He could not have picked better.

The iron master, as much as Telford, was the author of the bridge's success. The design called for sixteen chains, in four sets of four. Each of the wrought-iron links was just over nine feet long. Since the failure of any link in the chains could bring about the fall of the whole structure, and no one had experience of using such things in practice, they were made stronger than theory said was necessary and tested relentlessly.

The first sections of chain to reach Menai were hung across a gully on the Anglesey side, to try their strength. Telford also had a model made, from which measurements for the supporting rods holding up the roadway were drawn. At Hazledine's workshop in Shrewsbury, John Provis – William's younger brother – oversaw the operation of what was called a Proving Machine (surely a pun on the family name) which subjected each link to twice the strain it was anticipated ever to have to carry, 11 tons per square inch against a working load of 5½. Once made, each link was hit with a hammer, to check it did not shatter, and if it passed, was then heated, coated in varnish and linseed oil and dried in an oven to secure it against rust caused by the salty sea air.

By June 1824 the high towers on each shore had been finished. 'We understand that the stupendous chain bridge across the Menai is in a state of great forwardness,' *The Times* told its readers.[19] But those involved knew that the most difficult part was still to come. The towers were soon fitted with cast-iron saddles, to allow the chains to move over their tops as the bridge flexed with the weight of traffic or the wind. Now arose what Provis called 'the grand question of how are the main chains to be put up' – a source of 'much speculation and doubt'.

This was more than chatter. Even Telford was daunted. A man who had taken risks before, with no sign of worry, hoisting up the ironwork for Pontcysyllte, seems to have been close to a breakdown. Perhaps he knew that the bridge was to be the defining creation of his life and that if it failed, everything that had built his reputation before would be forgotten amid the shame. Normally he slept well at night, but as the pressure grew he found himself, for the first time in his life, lying awake worrying. 'A continuance of that condition must very soon have completely undermined his health,' Smiles records him admitting to a friend.[20]

And he had much to fear. The challenge was to haul the main chains aloft and join them to the shorter sections already fixed in the anchor tunnels on either shore. It was all or nothing. If the rope pulling them up broke, or if the chains snapped and splashed deep into the Menai, or (worse still) if the whole design for raising them proved impractical, or the chains too weak, Telford would have been left with an expensive skeleton of an incomplete and unbuildable bridge.

By the middle of April 1825 everything was ready for the lifting of the first chain. Telford hurried to North Wales from London. The day of the attempt, 26 April, Pring recalled, saw 'a scene our Ancestors never contemplated'. The weather was fine and the pressure was on to move fast since special powers had been gained to close the Menai to navigation for the day. At 2.30 p.m. a raft carrying the central section of chain was floated into the central part of the current. This chain was joined to the section slung down from the tower on the mainland side. Next 150 workmen, straining at capstans, began hauling the heavy metalwork into the air. 'Two fifers played several enlivening tunes,' recorded Pring, 'the chain rose majestically and the gratifying sight was enthusiastically enjoyed by all present in breathless silence.'

After a sweaty struggle the chain dangled near the top of the tower on the Anglesey side. Now all that had to be done was to join this section to the chain already in place leading from the top of the tower to the anchor point in the tunnels far below. It was a perilous task, linking two heavy pieces of iron under great strain, at the top of a thin stone tower. Other senior engineers might have watched from below. But Telford (by now almost sixty-eight years old) scrambled up the pillar to help, along with Provis, Wilson and Rhodes. All four were visible to the crowds below as small specks.

At last they waved their hats: the chain had been joined and 'general shouts of exaltation arose from the workmen and the numerous spectators who had assembled to witness this novel operation'.[21] Smiles claims that when his friends rushed to congratulate Telford they found this most material of men on his knees in thankful prayer.

There were more prosaic celebrations, too. The workmen were issued with a quart of 'Cwrw da' Welsh ale, paid for by the parliamentary commissioners. Emboldened by that, three men, a stonemason, a carpenter and a labourer, broke free from the ranks and walked across the strait on the steeply graded chain, only nine inches wide. A few weeks later one of them did it again. William Williams 'sat himself down quietly on the centre of the curved part of the upper Suspension Chain, with his feet resting on the one below it, and, in that position, actually went through the regular operation of making a pair of small Shoes', Provis remembered. 'He was led to suppose that the Shoes were purchased for Public Exhibition at the British Museum!'

Soon, such exploits became common. This was only the first of four sets. The second presented problems, as the raft drifted out of line in the strong current, damaging the links, but it was soon repaired and the remainder were raised so smoothly that by 19 July 1825, with the water full of pleasure boats, the last chain could be hoisted up. The workmen marched in single file on a platform resting on the lowest chains from Anglesey to Caernarfonshire and back, 'the altitude. . . giving them the imaginary appearance of "aerial beings" '.

By now Telford was a regular customer of the Holyhead Road as well as its engineer. He travelled it on journeys to Ireland, building business there. 'Last season I went to the land of sham-rocks and spent six weeks in the Metropolis of Pats Country,' he wrote to William Little in 1817; 'I confess it enables me to set a higher value upon our own.'[22] A return visit in 1826, to work on problematic plans for an Ulster Canal, was no happier. 'I made an extensive and rapid excursion in Ireland amidst extreme heat in every sense of the word,' he reported.[23]

He continued to work on canals, but road improvement was by now taking up more and more of his time. Even before the Holyhead scheme had got under way he had surveyed the Carlisle to Glasgow road, with Provis as his assistant, a task that included the building of four large stone bridges. As with the road through Wales, this route tied the constituent parts of the United Kingdom together.

His appointment in North Wales had also required him to improve the coastal route running from Bangor, near Menai, to Chester. Today this (rebuilt as the A55 expressway) has replaced Telford's inland road through Snowdonia as the main access to Holyhead. At the time, it was a secondary route intended to serve the growing industrial centres of Liverpool and Manchester but it threw up a series of engineering challenges nonetheless – most of all the need to cross the River Conwy at the edge of the walled medieval town of the same name.

Here, hard by Conwy Castle, Telford planned a second, shorter and lower but still dramatic suspension bridge, learn-ing lessons from Menai. Work on this was finished in 1826, not long after its predecessor. Today, squeezed between later road and rail bridges, its impact has been diminished. At the time it was spectacular. It stands as the only one of Telford's suspension bridges to retain its original chains, forged, of course, by Hazledine and lifted in place by ropes and tackle brought over from Menai once the work there had been done. On the day it opened, Provis reported to Telford 'the Chester Mail [crossed] with as many passengers as could possibly find a place about it that they could hold by'.[24] It was a characteristic event, in that the principal author of the structure was not there to witness it. As at the later opening of the Menai, Provis had to write to Telford, who was in Dublin, on his latest Irish expedition. It was characteristic, too, in that it was a success. 'The Horses went steadily over which was more than I expected they would, as the people were shouting and running by the side of them from the embankment to the pier, the passengers at the same time singing God Save the King as loud as they could.' The only damage in the excitement, he added, was 'a few broken windows of the public houses'.

Back at Menai there were growing fears of damage of a much more serious kind. Work on the final stages of the bridge had seemed to go well. The bars from which the roadway was to hang were in place by the end of September 1825 and soon workmen were hammering the deck in place, layers of wooden planking, coated in tar. 'We completed the third or top tier of plank on one roadway yesterday and commenced the other one,' Provis told Telford on 30 December 1825.

It should have been a great moment. But something was worrying the engineers that Christmas. 'It is much better that the time can be prolonged for opening the bridge and the longer this can be done the better,' Rhodes advised Telford, in tidy copperplate writing on thick paper that still bears the stamp of the mail coach which carried it urgently to London.[25] 'We had a severe gale of wind on the 19th from the SE and SW which came about 12 o'clock and lasted several hours. It was very sudden and unexpected and caused the bridge to undulate more than I have seen and at times [it] would be impassable.' A few days later, he added: 'When the bridge is undulating it generally has a cross twist.'[26]

Nor were the engineers the only ones to worry. 'Strange rumours were sent abroad, and strange fears were conjured up by persons whose *span* of mind was not sufficient to embrace and comprehend a work so novel and so mighty,' it was reported.[27] 'The presiding genius of Telford and the events of the week proved how utterly baseless were all these tales and terrors.' Yet that scorn for the sceptics came afterwards. As the opening of the bridge approached it seemed to be at the mercy of strong winds.

More than a century after the Menai opened another newly built suspension bridge – at the Tacoma Narrows on the Pacific coast of the United States – was caught on early colour film twisting and bowing with the wind in a similar way, although even more violently so that the deck finally gave way in a famous shower of concrete and spray. That was in 1940, when civil engineers had huge experience of suspension bridge design, access to advanced mathematics, wind tunnels and reinforced concrete. Telford was attempting to build a similar structure in 1826 without any of these things.

All that protected the Menai from a similar disaster was the strength and quantity of Hazledine's great chains. And whatever their fears, the bridge's builders had no choice but to press on. 'I expect the guides on one carriageway will be completed tomorrow,' Rhodes told Telford in mid-January.[28] The bridge, he said, would 'be passable on the 25th'.

The final preparations were made. Eleven powerful copper and brass lamps (and eight gallons of sperm whale oil to fuel them) were put in place to light up the first coach to cross.

Parnell wrote, asking to join Telford in Wales 'not only to see the bridge opened but to examine the whole of the road to London'.[29] On 29 January Telford inspected his bridge one final time. It was, ready, he decided, to open – as it did, in a hurry that wet January night, with none of the formal celebrations one might have expected for such a great and costly project. There were no special invitations. No fireworks. Never talk of a royal visit. Just the winter winds roaring through the heavy chains and a narrow, slippery roadway heading out into the blackness.

It was an immediate success – 'the most stupendous monument of human art', exclaimed the Sunday Times.[30] The hated ferry across the Menai could be avoided, 'an event that deserves to be recorded in letters of gold', Pring announced happily. Congratulations poured in. 'I feel a satisfaction in having had a connection with even a paving stone that was to belong to a work so stupendous and useful as that of the Menai Bridge,' wrote one workman from Ireland.[31] Yet even now not everything was right. On the day it opened 'the wind caused a trifling, though scarcely perceptible, undulating motion about the centre of the bridge' and Telford stopped the crowds waiting to cross.[32] After that, the bridge continued to sway alarmingly in the winds and did so for years. It needed repeated and heavy repair. Built today, it would almost certainly be condemned as unfit for use.

'Since you left here we have had a succession of gales of wind,' Provis told Telford on 8 February, 'no material damage was done, only part of the side railing which was only temporary fastened and a few of the suspending rods on Anglesey side shook loose'.[33] But this was worrying in itself. 'I was in attendance when the coaches passed . . . the Head of the Oxonian took the horses out and refused going over, but after prevailing with them a little they started and went over safely,' he wrote, 'the

Bridge certainly laboured very hard and the night being dark and the wind whistling through the railing and chains made it appear rather terrific.'

Terror was not what Telford had hoped for: in 1819 he had reported to the commission that the bridge would be designed to limit 'undulation and vibration'.[34] But soon after this storm Provis and Rhodes found that twenty-four of the iron bars holding up the roadway had broken. Telford sketched out and sent a new way of strengthening them which, as Rhodes pointed out, was 'rather impractical on account of getting out the pin'.[35] He devised his own solution. But even when the repairs were done, bad weather still caused problems. 'All that has happened,' wrote Rhodes, might naturally be expected in a work of this magnitude and novelty.'[36] That was true, but it was not reassuring. 'I think it would be advisable to have another set of rods next to the roadway,' Rhodes added.

Soon after, the first accident occurred. A coachman reported that 'the Bridge is in great motion which caused the Horses to fall all down together, and so tangled in each other's harnesses had to set them at liberty one at a time by cutting the harnesses'.[37]

'I should have advised you of this before,' Rhodes told Telford, as news of the incident spread, 'but this being such a trifling occurrence that you might think me too tedious and meddling with a thing that did not directly concern me.[38] There is a sense in his correspondence of wanting to spare Telford the worry, and in this case he was right. The coachman had lied. A few days later, after investigating, he wrote again to Telford to explain the real cause. 'I have learnt from good authority that the coachman driving the Chester Mail on the morning of the 1st was intoxicated and that he drove against a wall,' he told Telford. 'This probably was the principal cause of the Horses stumbling on the bridge and not the effect of the "great motion" as it was termed.'[39]

Confidence in the bridge returned. Parnell took the *Prince of Wales* coach over it in late March 'with a full load of passengers and luggage and nothing could have been more satisfactory to the party than the pace and regularity with which we trotted over'.[40] He was reassured that 'we could not discern the slightest motion.' In good weather that remained the case, but severe storms continued to damage the bridge. In January 1836 the bridge keeper reported a sway of sixteen feet in a storm and in 1839 hurricane-strength winds wrecked the central section entirely.

By this time Provis and Rhodes were looking to the future. The bridge had been built, opened and stood up to use. That was what counted. In the summer of 1826, Parnell, on behalf of the commissioners, agreed to raise Provis's pay by £50 a year if he remained at Bangor a little longer, which he did, writing his book on its construction before moving on to work with Telford on his final canal schemes in the Midlands. Rhodes, meanwhile, wrote to Telford in May pointing out that as the Menai Bridge was finished and the one at Conwy about to open soon after: 'I trust you will not think it premature of me to ask if you have anything further for me to do.'[41] There was a sense of an ending for all of them.

Telford's glory was complete: he was, proclaimed Southey, the 'Pontifex Maximus' ('the Greatest Bridgemaker'). 'Crowds of people are daily coming to see the bridge, the steam packets... are now regularly crossing from Dublin. The pavement is excellent,' Parnell wrote.[42] Even its critics could not find fault, he added.

In the first two months, the tollkeepers at Menai collected £66 17s. 9½d. This was a start, at least, in recovering the vast cost of the Holyhead scheme. Telford's original estimate for building the bridge was around £100,000. In the end it cost £211,791. The bill for the whole route, including the side road from Bangor to Chester, was even greater: £200,000 was granted by Parliament and the rest was held as a loan, to be repaid from income, including twopence charged on all letters to Ireland. Perhaps the road's champions should have worried at the success of a new steam packet service, which began running from Liverpool to Dublin in 1824, cutting into their traffic. But there was, everyone believed, all the time in the world for tolls to repay the debt.

Menai had been an experiment and in his heart Telford always knew it. It was an experiment conducted not at leisure in laboratories but in the real world, in real time and real weather, risking real money – and real human lives too. Not everything had worked quite as envisaged. In the decades – and centuries – that followed, problems, some serious, have had to be rectified. But tested, the concept worked. The experiment had been a success; and the success was immense.

24 Abingdon Street

The last years of a life can be marked by a fading of the light; by retirement and isolation; a pretence at satisfaction even as irritation at the arrival of a new order grows. So it was with Telford: but only late on and not completely. The tremendous motor which powered him did not lose all its kick. He had companions. He had money. He had a reputation. He worked to the end. He was not notably unhappy. He still, even towards his death, built new things and new institutions. And one of the things he created at the late age of sixty-four was a home.

From early boyhood he had been famously rootless. As a child he was away working on farms or as an apprentice; as a young man he had temporary rooms in London, Portsmouth and Shrewsbury; and for the greater part of his working life he was on the move. At the point when most people settle he did not.

There had always been places he knew well: where he must have been able to leave possessions and clothes, kick off his boots and enjoy a feeling of pause, at least, in his perpetual journeying. In London he had rooms at the Salopian Coffee House in Charing Cross and in Shropshire he worked from the offices of the Ellesmere Canal Company, which he continued to serve as its engineer. Letters sent to him at these addresses would eventually catch up with him. But neither London nor Shropshire offered a front door with a lock and key of his own. It was not until 1821 that he bought a house: news which (it was said) shocked a newly arrived landlord of the Salopian who had expected that his income would be boosted by his tenant's celebrity. 'What! leave the house!' he exclaimed, 'Why, sir, I have just paid £750 for you.'[1] There is no proof of the story and records suggest that the business did not change hands at this date: but it is the sort of tale Telford enjoyed telling over dinner.

He was still writing letters addressed from the Salopian in April 1821 but by the end of the year he was paying rates on his new home, his name added late to the parish records in pencil. He did not move far. The house was no rural retirement villa. Opposite the old Palace of Westminster and just in front of Westminster Abbey, 24 Abingdon Street was a step away from John Rickman and the parliamentary committees before which Telford was a frequent witness. It was part of a four-storey terrace with a front door onto the pavement not so dissimilar to the one not far away at 10 Downing Street, though Telford's house was smaller and newer than that.

Today it does not exist. Abingdon Street was bombed in the Second World War and partly demolished before the conflict was over; the soot-stained, run-down remains were cleared away in the early 1960s as a precursor to a planned office block to house MPs. That horror was not built: instead the site gave way to an underground car park and a threadbare lawn, Abingdon (or 'College') Green, much used by television crews to interview MPs against a Palace of Westminster backdrop at moments of political drama and which during elections becomes a temporary media encampment. There is nowhere now to put the blue ceramic plaque[*] that once recorded the building's illustrious occupant.

Or perhaps occupants. He always claimed to be proud of the fact that the house had once been used by Charles Labelye,

---

[*] It can instead be seen just inside the entrance to the Institution of Civil Engineers.

the Swiss-born builder of Westminster Bridge, and later by Sir William Chambers, the architect for whom Telford had worked as a mason when he first arrived in London. Perhaps by buying it he was also buying confirmation of his status as their equal. Chambers he had disliked. Labelye he respected, though there is no certainty that the latter ever lived at Number 24. Telford may have been right, even so: he knew Thomas Gayfere who had worked on Labelye's bridge, who lived in the area until his death, aged ninety, in 1812 and who could have passed on the story.

In his stylish house he was able to indulge in something close to luxury for the first time. It was furnished (by the time its contents were auctioned after his death) with such comforts as a four-poster bed with a prime goose-feather mattress, mahogany bookcases and 'a remarkably strong-built travelling chariot'.[2] A few years after Telford moved in, James Little (a nephew of his old friend Andrew), living there as a pupil, wrote to tell his mother what life was like:

> We breakfast at 8 and dine at ½ past 5. . . now that I am accustomed to it I like it very well. Mr. T is very fond of the mutton ham which is only presented on great occasions, Mr Anderson [a fellow pupil] & I get a little now and then by way of a treat – we have large dinner parties sometimes, & a fine set out there is – but do not imagine that Mr. T. eats nothing but cake, & drinks nothing but water, if you do you are quite mistaken, we drink wine every day.[3]

In these congenial surroundings Telford settled. By the early 1820s, wrote Southey after Telford's death, 'he might be called a regular inhabitant of the Metropolis'. He also showed off. 'His house was handsome, quite suited to his requirements, and in accordance with his professional position and he was very proud of it,' remembered James Mitchell, who preceded Little as the first pupil to live there.[4] 'After having lived in lodgings and hotels all his life, he appeared gratified and delighted. . . when any friend called and complimented him on his new quarters he used to say "oh, you must see my house", taking them over the principal apartments, pointing out the solid mahogany doors and marble chimney-pieces.' The tour, Mitchell added, 'usually finished by pointing out that the house had been owned by the engineer of Westminster Bridge . . . "and there, in a panel above the chimney-piece in the dining room, is a picture of the bridge by Canaletto" '.

'Below,' he used to add, 'I have a perfect village' – by which he meant the pupils and assistants with whom he shared the build-ing. 'His business had so much increased that he had resolved to take up a fixed residence in London,' Telford had explained to Mitchell's father when he offered him a place, 'and as he did not like to live alone, he had decided on having two young lads to live with him, to act as his clerks and to study engineering.' The second place went to the son of John Gibb, his superintendent of harbour works.

These young residents were not the first engineers to gain experience on Telford's projects. He had long established the practice of picking out promising individuals and bringing them on. Thomas Brassey, for instance, began work as a teen-age surveyor on the Holyhead Road the decade before and went on to become one of the greatest self-made Victorians, build-ing a substantial proportion of the world's railways (employing 100,000 men at his peak). Brassey was famous for emulating Telford's peripatetic energy, scrawling dozens of letters each night and marching huge distances by day. William Provis, too, the engineer of the Menai Bridge, was an early product of the Telford school and so was Alexander Nimmo, who began as a surveyor on the Highland roads and became a celebrated engi-neer in Ireland. But only once he was in Abingdon Street could the process of training be formalised. The building became as much a college and a business headquarters as a home.

In Telford's view the way to learn was the way he had done it: from the bottom up. 'It is absolutely necessary to give early habits to young persons, in practical operations. . . it smooths their progress thro' Life – give them the means of procuring, by Industry, an honest livelihood then if they have talents, let them work up in the ranks of Society,' he urged, about the sons of a wayward Eskdale cousin.[5] 'If civil Engineering is, after all, still preferred,' he cautioned another Eskdale family that had asked for help, 'the way in which both the late Mr Rennie and myself proceeded was to serve a regular apprenticeship to some practical employment. In this way we secured the means, by hard labour, of earning a subsistence and in time by degrees, acquiring the confidence of our employers and the public, rose into what is denominated Civil Engineering.'[6]

'This,' he added, 'altho' forbidding to a young person who believes it possible to find a short and hasty path to distinction, has been shown to be otherwise, by the two examples which have been the most successful – one of whom lives on – Evidence "That steep is the Ascent and slippery is the way".'

He enjoyed such bluster about the merits of manual labour (he was said to have kept his old mason's tools and leather apron safe, in case his luck failed and he had to return to the way he had once made his living) but bluster it was. The truth was that the Telford of the 1820s was a sophisticated, literate and scientifically aware man who had reached his position by paths which were hardly open to a younger generation. The days when a mason with a keen eye could make his way to the top had gone and he must have known it. James Mitchell, who arrived in 1821, won his place after spending two months drawing a panorama of the Caledonian Canal to prove his skill as a draughtsman, rather than for any manual skills. Telford's sermons on the stony path to civil engineering greatness were more self-referential than a useful picture of changing times.

He could not – or chose not to – see that Britain was on the brink of a boom in civil construction, to be led by younger engineers and not him. 'The profession has more candidates than can be employed,' he warned, '<u>the Lottery is great</u> the <u>valuable Tickets few.</u>'[7] That was true, perhaps, after the financial crash in the mid-1820s and he must have found it hard to be a magnet for the appeals of out-of-work surveyors, such as George Peat, a Yorkshireman who wrote to Telford as late as September 1833 begging him to 'employ me, or suggest anything which you may suppose likely to serve me'. The poor man added that he had been out of work for eighteen months, had a family of five to feed and had been 'reduced to great distress'.[8] But by the 1840s, as railway-building spread, anyone with a half-plausible claim to engineering skill could make himself rich by laying out tracks. The valuable tickets turned out to be many and those he chose to join him in Abingdon Street were among the biggest winners.

They found it a cheerful home – Mitchell described 'a halo of happiness throughout our establishment' – and Telford could be as kindly personally as he was austere in his theoretical pontifications. He was generous to William Anderson, the son of one of Rennie's engineers, paying for him to travel to Italy when his health was bad – Anderson sending back long descriptions of his travels. When young Mitchell left Inverness for the first time and reached the capital by coaster, he sailed up the Thames past the remains of bodies hung in chains, Royal Navy sailors who had taken part in the infamous mutiny at the Nore anchorage in 1797. Abingdon Street, by contrast, was welcoming. 'I was assigned a room, and entered on my duties. Our working hours were from nine til seven; the evenings we could dispose of as we liked. The old gentleman treated us like sons. If he had friends, we dined in their company.'

'Telford did not go into what is called "Society", but he was always delighted to see his friends at home,' he continued, among them Parnell, Rickman, Southey and two of his fellow Eskdale boys who had done so well, Admiral Sir Pulteney Malcolm and General Charles Pasley. 'Telford was the soul of cheerfulness and used to keep his guests in a roar of laughter. . . he had a joke for every little circumstance and he was full of anecdote. His ordinary manner to a stranger was that of a happy, cheerful, clear-headed, upright man; but any attempt to impose on him, or any exhibition of meanness or unfair dealing, called forth expressions of stern indignation and severe invective.'

Mitchell's aside about Telford not going into 'what is called Society' is interesting. Invitations declined tend not to be recorded but we must assume Telford will have had no shortage of invitations of a hospitable kind: in 1828, for instance, he was asked (and refused) to become a steward of the Literary Fund, which held an elaborate anniversary dinner each year.[9] He had no wish to be the sort of diner-out that an eminent engineer in the late summer of his life could easily have become, nor grace society hostesses' tables as a trophy guest.

His attitude towards both 'Society' and celebrity seems oddly ambivalent. Status – even fame – of a professional kind was something he not only enjoyed but from the start had actively sought and sharpened his elbows to maintain. When strings had to be pulled or a forelock tugged to secure career advantage he seems to have shown neither shame nor hesitation – as he had so often boasted to Andrew Little. And he was no ascetic: he had always enjoyed company, conversation, food and drink. But now, as the opportunity to sail out into London society from a smart home of his own opened up to him, he showed little inclination to grasp it, preferring the company of men he could talk to about engineering, career-building and the lessons of life.

Perhaps he found society chitchat boring; or perhaps he was not as confident or relaxed in the salons of high society as might be supposed; or perhaps his youthful feelings of reproach towards an unequal and exploitative world had not been an entirely passing phase, but remained, buried somewhere, in a now comfortably established breast. Perhaps it was a combination of all of these. What is clear, though, is that in the most generous and least high-handed of ways, he enjoyed playing king of the kids.

'I experienced kind treatment,' Mitchell wrote, 'and enjoyed consideration as an intimate in the house of a man so distinguished' – although he remembered, too, Telford's fury when he went out one night with the key to the wine cellar in his pocket leaving his master without drink for the evening. James Little, too, looked back fondly at the two years he spent in Abingdon Street between 1825 and 1827, in which he served as much as a companion for the old man as a student. 'I was often alone with Mr. T. & we had many conversations about Eskdale, & his firm friends,' he recalled.[10] 'A day seldom passes without his telling stories of them,' he explained in a letter sent to his mother at the time, 'his principal heroes are Sandy Esplen, Watty Runshire, Bauldy Regan, James Veitch, Auld Dalbeth.'[11] Those are the evocative nicknames of men with whom he had roamed in his youth: probably all dead, even by the time their old friend sat by his marble fireplace in Westminster and talked about them with a smile of recollection.

Among the residents of 24 Abingdon Street, one in particular stood out: Henry Palmer, who had joined Telford's staff in 1818. He was an assistant rather than a pupil and resented for it. Mitchell regarded him as 'self-complacent and indolent. He imagined himself a genius but had no experience practically of engineering works.' That was unkind. Palmer was a creative man, spending his time sketching out an early design for a monorail (and later, before his early death, patenting corrugated iron). He was also an incorrigible self-promoter and it was this that led to the creation of the organisation which has done more than any other to keep the memory of Telford's name alive.

On 2 January 1818, in the Kendal Coffee House on Fleet Street, the twenty-three-year-old Palmer chaired the first meeting of the Institution of Civil Engineers: an organisation still thriving today. 'An engineer is a mediator between the Philosopher and the Working Mechanic,' the minutes record, implying a gathering of something grander than the six impecunious young men who turned up.[12] It was to be a fellowship for engineers aged under thirty-five and as such intended to differ from the Society of Civil Engineers (founded in 1771 by John Smeaton, one of the greatest civil engineers of his time and later called the Smeatonian Society in his honour). That was a dining club of important men, including Jessop, Rennie and Watt, but seemingly never Telford, meant more for conviviality than serious study.

At first, the group made little impression. The original members – many of them mechanical not civil engineers – were joined by only four more, among them William Provis. Then, at a meeting of the institution chaired by Palmer in January 1820, Provis moved a motion proposing 'to render its advantages more general both to the members and to the country at large'. The best way to do that, he said, was by 'the election of a President, whose extensive practice as a civil engineer has gained him the first-rate celebrity'.

A letter was sent to Telford on 3 February 1820. 'The cultivation of the profession of civil engineer, is a subject which does not appear to have met with that attention in this country which its importance deserves,' it read, which had 'induced a few young men of the profession to come forward'.[13] Then came an appeal – which has proved its worth over almost two centuries – asking him to accept 'the office of President, feeling conscious that your name would at once stamp their endeavours with that consequence which would ensure a great and lasting benefit to the profession and the Country at large'.

It is hard not to suppose that the ground for its reception had been prepared in advance by Palmer. By the end of the month Telford had attended his first meeting – the upper age limit for membership having been abandoned – and in March he replied formally. 'I am fully aware of the necessity for such an institution,' he wrote to the handful of members, 'a sense of duty and gratitude induce me to accept the office until a fitter person can be selected.'[14] He donated thirty-one books to begin its library and on 21 March made his first speech as the institution's president, a position he held until his death in 1834. 'The principles of the Institution rest more upon practical effort and unceasing perseverance, than upon any ill-judged attempts at eloquence,' he began, adding that he had 'no share, or even knowledge, of the original formation of the Institution', but hoped that it would attract 'ingenious men of every country'.[15] It did, from the start, with his Swedish friend von Platen joining in May (it is hard to imagine any society headed by Telford that the fretful count would not have wished to join).

Telford stressed that 'it is quite evident that civil engineering has increased to an extent and importance which urgently demand such a separate establishment' and praised the contrast between individual initiative, which had led to the British institution, and state organisation and control which was typical abroad. 'Talents and respectability are preferable to numbers,' he cautioned.

His energy and credibility lifted the body to a new level. Throughout the 1820s he was in the chair at most of its sessions. 'Every Tuesday, the meeting nights of the Institution of Engineers, he used to have two or three to dinner before going to the Institution,' recalled Mitchell, whose informal notes of the meetings led Telford to begin the practice of keeping minutes.

Then, as now, the institution was a practical one: in Telford's view it was not to become a club or membership a mark of status to be used for personal advantage. 'We have no wish for learned discussions,' he told von Platen. '<u>Facts</u> and <u>practical operations</u> are to compose our collections and we should leave <u>project</u> and <u>theory</u> to those who are disposed to create new systems.'[16] In March 1822 he told his Swedish friend that 'there are now above 50 members in England, Scotland, Ireland, Holland, France, Germany, Sweden and the East Indies. . . the rising generation will, for their instruction and benefit, be attracted to this establishment as a centre. We will admit only what has really been performed – no theoretic projects.' One can hear him hammering this point home; the voice of the mason with a chisel. He insisted every member read a paper once a year on a practical project (a rule that was never enforced, though poor von Platen wrote to apologise for breaking it).

There was more than a touch of a Telford fan club about the whole thing. Many of those who joined were his intimates and not all were strictly civil engineers. The wider profession was less enthusiastic. 'Very few rallied to the support of the first President,' recalled Charles Vignoles (who took charge of the institution in 1869, by which time it was flourishing, but who had first met Telford at a dinner in 1824 and well remembered its difficult early days). The minute book shows that the institution was soon attracting two dozen or more people to each session but a decade after its foundation it still only had 139 members and what Vignoles recalled as 'the very rudest accommodation in Buckingham Street, Adelphi'.

Nonetheless, it was taking root. In February 1823 it held a drunken anniversary dinner which, reported the *Morning Chronicle*, began with toasts to the royal family and the army and navy, and a speech from Telford in which he declared that 'all the merit he would claim was in the good will to his fellow engineers'.[17] He proposed a toast to Sir Henry Parnell who in turn gave a long speech on the virtues of his friend. Everyone drank to the memory of departed engineers, Smeaton, Milne, Bramah, Watt and Rennie: but it was the giant, present in the chair, who overshadowed everything.

In 1828 the institution gained a royal charter after Telford advanced £200 of the £300 that this cost. It was a sign that civil engineering was becoming a profession, not a trade; but it was a profession that was torn, increasingly, between the older generation of canal and road engineers and the rising interest in railways. In 1828 Telford proposed the railway engineer Robert Stephenson for membership but it was not until after Telford's death that the institution took an active role in the development of railways (and even then the divide remained, leading in part to the establishment of the separate Institution of Mechanical Engineers in the 1840s). 'The institution is flourishing beyond expectation, last meeting was attended by above 40 members,' Telford told Charles Pasley in 1830 – in a letter which also informed his friend that the title of 'Honorary Member, the highest the Institution of Civil Engineers can bestow, has been conferred on Lt Col. Pasley – may he live to enjoy it 1000 years'.[18]

That award was a small sign of the manner in which the institution became something like a family in Telford's old age. It was full of friends and his enemies stayed away. 'There was scarcely a Member of any rank belonging to the Institution who attended so regularly as did Mr Telford himself,' recalled George May.[19] His increasingly weak signature is a reliable feature of the minute book. He continued to attend meetings, chairing his last on 17 June 1834, not much more than three months before he died. But by the 1830s he was more beloved figurehead than force of progress; isolated by deafness and – though neither he nor his friends liked to say so – by the passing of the old way of doing things.

In these years, the last before the start of the Victorian age, Telford became established as the presiding authority over civil engineering projects of all sorts. Any attempt to describe it all risks turning biography into inventory. The range of his work was so immense as to be almost indescribable and the depth of his involvement in each project that he touched was varied and often unknowable. He went everywhere, knew everyone and tried everything. 'I am completely entangled,' he explained to the engineer of the Edinburgh waterworks company in September 1820; 'I am now working from 6 in the morning to 10 at night.'[20]

He maintained, as old age approached, the obsessive schedule he had begun in his youth, chasing around the country, sleeping in coaches, living in inns, sending letters to his staff and contractors inked out by candlelight. He was surely the most mobile individual in the country at the time; and quite possibly the most mobile in history before the coming of the railways. As he had written to William Little in the winter of 1816, approaching his seventh decade when others his age would have been settling at home by the fire:

I have been and shall still, for some time, be much hurried. After Parliament was prorogued, I went thro. North Wales where about 500 Men are employed, & from thence into and along all the Eastern side of Ireland from Waterford to Belfast & Donaghadee and cross by Portpatrick to Carlisle, from thence to Glasgow and back by Moffat to Edinburgh – From

thence St. Andrews & Dundee to Aberdeen then up to the
Western parts of that Country, then across it and Banffshire –
and to every Town on the Coast to Inverness – thence thro'
Ross & Sutherland & back to Inverness – from thence across
to Fort William on the West Sea and back to Inverness –
Then back to Fort William – From thence by Tyndrum &
Inverary down to the Crinan Canal in Cantyre, thence back
by Inverary – and Loch Lomond to Glasgow. – and again
still nearly the same route and back to Inverness – & west by
the Crinan Canal and Glasgow to perform over again before
I reach the Border.[21]

'This,' he concluded, 'will give you some notion of my restless life, and at this season a Post Chaise can scarcely render
it Comfortable – the Weather here more than half Rainy and
already much Snow on the Mountains.'

An expedition like this would be exhausting today in a car
let alone in Telford's cramped, two-wheeled private carriage,
exchanging horses at inns along the route. But it was this persistence which made him invaluable and all-present. He knew first
hand what contractors and investors were up to from Inverness
to Bristol and he could try techniques in one place, only to adapt
them for another. A man more focused on one task might have
been less flexible in his thinking. Note, though, the relish with
which he charted for his friend the range and pace of his travels.
Telford was proud of how hard he worked, how far he went, and
the privations he had to endure. Where others of his age might
have been boring their friends with accounts of their aches and
pains, Telford dazed them with odysseys of his endeavours.

It was this blend of engineering omnicompetence and cultivated connections at Westminster which led him to be appointed
engineer to the Exchequer Bill Loan Commission in 1817: a
boring name for an ambitious undertaking so downbeat in its title

that one almost suspects an attempt by its authors to camouflage their great project. This was a Georgian precursor to Roosevelt's New Deal, authorised by the Poor Employment Act of the same year to fund 'Public Works and Fisheries in the United Kingdom and Employment of the Poor in Great Britain'. The scheme was a brainchild of Nicholas Vansittart, a 'good man and real patriot' according to Telford, but by then a Chancellor struggling against the political and economic tide ('a man of unexceptional abilities', concludes the *Dictionary of National Biography* in an uncharacteristic misreading of the importance of politics and its consummate practitioners). He hoped that capital projects would lift the country out of the economic slump which followed peace with France.

The commission placed Telford in a powerful position. Soon there was barely a large civil engineering scheme in the country that did not involve him. On some he did no more than read the paperwork or send an assistant, but on most he at least examined the site and the proposal himself, such as for the grant of £250,000 that was made towards the completion of the Regent's Canal running around north London from Limehouse, or the £78,000 that went towards an iron bridge over the Thames at Southwark. On a few he became much more involved.

One was the Gloucester and Berkeley Canal, part of a long-running struggle to bypass a difficult section of the River Severn, which was given a grant of £160,000 by the commission. Telford found the business in a mess, all but evicting the previous engineer John Upton, who then wrote to him bitterly: 'You had condescended to meet me at the Bell. I was there to the minute – you was gone.'[22]

For poor Upton this scheme must have been everything. For Telford (who proposed relocating the westerly entrance to the sea and kept a regular eye on progress into the 1830s) it was just one of scores of projects competing for his time. Even Upton found himself redeemed and roped in to the frenzy, reporting on applications to the Exchequer Bill Loan Commission, among them the Thames and Medway Canal and both Folkestone and Shoreham harbours, before eventually being accused of fraud on a Midlands turnpike scheme and starting afresh as an engineer in the Crimea and a lieutenant-general in the Russian army.

That late career was not atypical: this was a time when British civil engineering and engineers were rising to be the best in the world. Telford, too, was consulted in Russia and his reach extended to the British Empire. In 1824, jointly with Alexander Nimmo, he endorsed plans for the Welland canal linking Lake Ontario with Lake Erie in Canada. He wrote to advise the Canadian businessman William Merritt that a proposed design for the route above the Niagara Falls was 'very favourably circumstanced' for 'a country so rapidly advancing'.[23]

Telford's opinion here was offered only as an inducement to investors but in India he played a practical role. In 1822 he was asked to judge between competing proposals for linking the islands of Bombay and Colaba. The shortest route crossed over insubstantial sands and some in the city feared it would not be strong enough. 'This means of constructing mounds is by no means only theoretical or experimental,' he reassured them, in a report preserved among the India Office papers.[24] The route could, he added helpfully, carry iron pipes to supply water.

Back in Britain, Vansittart had given Telford temporal power through the commission, but spiritual strength came from another of their joint schemes in the glens and islands of the far north of Scotland. In late 1822 Vansittart was eased out of the Treasury in return for a peerage and large pension. It gave him time to indulge his evangelical Christianity. The end of the Napoleonic Wars had been followed by a bullish bout of government-funded Protestant building across Britain and the General Assembly of the Church of Scotland was keen to make sure the Highlands (much of it traditionally Catholic or without church buildings at all) would not be left out. In 1819 Vansittart had brought a bill to Parliament to fund church building in the far north and though he was defeated on the first attempt he succeeded in getting an Act passed in 1823. He turned, of course, to Telford, untroubled by that latter's apparent lack of faith.

The Commission for Building Highland Roads and Bridges, still active although most of the new roads had been finished, was the obvious vehicle for getting Vansittart's churches built. It knew the region, knew the landowners, had superintendents in place and in Telford had an engineer who could make things happen. In Rickman, who retired as secretary to the commission in 1828 but kept up his work on the new churches after that, it also had a bureaucrat of the first order.

The routine had been set long before on the Highland roads. Landowners applied for grants and the commissioners picked winners. In 1825 they chose the first thirty of thirty-two eventual sites for churches (forty-one manses were also built, some near existing churches). Telford designed (or asked three of his surveyors to design on his behalf) a plan, elevation and estimate for a standardised church and manse that could be built for £1,500 'particularly calculated to resist the effects of a stormy climate'.[25] From these Telford selected and improved one, submitted by his superintendent William Thompson, who had worked on the Crinan Canal and become one of John Mitchell's successors on the Highland roads following the latter's death in 1824, for a single-storey church with a T-shaped floor plan, which could be adapted to include a gallery where numbers justified it. 'Gibb and I worked the whole of yesterday on the Church and Manse plans,' Telford told Rickman in April 1825.

The basic pattern, rolled out on each site, had six wide Tudor windows and a stubby belfry. It was tidy but dull. It 'has the feel of Thomas Telford about it, trying rather unsuccessfully to create a set style', a historian of the project, Allan Maclean, writes.[26] All eyes and ears were focused on the pulpit. In winter, without a heater, congregations froze, and in summer, without windows which could open, they sweltered. Economy required standardisation and in eschewing anything more surprising, striking or florid, Telford may have been mindful of his countrymen's sober taste. These churches were, as Rickman said, 'solid Buildings [with an] inoffensive exterior affording the greatest possible accommodation at the least possible expense'. 'The Commissioners,' he added, 'could not afford to spend their own time or public money trying to conform to the dictates of taste.'

They were built under Joseph Mitchell's reliable supervision for just £750 each (and an additional £720 for a manse, where those were provided). Most have survived and some are still in use; others (as on the gentle Hebridean island of Berneray) have been rescued as homes; and a few stand as ruins in the heather, the victims of population decline or the separation of the Free Church of Scotland in 1843, which left them without a congregation.[27] After Telford's death the churchyard at Croick, at the end of a track leading up Strathcarron from Bonar Bridge on the east coast, sheltered families driven from their homes. 'Glencalvie people was in the churchyard here May 24 1845,' graffiti cut into the windows reads, a bitter record of the clearance of a nearby glen of that name. 'Glencalvie People the wicked generation Glencalvie,' they wrote, miserably after their landlord and his agent had emptied the place of its people. The Scotland in which Telford worked was an unsettled place.

And while he was busy with the Loans Commission and the Menai Bridge and the Caledonian Canal and his home in London and the new institution and the churches and the Glasgow road and waterworks, Telford found yet more to do in the waterlands of East Anglia.

Agricultural improvers had been busy in the fens for over a century. Now, armed with steam pumps, they were raising their game. In 1818 they appointed John Rennie to the post of engineer to the Commissioners for Drainage, and Telford was persuaded to accept the equivalent post to the Commissioners for Navigation. The interests of the two bodies did not align and neither did the personalities of their engineers. 'Mr Rennie I am informed is on the worst terms with Mr Telford,' noted the Earl of Eglinton (an investor in canals) at the start of the century.[28]

What caused the falling out? Telford claimed to be mystified. 'Altho' I never had any connection with him in business or ever intentionally did anything to injure or interfere with him,' he had told Watt in 1805, 'I, in every quarter, hear of him treating my character with a degree of illiberality not very becoming. This is so marked a part of his conduct that I really believe it does him a serious injury and proves serviceable to me. As I am desirous of not suffering in your good opinion, I mention this with a view to counteracting any insinuations which may be advanced to my disadvantage.'[29]

There is something disingenuous in this attack disguised (and only half-disguised) as self-defence, designed to undermine a rival. Nonetheless the pair seemed to tolerate each other until Rennie's death in 1821, overseeing work on what became known as the Eau Brink Cut near King's Lynn. Telford is scrupulously polite about his opposite number in his memoirs and after his death joined forces in the Fens with his son, John (later Sir John) Rennie in what became more than a decade's slog to align the complexities of large-scale drainage with the parochialism of local inhabitants. Reconstructing the Nene outfall (a six-mile channel a little further north, which flushed water out to sea in the Wash), the pair faced the anger of the townspeople of Wisbech, who feared trade at their port would be damaged.

Telford proposed a solution which they rejected. They were, he wrote, 'totally incapable of judging what would have been manifestly beneficial to themselves' – the cry of frustrated infrastructure planners through the ages.[30]

Telford the dredger lacked the glamour of Telford the bridge builder, but he was proud of this solid and useful work, which continued in the North Level of the fens long into the 1830s. He gave more space in his memoirs to his efforts here than to many of his greater schemes: they were, perhaps, fresher in the mind of an ageing man. 'I cannot, I confess, without some complacency, reflect upon the success of my drainage operations,' he wrote. The tone suggests he knew that others might not be so fascinated.

It took Samuel Smiles's extravagant pen to bring colour to this effort. 'The barbarous race of Fen-men has disappeared before the skill of the engineer,' he rejoiced, 'the half-starved fowlers and fen-roamers have subsided into the ranks of steady industry. . . wide watery wastes, formerly abounding in fish are now covered in crops.' The extinction of natural habitats must have been stupendous. Did Telford notice or care? He never wrote of it.

Extinction was creeping up on him in other ways, too, though he did not sense it at the time. Following the success of the Holyhead Road, he found himself busy surveying other routes which he hoped might be built along similar lavish lines. Some came into being. After a frenzied six weeks of surveying he laid out two new roads across Lanarkshire, one running east to west and the other north to south. In 1822, Cartland Bridge was built to his design just outside Lanark, a road viaduct 129 feet high which looks like something out of the railway age. In Derbyshire he laid out the climb from Glossop over the Snake Pass to Sheffield. The turnpike opened in 1821 and is still one of the main routes across the south Pennines. Freed from tolls in

1870 it is renowned today as a place of grisly crashes and regular traffic reports of its closure by snow.

Elsewhere in England he advised the Exchequer Bill Loan Commissioners on the condition of Westminster Bridge (which Labelye had built the century before) and also built a series of fine bridges over the River Severn. Mythe Bridge, built in 1826 just north of Tewkesbury in Worcestershire, was a classic Telford and Hazledine collaboration, constructed of 'the best Shropshire iron' in place of an original design by others which had shorter arches and (the trustees feared) weak foundations. It has a fine, wide arch, drawing on what had by now become a standard Telford design for iron bridges. The first, at Bonar, in Scotland, was followed by seven more by 1829, of which six are still in some sort of use. Its masonry abutments are pierced to allow floodwater through. 'I reckon this the most handsomest bridge which has been built under my direction,' Telford declared.[31] A little to the north, near Worcester, he oversaw the similar Holt Fleet Bridge. Like Mythe, it is still there.

Near Gloucester, at the insistence of local magistrates who disliked iron, he built Over Bridge in golden stone. It is a work of striking purity, its chamfered arch in vertical masonry blending into horizontal stonework above. In designing it he drew on the work of the French engineer Jean-Rodolphe Perronet, who had designed a similar structure over the River Seine late the century before. That Telford knew of Perronet's work is a sign of his dedication to research; but without having seen its French predecessor at first hand he was unable to avoid a fault which occurred in both structures. When the wooden supports for the arch's construction were removed it dropped by ten inches. That was less than on Perronet's precursor, but as Telford reflected in his memoirs, 'I much regret it, as I never have had occasion to state anything of the sort in any of the numerous other bridges described in this volume'. Visiting a few years later, Marc Brunel noted the error (perhaps with private glee) in his diary.

Telford blamed himself for economising on the construction of the abutments. It is telling that he has only good things to say about the contractor John Cargill, who had by then worked with him for over two decades. His old crew still remained loyal, and he to them. But their era was ending.

More and more, Telford found that his schemes, once planned, were left undone. In January 1823, for instance, he could be found at the Beaufort Arms in Monmouth, surveying a new road through South Wales to the port at Milford Haven which in other circumstances might have developed into a southerly pair to the Holyhead Road. By 1826 he had supplied detailed plans, but nothing followed. The same was true of his survey of the route from London to Liverpool and his efforts to shorten the Great North Road from London to Edinburgh, on which he estimated a better line would save thirty miles.

There was nothing wrong with these plans. Not long before they would have been funded and built. But by now everyone wanted railways.

These Railroad Projectors

'We are a dull plodding people,' Telford wrote to the French statistician and engineer Charles Dupin in 1832, 'always busy, seldom producing anything new or striking, though not infrequently riding a Hobby to excess; the application of Steam Power, seems to be at present the chief pursuit of ingenious persons.'[1] Dupin knew his correspondent's work: he was one of many European visitors who had toured the Caledonian Canal, Pontcysyllte and the Holyhead Road, and his account, published in Paris in 1824, helped spread Telford's reputation on the continent – and particularly in France, where Richard Tholoniat, Professor of British Civilisation at the Université du Maine, describes the national view of Telford as 'ever-admiring'. 'Nothing deserving your acceptance has lately occurred in this Country,' Telford grumbled, but Dupin will have known that beneath his weary tone lay his usual curiosity about new technology. As an old friend of James Watt and someone who led the use of mechanical pumps in all sorts of civil engineering projects, he was no enemy of steam.

Nonetheless its power was blasting apart Telford's world. The old ways with horses which he had taken to its limit – the sweating, soft-feathered dray tugging a canal barge, or the racing black stallion at the head of a heavy, rattling coach – could not compete with the pure hot energy created by coal and water and running on rails. The laws of physics could not be defied. Before the development of rail, Telford's specialisms, roads and canals, had been slugging it out for competitive advantage, each with innate strengths and weaknesses: canals offering a smooth, cheap, slow ride for bulk and weight; roads a faster, rougher ride for passengers. Rail then overcame both. It gave an evenness of ride at which greater speeds became possible and could carry weight and volume as well as passengers, and it did not suffer the energy losses that encumbered fast water-borne transport.

What did he make of this challenge? He was too acute, too audacious, not to realise what it would mean for the things he had built, though he tried to resist, perfecting roads and canals and studying the science which he hoped would prove their success. It was a time in which he reached the peak of his reputation and importance even as the foundations were collapsing.

In 1821 the Stockton and Darlington railway was authorised by Parliament. That was not news. Many other railways had been built before. What was new came next: George Stephenson's appointment as its engineer and his decision to use locomotive engines in place of horses to pull the wagons. He was not the first to try out steam engines on iron tracks and neither was his railway in itself revolutionary. But he made the technology famous.

Even before the line opened on 27 September 1825 its reputation helped fuel a craze for steam railways which spread terror through the indolent and prosperous world of canal proprietors. 'Is Inland Navigation to be ruined by these Rail Road projectors? If so, Canal Proprietors had better stop all further improvement; and then, what will be the state of the country?' the Chairman of the Trent and Mersey Canal scrawled in a postscript to a letter in June 1824.[2]

Telford, the recipient of this letter, was caught. On the one hand, in most things his habit was to back the technology which worked best. 'It is, I presume, at this day, needless to observe that the prosperity of a country is most perfectly promoted, by introducing perfect modes of intercourse in its several districts,' he had written in 1810, at the start of an essay on the proposed 125-mile-long railway from Glasgow to Berwick.[3] He had seen the advantages of horse-drawn lines for freight ever since he had visited the Peak Forest Tramway in Derbyshire in 1797. Writing in Shropshire in the 1790s he had described the advantages of railways over canals in some circumstances. Later he advised on plans for a horse-drawn line across the Cotswolds, and temporary industrial trackways had proved invaluable in building the Caledonian Canal.

Neither does he seem to have expressed hostility to the idea of steam locomotives, amongst the first of which were built in Shropshire with help from William Hazledine's brothers. In 1824 he sent his assistant Palmer to Darlington to see work on Stephenson's new line and a nearby colliery line that was already using locomotives.

On the other hand, he was, in his heart, a canals and roads man. By the time steam locomotion had proved its worth he was too tired to change his habits. 'Mr Telford often expressed an opinion that the wear and tear of Engines and Rails would be so considerable that the expense of carrying coals & heavy materials would be much greater upon Railways than upon Canals,' one friend recorded.[4] He expressed a loyalty to his old schemes even as technology changed and he fought to defend them against the onslaught as best he could. His papers contain a rant against the risks, *Observations on Railways with Locomotive high-pressure Steam engines*, published in 1825, which pointed out all sorts of disasters from horses tripping on the rails at crossings to a 'sudden explosion' that 'unkennels the passengers, parboils the pilot and attendants and scatters the luggage abroad like that of a vanquished army'.[5]

Even if Telford did not believe this stuff, it is telling that he read it and kept it. And as early as 1824 he seems to have recognised and resented the fact that railways could only be operated by a monopoly company, in control of the track, in contrast to roads and canals which were open to all comers prepared to pay a toll for using the infrastructure. Canal companies were legally barred from operating barges on their own waterways. Railway companies barred everyone but themselves. It was a fundamental challenge to liberal economic thinking, and seemed a disturbing move towards monopoly control.

After Telford's death John Rickman wrote to Robert Southey recalling their old friend's 'dislike of Steam Conveyance'.[6] Southey, in return, noted that 'The Railway Speculators have sometimes been disposed to disparage Telford as a person who was far behind them in intellect (it was never a Match!).' Telford had been accused of 'undue slight of Rlys', Rickman argued, but the truth was that enthusiasm for the new technology had placed him in a difficult position.

After the Stockton and Darlington opened, all sorts of promoters came forward, some of them seeking money from the Exchequer Bill Loan Commission. Telford – cast as a friend of canals and an enemy of rail – was caught. His role as engineer to the commission meant he had a duty to judge these, but the promoters resented being examined by a man they saw as utterly opposed to their success. Among the projects which applied to the commission were the Newcastle and Carlisle line and the Liverpool and Manchester railway. Telford's papers contain a prospectus from 1824 for the latter, a 'projected Rail Road' which would allow 'the transit of merchandise between Liverpool and Manchester in four or five hours'.[7]

In his memoirs his pupil Joseph Mitchell claims that Telford was asked to engineer it. 'To the great disappointment of his staff, he felt it his duty to decline,' he wrote.[8] 'He said he had been so long and intimately connected with canal works, and the several companies had placed such confidence in his judgement that he felt it unfair to his friends and the canal interests to give the benefit of his great experience to this novel form of locomotion.'

That is a generous way to describe what actually happened. The Liverpool to Manchester line was a hotbed of competing engineers (including, at different points, Stephenson, Vignoles and the two younger Rennie brothers) out of which Stephenson emerged pre-eminent. He sought to frustrate Telford, who 'represented exactly the school of London engineers', writes the historian David Ross, which Stephenson 'actively despised'.[9] That the ever-mobile Telford could, by now, be seen as a 'London engineer' was in itself a sign he was getting old. He requested information on the line. The company failed to send it. When one of his assistants was sent north to see what was happening on the ground, he was ignored. Telford, who had been asked by the commission to 'survey not only the works in progress but the whole of the intended route', eventually had to make his way there himself.[10]

He inspected it in February 1829 in Stephenson's company. It must have been a painful encounter for both men. 'To accomplish this upon a line of 30 miles in the present unfinished and complex state of the works was a tedious and laborious task,' he wrote – adding that 'I am also at a loss. . . as to when this communication will be open to the public or how it is to be worked.'[11] In turn, the Liverpool and Manchester Company denounced Telford afterwards in a printed statement. His report, it declared, was 'an extraordinary document. . . one more abounding with inaccuracies and erroneous statements can hardly be described'.[12]

Telford fired back – he could hardly do otherwise, given the slur – with a full-throated attack not only on the lack of help given to his assistants but also on the design of the scheme; though in the end it secured £100,000 in commission loans.

At the start the company had planned to use a series of fixed steam engines hauling wagons up inclined planes on ropes and some of its supporters remained sceptical about the use of locomotives. What emerged was a compromise, with cable haulage in the two steepest parts. It opened successfully eighteen months after Telford's visit, the world's first intercity passenger line and as such a huge advance on the previous use of railways for local freight. It was a wonder of the modern world and a younger, curious Telford would have rushed to visit and promote it. But writing to the company in 1832 he resisted further involvement. 'I had a violent attack from which I am not yet quite recovered,' he declared, 'so that it will be quite unadvised my attempting a journey.'[13] Excuse or not, he had the same message for the Newcastle to Carlisle line, whose promoter at least had the grace to send his 'best wishes for a speedy return to health'.

Not long after dawn, on a wet morning at the start of November 1833, Telford climbed on board a noisy, smoky contraption outside a factory on the Gray's Inn Road in London. Ageing and frail, with less than a year to live, he was determined to prove that there was a modern alternative to railways. The vehicle he chose was a steam carriage, owned by the pioneering 'motorist', Sir Charles Dance, who was among those who joined the expedition; so were Henry Parnell and John Rickman, loyal to Telford's roads to the end. Expert advice came from another member of the party, the Irish-born engineer John Macneil. He had joined Telford's team in late 1826 and was soon promoted to be the resident engineer of the section of the Holyhead Road from London to Shrewsbury (his loyalty to his master was such that he gave his first son the forename Telford).

In May 1833 an Act authorising the building of the London to Birmingham railway had been passed (on the second attempt, after objections from landowners brought down the first bill). With this came the certainty that trains would soon provide a competitor not just to canals but to the Holyhead Road. Now Telford made one final bid to save his creation.

Today, the suggestion that steam carriages could have substituted for trains in the 1830s sounds fanciful. But at the time many people believed they were the future. Among them, Dance had run a short-lived service between Cheltenham and Gloucester using three vehicles he bought from the engineer and surgeon Goldsworthy Gurney, who himself had learned from his fellow Cornish steam pioneer Richard Trevithick.

These were lumbering, hissing vehicles but they had attracted full loads towing something like a steam bus, outpacing horses. Critics pointed to the tendency of their boilers to explode but the greater objection came from reactionary turnpike trusts, whose income depended on stagecoaches. They were uniformly hostile to change, charging steam carriages unsustainable fees. Advocates of the vehicles argued that high toll roads were all that stood in the way of a new age of express road travel. That was optimistic: engineers struggled to make the machines work and steam technology, at least in its state of development at the time, always seemed better suited to long trains, guided by rails. Some now argue that high tolls were a useful excuse for failure.

Whatever the reality, in 1831 a parliamentary committee investigated the matter, taking evidence from among others Macneil and Telford. 'The injury roads will sustain from the introduction of steam carriages will be much less than is commonly supposed,' the latter assured MPs, arguing for a fairer system of tolling to promote their development.[14] The committee came down in favour of steam vehicles as 'perfectly safe for passengers', but by then it was too late. Parliament failed to pass the hoped-for bill that would have imposed a national reduction in tolls. When steam locomotion came, it ran on rails instead.

In 1831, Telford was cautious: he told the committee that he had no experience of steam carriages, though he did not object in principle. But the threat of the railway to Birmingham soon helped firm up his mind. In late 1833 – as building work started on the London to Birmingham railway line – he set out from London on his steam expedition, an unashamed act of showmanship.

The ground was laid by Macneil, who wrote to him on 20 October to assure him that supplies of fire and water were in place along the route. 'Every person I have spoken to is most anxious to have the trial made,' he explained, 'as they think it will be a <u>clincher</u> for the present tottering state of the rail-way.'[15] The share price of the new rival company was falling fast, amid reports that a fresh bill would have to be taken through Parliament to adjust its course. A few days later he wrote again to add that 'the railway people are at loggerheads with one another'.[16] Telford saw his chance. He wrote to invite fellow engineers to join him, many of them his friends, including Bryan Donkin, a renowned innovator; he was one of the first to find a way of canning food, as well as a contractor on the Caledonian Canal and an early supporter of the Institution of Civil Engineers. 'Being assured of the pleasure of your company I am determined to risk all consequences,' Donkin replied.[17] 'May fire and water befriend us.'

On 26 October, reported *The Man*, 'the long-talked of experiment of a steam carriage between Birmingham and the Metropolis will be at length put to the test. It is said that a number of engineers and gentlemen interested in the improvement of the means of conveyance on the Holyhead Road have hired Sir Charles Dance's carriage.'[18] The next day the *Observer* reported that the trip was to be delayed, but on 31 October, Dance's carriage puffed its way from its depot south of the river, crossed Westminster Bridge and travelled up Whitehall at the busiest part of the day, ready for an early start the next morning.

On 1 November, with Telford on board, the vehicle made its way north along the Archway Road, after which, reported the *Observer*, 'one of the tubes, of which the boiler is composed, burst and although the fracture was small, it was sufficient to stop the progress of the carriage'.[19] This delay meant the vehicle could not reach Birmingham before nightfall: and when it stopped at Stony Stratford – after averaging seven miles an hour on a fifty-one-and-a-half-mile journey – it became clear that another fault needed repair. 'The horrible state of the roads,' the paper reported generously, meant it could not 'reach Birmingham with anything like éclat' – although Telford and his team 'expressed themselves thoroughly satisfied with the principle and vast importance of the invention, as well as its easy application'. Given it had failed twice, that was hardly honest.

In the end, deserted by its celebrity passengers, the vehicle did steam into Birmingham, and the same month Telford put his name to a public declaration stating that 'there can be no doubt that, with a well-constructed engine, a steam-carriage conveyance between London and Birmingham, at a velocity unattainable by horses, and limited only by safety, may be maintained'.[20] He proposed special fast routes for steam carriages and soon after was listed as the consulting engineer to the London, Holyhead and Liverpool Steam Coach and Road Company, chaired by Parnell. It sought £350,000 in capital from shares of £20 each, proposing to run thirty carriages a day.

If anyone invested, they must have lost their money. 'Steam carriages will in no distant time be quite as general use on our roads as steam vessels are on our rivers,' asserted one lobby group optimistically, arguing that the saving on fodder for horses would 'feed sixteen million of our starving fellow-countrymen'.[21] But though steam carriages were considered dutifully by a second parliamentary committee in 1834, by then enthusiasm was falling away and everyone was putting their effort into railways. Telford's plan won support only from dyed-in-the-wool opponents to trains, such as the author of *A Plea for Ancient Towns Against Railways*, a pamphlet which stood in vain against the sort of thoughtless construction which, later, saw Parliament presented with a plan to demolish Stonehenge to make way for a new line.

So instead of carrying steam buses the story of Telford's road became the story of all transport at the time. For a brief moment, the Holyhead Mail beat all records. In February 1834 Macneil wrote to tell Telford that it had carried the text of the King's Speech from Parliament to Holyhead in just fifteen hours, with 'one of the fastest Steamers' on standby to rush it on to Dublin.[22] In 1836 the service was accelerated one last time, travelling from London in just twenty-six hours and fifty-five minutes. It was a glorious final fling for express horsepower. But when the London to Birmingham railway opened on 17 September 1837 passengers switched from the road overnight. They were attracted not just by the speed and comparative comfort but by the space. Trains could carry many more people at lower cost. Stick-in-the-mud travellers continued to use Telford's road for a few months more, but the last stagecoaches to Holyhead ran in early 1839. By then the ferry trade to Ireland had moved to Liverpool (connected to London by rail) and Telford's road, and the Menai Bridge, though barely a dozen years old and built to last for centuries, was without traffic.

The parliamentary commissioners for the road struggled on, maintaining the route and issuing mournful annual reports about their declining traffic. In 1850, however, when Robert Stephenson's rail route to Holyhead opened, and along with it his Britannia tubular rail bridge over the Menai, they gave up. 'We are of the opinion that the road is no longer of such national importance as to justify in applying to Parliament for a grant of public money for its future maintenance,' they decided. Soon after, the Victorian traveller George Borrow walked thirty-three miles along Telford's old road and found it an empty museum piece, grass growing where not long before stagecoaches had raced. He met almost no one.

That was almost – but not quite – the end. The Menai Bridge was (narrowly) spared the indignity of being converted to carry trains (an ageing Provis was called before Parliament to explain whether it could be done) and Telford's road lingered like a ghost, the last tollgate not closing on Anglesey until 1892. The bridge itself remained open for local travellers. Its deck was strengthened in the 1890s, to take heavy loads Telford had never intended his chains to carry, and it found an unexpected new life after the First World War. It was rebuilt in the late 1930s by the engineer – and Telford's biographer – Sir Alexander Gibb, the great-grandson of one of his subject's assistants, John Gibb. He changed the bridge's appearance and replaced the chains but he also assured its future. Today it is busy again, with cars and the green buses of Arriva Wales's route 4A. The main road now crosses Stephenson's rail bridge, itself rebuilt after a fire in 1970 to carry cars as well as trains. But you can still drive to Holyhead by the Menai Bridge, if you wish.

And in that lies a truth: Telford was unlucky in his timing. Had he lived and died a little earlier, he would never have known his roads and canals to be anything other than an assured success. Had he lived later, he might have known enough about the internal combustion engine to see that his roads, at least, would fight back. Instead he died on the brink of change, his vast, solid schemes a victim of steam.

The new railways were yapping terriers; the old canal companies sleepy pedigrees. It took time for them to do more than stare at the threat. Most of their routes had been built before the end of the eighteenth century and since then they had done little other than collect profits. Between 1795 and 1825 only three significant canal schemes were built in England (two of which, the Grand Union and Regent's Canals, indirectly involved Telford). He was also engineer of the Weaver Navigation Trusts and assisted on small schemes in Scotland, but it was not until the mid-1820s that the industry began to panic and invest on a large scale once again. A recovering economy and rumours about railways gave birth to a late, last bout of canal building. Telford, of course, was at the centre of it.

One consequence was a request for him to assess the potential of 'an additional tunnel through Harecastle' in Staffordshire.[23] The original route under Harecastle Hill had been built in the 1770s by the pioneering canal engineer James Brindley. Then it had been a source of wonder, nearly 3,000 yards long. By the time John Rennie came to inspect it in 1820 it had subsided into a congested, lethal wreck, its roof only six feet above the water in places and its rotten bricks half rubbed away by boats. He proposed closing and rebuilding it, but his death and the lacka-daisical habits of canal owners meant nothing happened. Only the far-off rumble of railway competition prompted them to act.

In January 1822 they asked Telford for proposals to repair not only the tunnel but also the water supply to the route. He prom-ised to look in 'on my way to Ireland, on some public business' and by March had settled at the Red Bull Inn at Lawton, nearby.[24] His draft report records that the old tunnel was 'tolerable good', but that the 'only effectual remedy, simple and practically useful, is another tunnel'.[25] The bad news, he added, was that 'in work of this nature, an accurate estimate is what no engineer can be expected to give'. It was a get-out clause he would have done well to have used when first making plans for the Caledonian Canal, but it did not deter the proprietors at Harecastle. Two years after his survey he was appointed consulting engineer and by the following year construction was under way.

This was a typical Telford project, done under his instruction by others on site, in particular the young resident engineer James Potter, with their superior regularly inspecting progress and reporting to the canal owners. Tunnelling was done at what, even today, would be an astonishing speed, the whole route built in little over two years, with hundreds of men working from fifteen shafts sunk down from above and sixteen cross-passages dug from the old tunnel next door. Nothing on this scale had been tried before: the kilns alone had to bake seven million bricks to line it. By November 1827 Telford was able to tell the proprietors that he had 'passed through and examined the New Harecastle tunnel'.[26] 'I have to congratulate the company on having a tunnel more perfect than any other which has hitherto been constructed,' he added. His involvement continued in the months that followed, discussing the tricky issue of water supply with, among others, the second Josiah Wedgwood, who wrote to him from his factory at Etruria, in Staffordshire.

Although the Harecastle tunnel was built fast, well and usefully, so that today it is still open to traffic, it brought with it a flavour of changing times. Some other routes, examined by Telford in response to the railway threat, came to pass. He surveyed the course of the Macclesfield Canal in 1825, for instance, a late proposal for a waterway accepted partly because landowners disapproved of the alternative, 'a railroad transversed by smoking steam carriages'.[27] Even here, however, canals were under pressure. 'We heard of the appearance of a gentleman deeply interested in the success of railroads,' reported a local paper, 'but as soon as he was shown Mr Telford's report he was off again like a shot, leaving canal triumphant in possession of the field.' Telford returned in 1829 to inspect the work and it opened in November 1831.

Canals could still be good business. Telford's surviving business archive[28] contains a mass of paperwork on waterways, among them the Paisley and Ardrossan, the Edinburgh and Glasgow Union and the Weaver Navigation. He remained a hero to the trade: Joseph Priestley's definitive guide to the navigable waterway system at its peak, published in 1831, is dedicated to him. He had far more work than one man could deal with; but even as this continued, other fresh proposals for canals began to fail in the face of the railway revolution.

The young Joseph Mitchell joined him on an expedition to Devon and Cornwall in 1822 ('TT had his own carriage but posted,' he noted, which meant exchanging horses along the way) as part of a survey of a putative south-westerly version of the Caledonian Canal, a scheme which would have cut across the peninsula and shortened the sailing route via Lands End. But the route, which had first been surveyed by Rennie in 1811, came to nothing. Telford proposed a straighter, wider path in 1824, but by now, as he recognised reluctantly, steamboats rendered the project pointless. As late as 1831 Priestley's canal handbook described hopes for the route, whose 'great advantage [would be] sufficient to tempt the cupidity of the most sceptical', but it went on to admit that work had not started, having been 'drawn up in a time of high expectations' which had passed.[29]

It would have been an English Panama Canal, had it been built; and the ghost of that latter scheme lingered in Telford's papers, too, in the form of maps and charts of the isthmus of Darien in Colombia. The country had, in 1818, appointed English naval officers to survey the route and sought out English engineers to assist. If Telford was involved, no record seems to survive. Rickman noted that he never talked of it and the scheme might not have been real; but what a wonderful thing a Central American crossing might have been in his hands.

Faced with the rise of railways, Telford responded in the only way that seemed rational: to test their worth against his favoured canals. In the 1820s he paid Palmer a substantial sum for a series of experiments 'investigating the comparison of canals and railways'.[30] He compared horse-drawn rail traffic with boats on the Ellesmere Canal; tests which later spread across the country with the help of a wide team of engineers. 'With small velocities the force of traction on Canals is less than on railways,' they concluded, 'when the velocity is equal to four miles an hour, the forces are equal. Beyond this, the advantage is in favour of railways.' Telford pointed out that a single horse often pulled a dozen seven-ton barges on a canal. No horse-drawn railway, he believed, could compete with that. But that, of course, was before steam locomotion.

Still, he did not surrender. In 1829 the engineer George Buck reported to him on the operation of the Stockton and Darlington line, which had now been open four years, and also attended the Rainhill trials in Merseyside which proved the worth of Stephenson's *Rocket*. A year later Telford was present when an experimental, streamlined canal boat was pulled from Paddington basin by a pair of galloping hunters, reaching thirteen miles an hour, three times the usual speed, before the horses began to tire. And three years after that he commissioned Macneil to find ways of reducing the drag on fast barges.

Macneil was something of a one-man research institute, inventing and using a dynamometer to measure the effects of surface roughness and steepness on pulling power on roads. He tested model boats made of sheet copper in a seventy-foot-long tank in the Adelaide Gallery on the Strand – 'the National Gallery of Practical Science, Blending Instruction with Amusement' – a place popular with Georgian parents hoping to direct their sons towards improving, modern thoughts.

Visiting the pair at home one night while the tests were going on, Rickman found them sitting at the tea table talking of the experiments they had carried out 'all last week, from daylight to dark'.[31] 'I am convinced that such a sensation has been created, as to overpower the Railway models with which the Gallery is encumbered,' Telford declared.[32] Hopeful, but not realistic. The model trains were winning and the lesson was clear. To survive, canals had to become bigger and faster. Now, even the proprietors could see it. Harecastle had been a start, but Brindley's old route still ambled through the Staffordshire countryside. In the late 1820s, however, a last bout of canal building began, an attempt to build waterborne expressways, with Telford at the heart of it.

In what Telford called 'the great and flourishing town' of Birmingham he surveyed the overcrowded and immensely profitable system which served the Black Country.[33] It was 'little better than a crocked ditch. . . the horses frequently sliding and staggering into the water'. He proposed a new line which would remove the need to lift boats and water up to the summit at Smethwick, between Birmingham and Wolverhampton, with a wider route in a seventy-foot-deep cutting spanned by a series of iron bridges including Galton Bridge, which when it opened in 1829 was the highest single-span structure anywhere. Isambard Kingdom Brunel described it as 'prodigious' when he came to see it a year later. Just as modern were the complex of piers, basins and arched, multi-storey warehouses which Telford designed for the docks at Ellesmere Port, at the end of the route to Chester and (indirectly) Pontycysllte on which he had begun his canal-building career almost four decades before. Their destruction, in a fire in 1970, was the greatest loss of any of his creations.

His other project in central England was even more massive: the last of the English narrow canals to be built and his last collaboration, too, with what remained of his team: John Wilson (who had worked at Pontcysyllte), Alexander Easton (who had kept the faith at the westerly end of the Caledonian Canal) and William Provis (who had built the Menai Bridge).

The Birmingham & Liverpool Junction Canal was an alternative to the roundabout Trent and Mersey Canal: a fast, straight and wide waterway through Staffordshire and Cheshire dug into deep cuttings or raised on embankments to keep it as level as possible. It was built with a swagger equal to the first railways and intended to scare them away from its territory. It scared his employers at Harecastle, too, on the Trent and Mersey, who bridled at news that their engineer intended to assist a competitor. He wrote back stiffly, explaining that he had never thought himself 'exclusively attached to this service'.[34] 'If the Company should continue to disapprove', he added, 'altho' I have no wish to decline the management of the Tunnel Works, I shall await the decision of the Select Committee previous to taking any further active steps respecting them.'

That was a pompous way of saying he would go ahead whatever they thought: an attraction of the project, other than its scale, being that it would at last connect the Shrewsbury and Ellesmere Canals on which he had started his career (and with them the aqueduct at Pontcysyllte) with the rest of the trunk network. He might have answered differently had he known what trouble this last canal was to bring him.

It was designed and built in three sections. In the north it required a great flight of fifteen locks at Audlem, but that challenge was as nothing to the deep cuttings and high embankments required on the middle section near what became Norbury Junction. In the middle of the lush, grassy country of flat plains and low hills which makes up the borders of Staffordshire, Cheshire and north Shropshire, this was, on the face of it, easy country in which to build, compared with the mountain lands which had taken up so much of Telford's life. In the summer of 1827 he inspected the works, which seemed to be going well under Provis's intelligent direction. But wet weather and slithering, rich marly soil soon turned this into a cursed project.

On 9 January 1831 Telford's trusted friend, the contractor on the southern portion of the route, John Wilson, died suddenly aged fifty-eight; but the effect of that on morale was as nothing against the hellish task of supporting the great embankments on the route further north, especially at Shelmore. Telford had not wanted to build this: it was made necessary by the refusal of Lord Anson at Norbury Park to allow the canal to cut through the estate where he kept his game birds. Huge quantities of earth had to be tipped onto this unnecessary structure, year after year, only for it to slip to the bottom. In July 1832 Telford assured the canal company that it was 'gradually assuming a more shapely and consolidated state', only for a great part of it to collapse a month later.[35] The fault lay in the nature of the local soil: no one should ever try again to use it on such a high embankment, he advised in his memoirs, but with work so advanced and the route fixed there was no alternative but to plug away until the thing was done.

In September 1832, unwell and very deaf, Telford gathered his strength to travel to Shropshire with John Rickman to assess the situation. They galloped up his improved road to Birmingham, covering 110 miles without leaving their seats in the carriage, and then travelled up the splendid new line of the Birmingham Canal. In Wolverhampton they defied the threat of 'Bilston Cholera', which had shut the town down, and travelled through Ironbridge, where Telford showed Rickman the inclined plane at Coalport, which had so impressed him as a young man. In Shrewsbury, where even now he held on to his post as county surveyor (Telford never gave up work, though in this case most of the county tasks had been handed to assistants working from the canal office at Ellesmere), he found himself as 'the appointed Doctor', called upon to advise on repairs to Shrewsbury's official buildings. They then set off to see the canal works, a journey interrupted when Telford's health collapsed, 'apparently attacked by jaundice'. Too weak to see the rest of the route, he remained in Market Drayton, sending Rickman off with Easton as his guide, 'being anxious that his own Illness should not prevent me seeing the entire line of Canal'.

It must have been a miserable return to the county that had made him, all the worse because of the perpetual disaster at the Shelmore embankment. In February 1833 the grandees behind the canal company – including the Marquess of Stafford and Viscount Clive – came to Abingdon Street for a crisis meeting with their ailing engineer. Telford had no choice but to agree that William Cubitt, a promising canal engineer, should take over the work. Under his command, the piling up of mud continued at Shelmore with no end in sight. 'It certainly takes more earth to make good than even I anticipated,' Cubitt admitted at the end of the year.

In March 1834, Telford came back to Shropshire for the last time, enfeebled and very deaf, to see the calamitous embankment on his still unfinished canal. He met Cubitt in Shrewsbury and they travelled together to Shelmore which had 'caused so much trouble, expense and procrastination'. He can have done nothing other than stare bleakly at the slippery slopes: he was no help, now; it was over; his work was done. He must have hoped to live long enough to know that Shelmore was open, but in May that awful soil slid again and it was not until 2 March 1835, six months to the day after his death, that the first boat inched across the precarious structure, which today, covered in trees in the quiet countryside, is still a perturbing sight.

Not long before he died Telford confessed to a correspondent that 'suspicious Symptoms occasionally remind your humble Servant, that he is also mortal'.[36] His health had held out well until his seventieth year when he collapsed in Cambridge with 'a severe and painful disorder' as he was returning from his work in the Fens. The shock was intense: even his script took on the shaky hand of an aged man. 'I had much to suffer after your kind visit,' he told John Rickman in September 1827.[37]

He was not, however, one to lie long in bed. 'I continue to get stronger,' he noted by the end of the month; 'I am getting about again,' he added a week later.[38] 'Being obliged to be in Gloucester on Monday and being resolved to make easy journeys, I leave town tomorrow (Sunday).' That was not much of an indulgence. Others would have found such a pace hard, even when fit.

He claimed, now, to want to wind down his work. 'I have now recovered my usual state of health, but have, of late, declined undertaking any new work, tho' several have offered, I mean to be content with completing those in hand,' he had told Rickman in 1827. 'I intend only to construct one more Canal in the Centre of England. . . and then have finally done with them,' he added, but there was another lure to work, too, 'the pressing application of the two principal Cities of my native Country to furnish designs for their principal Bridges'.

In Edinburgh this gave rise to Dean Bridge, a four-arched stone structure high above the Water of Leith, built between 1829 and 1831. In Glasgow it led, a little later, to a maturely tasteful seven-arched stone crossing over the River Clyde, part of the urban swagger of a fast-growing city. It was completed, just after Telford's death, by John Gibb, who had earlier worked with him on improvements to Aberdeen harbour and oversaw the building of Dean Bridge, too. Both these Scottish finales survive: the Glasgow bridge was raised and widened in the 1890s, but rebuilt using stonework and the balustrades from Telford's design at public request.

He was pushing himself to the last. 'I have received some distinct intimations that, even in this favourable Climate, existence is precarious; under that impression I have been endeavouring to contract my engagements,' he told James Little in 1829, 'but have hitherto only imperfectly succeeded, I have this Year been unavoidably led to extend them.'[39] Typical was the advert in a London newspaper in 1828 calling for suggestions to be sent to him in Abingdon Street in conjunction with an inquiry into the supply of fresh water to the capital, a sideline he could easily have declined. More demanding still was his role at the head of a scheme already under way, the construction of St Katharine Docks, just down the Thames from the Tower of London.

Until 1800, ships in the capital had tied up and unloaded in the river itself, affected by its tides. The first modern docks had been built at the start of the century, chiefly by his old acquaintances William Jessop and John Rennie. With both of them dead, Telford was the obvious choice to engineer one of the next big schemes, a dock to be built on an area of dense slum settlement around the medieval church and hospital of St Katharine (which was demolished and moved to Regent's Park). He submitted plans for a new site against strong opposition from established docks in 1824, together with the architect Philip Hardwick (a fellow member of the Institution of Civil Engineers who later designed the Doric arch which stood before Euston station until its miserable and unnecessary demolition in 1961–2). Telford's design was ingenious, maximising warehouse space on the cramped site. Two docks each covering about four acres were joined by a shared central basin which allowed them to be used independently of each other and entered at low water. Hardwick added massive warehouses on the water's edge, supported by hollow cast-iron Tuscan columns. History has not been kind to them: of the six that he and Telford built, the three to the east were bombed in 1940 and the three to the west were demolished in the 1970s. (The surviving basin is now a yacht marina.)

An Act authorising construction was passed in 1825, rendering homeless, in many cases without compensation, the 11,000 people who lived on the site. By 1826 work was well under way, overseen by Thomas Rhodes, the superintendent who had just proved his worth on the Menai Bridge. Telford, who was paid £500 a year (with the same going to Hardwick) visited the site regularly, even as he recovered his health in the autumn of 1827. The first vessel sailed into St Katharine Docks in October 1828, and by 1830 they were complete.

By now he was allowing others to take much of the strain. 'The several works in progress, under my management are all in such regular train, with able and faithful Resident Engineers,' he told Rickman gratefully, 'that only the monthly accounts remain to be examined and entered here.' The spring was winding down and the clock was slowing. He could see the end was coming.

A Blessing on Him

On 11 June 1834 the artist William Brockenden, a portrait-ist attracted to the intricacies of engineering, sat down to sketch Thomas Telford. In pen and ink he drew a dying man: cheeks pinched; hair stripped back; eyes sunk; mouth turned down. He showed Telford in his usual heavy dark coat, but his body no longer filled it and the swagger had gone. The lines of the picture drift into the page, tailing off to nothing.

The drawing, suggests Charles Hadfield, the debunker of Telford's reputation, is 'infinitely sad: an empty man haunted by emptiness'.[1] Is Hadfield reading too much into the melancholy picture that any once-great figure presents when enfeeble-ment and death are close at hand? Most endings are untidy and Telford had tied up the threads of his life as neatly as any man can. But there was a failure late on about which he never spoke or wrote and which cannot but have hurt: his strange, late defeat by the twenty-four-year-old Isambard Kingdom Brunel. It marked the passing of generations: the greatest civil engineer of the Georgian age falling before the greatest of the Victorians. For the old man it was, if not quite the end, then the moment his retreat from active life became certain. The strengths of char-acter – the certainty, the self-assurance, the boldness – twisted into weaknesses.

The roots of the story lie in the early 1820s, a time when Telford came to know Isambard Kingdom's father, Marc

Brunel, a French engineer who had arrived in London with an extraordinary story behind him. He had met his future wife, an Englishwoman, Sophia Kingdom, in France before the revolution. As the Terror worsened he had fled first to America (where, at the age of twenty-seven, he was appointed chief engineer of New York), before sailing for London in 1799. In the capital he sought out and married Sophia (who had in the meantime narrowly avoided execution in France as a spy). Their son, Isambard Kingdom, was born in April 1806 and in the same year Marc joined the ranks of respected British engineers. He became a fellow of the Royal Society and later of Telford's Institution of Civil Engineers. He might easily have become one of Telford's intimates but though there was no public falling-out, the suspicion must be that the pair respected each other; nothing more.

Unusually, for instance, Telford was not present at the institution on the evening Brunel read a paper on his plan for a radical Thames tunnel, the first ever under a river and the great engineering effort of the time. Nor is there any record in Brunel's diary of his visiting the site, although it was only a short cab ride from Abingdon Street and many others made the trip. Brunel did head the list of those who proposed Telford's elevation to the Royal Society in May 1827 ('Went. . . for Mr Telford's certificate,' his diary reads – the only mention in that decade of the country's leading civil engineer); but he was not a regular attender at the Tuesday dinners in Abingdon Street.[2] Isambard Kingdom's papers show that his father did call on Telford, but infrequently.

Perhaps Telford had been influenced by his friend, the military engineer Charles Pasley, who had been on the committee that approved the tunnel in 1819, or by his fellow member of the institution, Charles Vignoles, who was cast out as the tunnel's engineer in favour of young Isambard Kingdom. Whatever the truth, it was the tunnel which led to the next chapter in this tale.

In January 1828 Isambard Kingdom was overseeing work under the river when it flooded. He was nearly drowned in a disaster so serious that it was seven years before work began again on the project; and his injuries nearly killed him. Isambard was sent to the sedate Bristol suburb of Clifton to recover. Here he learned of a long-standing plan to bridge the deep Avon gorge. In 1754 a local merchant had left £1,000 to fund the scheme, a sum which, invested well, had risen to about £8,000 – the money still unspent.

That was about to change. In October 1829 the Bristol papers announced a competition. 'ANY persons willing to submit DESIGNS for the ERECTION of an IRON SUSPENSION BRIDGE at CLIFTON DOWN over the RIVER AVON, to the consideration of the Committee appointed to arrange proceedings for carrying the measure into execution, are requested to forward the same, accompanied by an Estimate of the probable expense.'[3]

There were more than twenty entries from, among others, Telford's old ally William Hazledine and Isambard Kingdom, who submitted four designs with spans varying from 760 to 1,160 feet, all wider than the Menai Bridge's 580 feet. Isambard had continued to develop as an engineer, joining the Institution of Civil Engineers as an associate member in 1829 and rising within two years to full membership. But now he was about to face its president head on. A committee announced that the Bristol designs would be put 'to the further consideration of Mr Telford': he was, after all, the most respected suspension bridge engineer in the country.[4]

Brunel was among the finalists (Telford is reputed to have declared the design 'pretty and ingenious') and the outcome cannot have been what the merchants of the city expected. Within days their respected judge had rejected all the entries as unbuildable, given high winds in the gorge, awarded part of the £100 prize money to Hazledine anyway and submitted his own (unsolicited) design for a hybrid suspension bridge with three narrower spans, held up by a pair of pointed Gothic stone towers rising from the bottom of the gorge.

It was an ugly thing with none of the lightness which characterised Telford's best work: as if he had swallowed a precursor to high Victorian style and half digested it with the remains of his preferred Greek Revival elegance. Even so, for a moment his reputation carried the day. Visiting his friend in Abingdon Street, Southey announced, loyally, that he was 'greatly pleased with the design'.[5] In January 1830 a bill was put before Parliament to build Telford's bridge, the Bristol committee putting a brave face on it by praising the design for combining 'every essential quality of adoption to the spot, symmetry and harmony of parts, beauty, lightness and strength'.[6] A more honest judgement is that it looked like a mad wire washing-line, strung from the worst bits of the future Palace of Westminster, and within a week the committee turned tail. They were, they announced, 'not at all surprised to find that the Design submitted to the public should have been the subject of criticism'.

Telford, so bold at the Menai Bridge, had lost his nerve, spooked by the repeated damage to that bridge caused by twisting in storms. Not long before, he had, according to Smiles, replied pugnaciously to the Austrian government after it consulted him about a suspension bridge across the Danube between Buda and Pest. 'We do not consider anything to be impossible. Impossibilities exist chiefly in the prejudices of mankind, to which some are slaves, and from which few are able to emancipate themselves and enter on the path of truth.'[7] If he once believed this (and the words attributed by Smiles make him sound more spiritual healer than civil engineer) he did so no longer.

Isambard Kingdom in particular did nothing to hide his scorn for Telford's design. 'As the distance between the opposite rocks was considerably less than what had always been considered as within the limits to which Suspension Bridges might be carried,' he wrote to the committee after his rejection, 'the idea of going to the bottom of such a valley for the purposes of raising at great expense two intermediate supporters hardly occurred to me.'[8] Brunel could be a bully, but in this he was right. He had done his homework, making five visits to what he always called the 'Menia Bridge' (the repeated misspelling in his papers is curious). 'I endeavoured to collect all that was good and to remedy the defects which had been discovered,' he reported.

Now Telford's humiliation began. In December 1830 a second competition was announced, judged this time by, among others, Davies Gilbert, the Cornish MP whose advice had helped make the crossing at Menai possible. One entry came, Gilbert admitted, from 'the highly respectable name of Mr Telford to whom this country is indebted for the Suspension Bridge extended over the Menai Straits'.[9] Delicately, he then excluded him. 'We are compelled to forego the advantages now to be derived from his abilities and eminence on account of the inadequacy of the funds requisite for meeting the cost of such high and massive Towers as are essential to the plan which that distinguished individual proposed.'

That left Brunel's design in the race, alongside four others. All of them, Gilbert decided, had flaws but 'Mr Brunel's plan has the important advantage of being adequately <u>strong in the most essential part, namely in the great chains</u>'. Brunel (after lobbying Gilbert) was chosen as the winner.

In London, his father was triumphant. 'Isambard is appointed Engineer to the Clifton Bridge,' Marc's diary for 19 March

1831 reads. 'The most gratifying thing is that Mr T. . . d Captn. Brown & Mr T Clark were his competitors and that, for my own part, I have not influenced any of the Bristol people in his behalf either by letter or by any interviews with any of them.'

This entry is doubly revealing. When he first wrote it Brunel mentioned only Telford's defeat; he inked in the other names later. And in the whole long diary this is the only name, on any page, which is not spelled out in full. Telford was his target.

No exoneration for the folly of his bridge is available. The kindest thing we could say is that Telford was psychologically unable to accept that his talents were failing; more unkindly we might speculate that he was overcome by jealousy. He makes no mention of either Brunel or the Bristol bridge in his autobiography, and only a few bills for the cost of drawings from his clerks seem to survive among his papers. If it is true that he filleted his records before his death to shape his reputation (as Hadfield suggests) then this would have been one of the errors which he wanted to bury. It is surprising that as realistic a man as Telford could have supposed that by omitting to mention the defeat he could do other than diminish himself: this was too good a story to be suppressible. Perhaps he knew that everyone knew, but felt so wounded by the humiliation that he simply could not bring himself to think or write about it.

The pain may have been made all the greater by the fact that Marc Brunel seems to have done much of the early design for the Clifton bridge on behalf of his son. The claim in his diary not to have swung things was untrue. 'I met the Deputation of Gentlemen applying for the Act of Parliament for the Bristol Bridge,' he noted the year before. 'Had a great deal of conversation with them on the Subject of the Bridge, and most particularly with one single span. Explained to them how the lateral agitation may be prevented and how the effects of the wind may be counteracted.'[10]

Nor was Telford to know that the Clifton bridge would not be finished until 1864, after the death of both Marc and Isambard Kingdom Brunel and in a different style, with stronger chains. Telford's doubts about the original Brunel design had some foundation. But his own proposal is beyond rescue by posterity. 'However beautiful Mr Telford's design might be,' a pamphlet celebrating the opening explained politely, 'the Trustees exercised a sound judgement in adopting the present design of the late Mr Brunel.'[11]

If only Telford had done as he had promised and stopped work sooner. 'Having come to the resolution, to, as much as possible, avoid engaging in any New Undertaking,' he announced in 1830, 'I intend dedicating the short time I can expect to remain amongst you, to the giving of some account of the manner in which I have been employed since issuing from an Eskdale Cottage.'[12] It was a familiar theme, repeated in many letters over several years: the soldier-like desire to stand down from duty. 'Having now shared in 75 Years incessant exertions,' he told James Little in 1832, 'I have for some Years past, proposed to decline the contest, but the numerous Works in which I am engaged, have hitherto prevented my succeeding. In the meantime I occasionally amuse myself in stating what manner a long life has been incessantly employed.'[13]

By that he meant his sprawling efforts to write an autobiography, a book which presents Telford's admirers with a problem. On the one hand it is, in places, a reasonable record of his work, most of all through the finely engraved plans and drawings in the *Atlas* which accompanies it. On the other it is largely an unreadable lump, lacking, after the first few pages, warmth or personality. Its different sections are badly out of proportion, 'Mr

Telford not being good at that kind of Engineering,' exclaimed Rickman as he sweated to edit the 'perplexing' text.[14] Telford fiddled about with the words during his lifetime, but left his friend to complete it. Two drafts in Telford's hand survive in the archive at the Ironbridge Gorge Museum, the latter full of corrections and editorialising by Rickman, who wrote to their close companions asking for information with which to fill the many gaps. As a monument, it is less than worthy.

It also lacks the fluency of Telford's previous writing, which, even excluding his poetry, amounted by the end of his life to the contents of a small bookcase: printed plans and surveys for around eighty separate engineering projects, some of them (such as his detailed reports to the commissioners on the Highland and Holyhead roads and the Caledonian Canal) rewritten annually over decades. Even if Telford did not produce all of this himself – his team of assistants was also involved – he will have read and edited it, along with a vast personal correspondence.

On top of this (and finding the time for his travelling and personal reading) he was a leading contributor to, and financial supporter of, the *Edinburgh Encyclopaedia*. When a collected edition of the *Encyclopaedia* was published in 1830, his entries written since the series began in 1812 amounted to more than 300 pages, on subjects including architecture, bridges and inland navigation (the latter the length of a book in itself). His piece on architecture displays his intellectual depth – as one of the first writers to see equal merit in works from lands far from Europe, as well as the classical tradition – and his respect, as an engineer, for the importance of aesthetics, as set out by his friend Archibald Alison.

He found it harder, though, to write about himself. At home in Abingdon Street, as he rifled fretfully through his papers and tried to sum up his life's work, his mind was drawn back to his birthplace. 'I fancy that few of my standing, are now to be found in the course of the Esk,' he wrote to William Little's widow in 1832.[15] 'I shall be glad to learn the Country News,' he added at Christmas the same year.[16]

That was the first in which he did not go back to Scotland, ending a series of annual or biannual visits which had made up the rhythm of his life since 1801. He last saw the Highlands in 1829. His health was worsening and travel was becoming hard, although when George Turnbull moved into Abingdon Street in 1830 as the last of Telford's young clerks he found his master remained 'a fine, hale, hearty old man', still able to hold dinners for his friends. 'Mr Telford was of a most genial disposition and a delightful companion, his laugh was the heartiest I have heard; it was a pleasure to be in his society,' he recalled.[17]

Soon, however, decline set in. Telford was seriously ill in January 1833 and again in August. 'During the last 12 months I have had several severe Rubs, and at 77 – they tell more seriously than formerly and require less exertions, and more precautions,' he wrote after the second attack.[18] His growing deafness, Rickman recalled, left him 'uncomfortable in any mixed company'.[19]

He kept up a regular correspondence with Eskdale until the end, even so, sending funds as he always had to be handed out to those thought to be in need. 'Let me know any other case which you think should be attended to,' he had told William Little in 1816, enclosing a banker's draft for £50 as well as £1 each as a Christmas gift for Little's wife and children. 'This very severe weather reminds me that a small aid may not be unacceptable,' he wrote at Christmas 1830 – adding alongside it a draft for £25 to replenish his Eskdale charitable fund, 'destroyed' by his ne'er-do-well cousin James Jackson, whose behaviour was a long-standing source of grief to Telford.

Generosity and compassion were his most endearing characteristics. George May remembered 'his active benevolence in every case of misfortune that was presented to him'.[20] 'I never knew an instance of unkind rejection,' he added, 'the possession of any talent, literary, scientific or mechanical, I always observed, was an irresistible passport to his bounty.' At a time when workers were easily replaceable (the death rate in Brunel's Box railway tunnel near Bath, built in the late 1830s, rivalled some of the trenches in the Somme) he took care of his men's safety. Even in his latter years, living in the moderate comforts of Abingdon Street, he remained rooted, perhaps ostentatiously so, darning his stockings before retiring to bed, for instance, rather than letting his servants do it. He had been paid well but not excessively, charging a fee of around five guineas a day in 1810, rising to seven by 1824 (the latter a little more than £300 in real terms today, although the purchasing power of the sum was greater than that implies).

He invested his money cautiously. But wealth was never the draw. 'I admire commercial enterprise, it is the vigorous outgrowth of our industrial life,' Smiles records him saying (in another of those quotes which is difficult to source and which may have been adapted). 'I admire everything that gives it free scope: as, wherever it goes, activity, energy, intelligence – all that we call civilization – accompany it; but I hold that the aim and the end of it all ought not to be a mere bag, of money, but something far higher and far better.'

He lived up to this, giving much of his wealth away. It was characteristic that he agreed to underwrite a £1,000 bond to allow a young member of the Little family to open Langholm's first bank (and then watched its growth with pride) even though he initially doubted the business's prospects. In return he was sent 'the magnificent present of half a Dozen fine Mutton Hams', which arrived in perfect condition. 'They have,' he added, 'already improved my morning meal.'

The new bank was part of a land much changed from his youth. 'All the great towns are greatly enlarged, and there is scarcely a country village that has not been renovated and extended,' he wrote after one of his last Scottish tours.[21] In 1829 he was in Eskdale on a visit which must have been bittersweet. He had been asked by Sir Pulteney Malcolm (the Westerkirk-born naval officer and one of the four brother-knights to emerge from the remarkable Malcolm family) to survey a better road from Carlisle to Glasgow, and among the options he looked at was a route 'passing up the Meggot by Glendinning &c.'[22] That was the farm where he had been born more than seventy years before; but he was too professional to overstate the case. The route would, he noted, 'only save 2½ miles more, and this thro' a thinly inhabited country and to be acquired at a great Expense'.

Did he think back as he tramped up the valley, to his time there as a shepherd and to the changes a road might bring? Did he even want to leave it untouched, in a life which changed so much else?

There were a few warm tokens of remembrance. While still working on the Caledonian Canal he had ordered glass goblets to be made and gave them to his closest colleagues. One went to Matthew Davidson, engraved with his name, who was said to have cherished it (as did his son, until it was smashed many years later). There were pictures, too. In 1829 he sent to Eskdale an engraving adapted from a portrait done in about 1822 by the artist Samuel Lane, who painted many of Telford's friends, including Rickman.

This shows him in his sixties, hair still dark, with Pont-cysyllte in the background: a man near the end of his pomp, slumped a little in a soft chair. It has neither the mop-haired insouciance of Sir Henry Raeburn's earlier portrait (from around 1801) in which the sitter seems ready to punch anyone who doubts him, nor the last-gasp vacancy of Brockenden's final drawing. As a series they record a life which was coming towards its close. The visits to Eskdale – 'laughing with his Langholm cronies', as James Little remembered – were over.

———

What would it have been like to meet Telford in his final years? On the face of it he might have appeared a relic, harrumphing about the modern world. He was, by his old age, a confirmed reactionary: he had spent too long in Rickman's pedantic company to be anything else. When, for instance, young James Mitchell left Inverness to live in Abingdon Street, his parents warned him to tone down his views. 'As I had sometimes expressed Radical opinions, which were then held in abhorrence, I was warned never to give any indication of them where I was going,' he recalled. Not long after this another established engineer came asking for a job. 'I should be happy to engage you, but I am afraid you are such a Radical,' Telford exclaimed, 'you make speeches at meetings; you are dangerous.' The man listened in silence, then cried out: 'Stop sir, stop; you should take a pill – that irritation proceeds from the stomach.'

Stories like this imply that Telford had become a man of fixed, even bitter, opinions. But his writing suggests that under the skin he was still open to new experiences. In the 1830s, for instance, he kept up a happy correspondence with James Little, by now trying his luck in Bombay. Like many at the time, Telford was curious about India, 'whose comfort and happiness, must be much greater under the mild British Government, than when harassed by the contentions of their remorseless Chiefs'.

His opinion was reinforced by the size of 'the Eskdale Corps' involved in ruling Bombay. He was fascinated by Little's report of 'a journey to the Mountains (Ghats) like other Indian

Pilgrims in quest of cool air, clean Streams, and the Source of the Sacred River (Kristna)' and laughed at his 'devotion at the Temple containing the Image of Indian adoration' which he mocked as 'a pretty account of a disciple of John Knox'.[23] 'The Parsee Tomb is a novelty,' he told Little; 'it shows what strange customs prevail in different Countries. . . the Vulture is certainly an uncouth Angelic Judge.'

He delighted, too, in the '<u>young sharks</u>' which Little sent him. 'From the shape of their jaws and the sharpness of their teeth they would depopulate a whole Region of the Ocean,' he noted, adding 'you should have accompanied the Fish with directions in what manner they are treated at Table, in their native country; I can only recommend that they may be peeled by the fingers like a dried spalding – they are however a Novelty and I thank you for them.' There's more than half a smile behind that comment; the old man still had his humour and curiosity.

Nor, at home, was his mind wholly unbending. As an early supporter of decimalisation he hardly seems the stereotype of a crusty old Tory. When in 1831 Charles Pasley wrote a paper proposing that Britain should adopt decimal measurements, Telford stepped forward in support. A year later the radical (though admittedly by this point much-mellowed) MP Sir Francis Burdett, who had always opposed the politics expressed by Telford's high Tory friends, picked him as a sole commissioner appointed to resolve the chaotic and polluted state of the capital's drinking water. (The previous commission in 1828, which also involved Telford, had achieved nothing.) Burdett's notorious past as a landed, fox-hunting but reformist enemy of the old corruption embodied by some members of Telford's circle did not stand in the way of their late alliance. Burdett went on to campaign for Telford to be put in charge of a better sewerage system; but this, like his efforts with water, went nowhere. He began work in 1833, but his proposal to stop taking supplies from the Thames and draw water instead from the smaller Rivers Lea, Wandle and Ver ran into the sands of his old age and political confusion.

The confusion, most of all, came from the battle for parliamentary reform which swept up MPs, their debates almost in hearing distance of his front room. Telford was no enthusiast for reform. But neither did he stand in its way. 'What the New Charter will accomplish, remains to be seen, and not being a politician, I pretend not to conjecture,' he wrote when the Reform Act was passed in 1832.[24] He did not sound upset by the measure. When Sir John Malcolm, the fourth son from that remarkable Westerkirk family, returned from India and announced he planned to stand as a candidate for Carlisle as an opponent of reform in the election that followed the Act, Telford cautioned against. 'I fear that our friend will not, in the present temper of the Country, be successful – my advice is to give himself no trouble, and go to no Expense.' It was good advice. Sir John lost.

Sir John Malcolm died, after a bout of flu, less than a year later. The era of the Eskdale knights was ending. So was Telford's. In early 1833 he carried out his final task as a civil engineer: a short study of Dover Harbour, which had been damaged in winter storms. He did it at the request of the Duke of Wellington, the Warden of the Cinque Ports. It was an insignificant job on which to end a great life. With the exception of the unfinished embankment on the Birmingham and Liverpool Junction Canal, his work was done.

Apart from one thing. And in this life, of all lives, it is appropriate that Eskdale returns in Telford's closing scene. The immediate cause was Sir John's death. 'We have all been sadly deranged here by the loss,' Telford wrote in October 1833, describing a plan to 'express a public Sense of his talents'.[25] This led to the tall obelisk which now stands on the top of Whita

Hill above Langholm. Intended as a memorial to the late imperial adventurer, it must also have seemed to Telford a chance to leave his mark on the valley. It was a plan, he explained, which 'Pasley. . . and I proposed for an Eskdale Monument.' The 'and I' here is significant: Pasley's papers suggest that Telford was not involved directly in its instigation. He was clinging on to importance.

On 19 May 1834, Pasley and Sir Pulteney Malcolm (one of Sir John's older brothers and the Royal Navy officer who had led the blockade that guarded Napoleon on St Helena) came to the door of 24 Abingdon Street. The man they met inside was the man drawn by Brockenden a few weeks later: in pain, barely able to hear and with only a few months left to live. It must have been as much as a kindness as anything else that they asked him to draw up plans for the monument. Telford worked quickly, even so. Sir Pulteney took them to Eskdale the following week.

Over the summer what Telford called the 'bilious attacks' from which he suffered grew worse. On 23 August he was brought down by one. Two doctors were summoned to his bedside, including Anthony White, one of the most famous surgeons of the time, who lived not far away in Parliament Street. They called on him twice a day for the next week but there was no improvement. At around five in the afternoon on 2 September he died, 'very peacefully' in his bed at home in Abingdon Street.[26] Only his clerk, George Turnbull, and his long-standing servant, James Handscombe, were in the house. 'I lost my kind, old master, whom I loved and respected sincerely,' Turnbull recalled.[27]

The news was rushed to Rickman, who was away in Ellesmere Port, and to Langholm by the racing Royal Mail coach along the roads he had improved. Pasley and Sir Pulteney were at the Malcolm stronghold at Burnfoot on the banks of the Esk.

'News arrives of Telford's death, which we all exceedingly regret,' Pasley recorded crisply in his diary before adding, with the matter-of-fact tone of a military engineer: 'he sent a second design for Sir John's column which is not a good one, to save expense it is too weak'.[28] The column was adapted and finished the following year: a solitary memorial overlooking the Esk marking the end of a working life that began when a teenage apprentice mason carved his own father's gravestone and set off from his valley to change his world. Both Telford's Eskdale memorials, then, were to others.

The aftermath was dispiriting. News of his death made the front page of *The Times*, but Telford's friends were getting old and the day had long gone when he could be counted as the country's leading civil engineer. The dean of Westminster Abbey agreed he should be buried in the church (Telford having himself asked for the more modest St Margaret's Church, just outside), 'on account of the eminence of his character' but 'neither', he added, 'in Poets' Corner nor in the North Transept'.[29] On 10 September the funeral took place. It was a quiet affair: only thirty-six people were invited. A procession including Sir Henry Parnell and members of the Institution of Civil Engineers walked to the service from 24 Abingdon Street, but there was no family to mourn him and the contents of the house were auctioned off soon afterwards.

It was left to Rickman, Southey and the institution to do what they could to preserve his memory. The institution paid generous tribute to 'the great loss they have sustained by the death of their venerable President'. As its members recalled, 'the boldness and the originality of thought in which his designs were conceived has only been equalled by the success with which they have been executed. . . members of that profession can never forget the liberality with which he patronised and encouraged young men, his ready accessibility, and the uniform kindness of feeling and urbanity of manners evinced in his intercourse with everyone.'[30]

Telford left the institution his papers and drawings as well as a legacy to fund a prize for engineering: among the early winners of a Telford Medal, in 1838, was Isambard Kingdom Brunel. The prize continues to be awarded. Both men would have been amazed by the research of one winner from 2011: 'the accelerator-driven thorium reactor power station'.

The rest of his estate was shared out in his will among his friends, names that had run through his life. Among the bequests were £800 for the children of Archibald Alison, the cleric who had employed him as a young mason to fix his house, and gifts for Charles Pasley, John Rickman, Joseph Mitchell and, separately, the rest of the family of the late John Mitchell – 'Telford's Tatar'. There were also bequests of £1,000 each to the libraries in Langholm and Westerkirk, the former probably a descendant of a collection begun by the town barber, from which he and Matthew Davidson may have borrowed books as young men. 'Does the circulating Library. . . still continue to flourish,' he had once asked Andrew Little, 'in what state is it?'[31] Now his support led to an unseemly legal dispute, in which a newer Langholm library sought to grab a share of his gift. The squabble went on until it was resolved in favour of the older of the two a decade after Telford's death. After that the institution grew fast, with a new building, but its minute books show that it had spent all but £100 of the bequest by 1900. When it finally ran out of funds in the 1960s the local council tried to throw the collection in a skip, from which it was rescued by locals, returned to the shelves and is slowly being restored. 'Telford's Legacy' is once again being stamped in gold on the books' spines.

There was also the question of the papers that made up his unfinished autobiography. It fell to ever-dutiful Rickman to complete it. 'I had almost placed myself under a vow, to write no Letters, nor do any superfluous or not indispensable act,' he wrote afterwards, 'til I had put into tolerable shape the posthumous work of Telford which he commenced in 1830 under assurance that dying in the Midst of his labour should not be lost, nor the memory of his work perish.'[32]

It was a bleak task; Rickman edited and drew together what survived, adding comments of his own, biased by his own deep Toryism. Thomas Paine's work, which Telford had posted from Shropshire to Langholm in the 1790s, was dismissed, for instance, as 'some of the political trash of the day'; 'after the horrors of the French Revolution,' he claimed, 'Telford silently abandoned active politics to the care of those citizens who spend their time discussing what they rarely understand'. The reality, given Telford's sustained involvement with politicians, was always more nuanced.

When the book was published in 1838 Rickman added a foreword – almost an apologia – explaining why it might disappoint. He pointed to Telford's desire to control publication even after his death, appointing an expensive engraver to produce the illustrative plates. The *Atlas* of pictures which eventually resulted was enormous: 'Mr Telford was always favourable to a large scale,' Rickman noted mournfully. And once published, there was the question of selling a book that was extremely expensive to print. The cost of production came out of Telford's estate, which left Rickman and Southey in an awkward position. Both had been left money from the residue. 'We Legatees must consent to lose by the Publication of every copy sold – But as Telford himself expected this result & cared little about it, his Extors & Legatees must not quite flout his liberal intention,' the former wrote to the latter.[33]

In the event Telford's estate was large enough to pay bequests one and a half times greater than listed in the will. 'This is truly a Godsend and I am most grateful for it,' the impoverished Southey wrote when the money came to him.[34] He deserved his reward for standing firm as one of the Telford loyalists – friend to the man he would rather meet than 'all the statesmen in Europe', he said. 'I owe Telford every kind of friendly attention and like him heartily.'[35] He had plugged away to sustain the engineer's reputation and confidence. 'Tell Mr Telford, when you see him, that Sir Walter S[cott] is full of admiration,' he wrote to Rickman after the author had visited Pontcysyllte and called it 'the most impressive work of art he had ever seen'.

But following his death both men knew that their friend's place in history was insecure. 'Three years have now past, and Mr Telford and his merit are passing into oblivion,' the faithful Rickman observed as he struggled with the book.[36] He encouraged Southey to review it favourably (and anonymously) in the hope of promoting sales, warning him not to give praise beyond that which the inferior material would bear: 'You will easily give well-flavoured pudding though not ornamental pastry – especially not Puff Paste.'

'Few men,' said Southey's review, 'have been so fortunate in all the circumstances of life as Mr Telford.'[37] His concluding words stand today as an honest memorial to his friend:

In the prime of life he found his proper place in the world, and he retained it to a good old age; retaining also his temperate habits, his equal temper, his cheerfulness, the love of his profession, and a benignity by which his fine countenance was characterised as strongly as by the intelligence which marked it. . . A pleasant companion, a constant and considerate as well as kind friend, it is seldom that one individual has rendered essential services to so many; and it was honourable to himself and useful to his country. He never seems to have been visited by any calamity – there was a blessing on

him, his days were long in the land and his good name will
be as durable as those great and numerous works which will
perpetuate it.

Dear old Southey, with his homely instinct to make things rhyme,
is trying to make his friend's life rhyme. It is in the nature of
epitaphs to be composed over the embers of a life. But those words
of Southey's frame the hearth yet miss the fire. Telford's distinction, his singularity, his raging energy, are hard to catch in mellow
retrospect, and his was no marbled example of a well-balanced life.
There is no equilibrium here. It was a life lived almost on the run,
though from what, or to what, I doubt he ever paused to think.

A simple inventory of his work could have filled a book of this
size, and a full study of the archive he left behind could take a lifetime, but even his great structures do not define him. To capture
what does we would find ourselves on lonely roads in freezing
carriages on dark winter nights, in Swedish bog and forest, or
striding across Scottish hillsides, his juniors in tow. We'd be with
him in rough roadside inns, burning the midnight oil over plans
and specifications, trying to keep abreast of his ceaseless correspondence. We'd stand over him in country inns where he penned
before bed his frank and affectionate letters to his blind friend
Andrew Little. We'd hurry away with him from opening celebrations for great projects where, his work done, he shrank from the
ceremonials. We'd see a youth on an Eskdale hillside, trying to
be a Robert Burns. We'd see a man who made a difference, who
through the force of his personal intelligence and drive shaped the
successes of the summit of British achievement: a figure whose
role in history could not have been replaced by another and who
shaped our world today. History should be about leaders as well as
the wider forces of change – the individuals who nudge things in
new directions – and Telford was one of them. He was a designer
of solutions more than an inventor of new ideas. He brought

things together; made them happen; saw beyond the tasks of the moment a pattern of modernity that could improve his country; technology deployed as a kind of art.

But none of this is simple. None of it is easy to define. No homilies, then, no respectful round-ups of a rounded life. This was not a rounded life. Gigantic as his achievement was, something dissatisfied runs through it all. It is not possible to detach from Telford's immense work and towering life the half-glimpsed sense of a thorn in his side; the moving on, the lack of family, the days and nights of work, the desire to prove that what he did mattered, the lonely repetition of his love and memory of Eskdale. There is no easy epitaph for a driven man.

APPENDIX: SELECT LIST OF THOMAS TELFORD'S CIVIL ENGINEERING WORKS

Bridges and aqueducts

DATE	PROJECT	RESIDENT ENGINEER
1775–1778	Langholm	
1790–1792	Montford	M Davidson
1793	Longbridge, Salop	M Davidson
1793–1794	Chirk, Denbigh	M Davidson
1794–1805*	Pontcysyllte Aqueduct	M Davidson
1795–1796	Longdon on Tern Aqueduct*	M Davidson
1795	Bridgnorth, modifications	M Davidson
1795–1796	Buildwas*	M Davidson
1795	Bolas, Salop	M Davidson
1796–1801	Chirk Aqueduct*	M Davidson
1795–1799	Bewdley	M Davidson
1805–1808	Tongland, Kirkcudbright (with A Nasmyth)	A Blane
1805–1807	Wick (demolished)	
1805–1811	Ballochindrain, Argyll	
1805–1809	Dunkeld	
1805–1815	Calder, Invernesshire (demolished)	
1805–1815	Easterfearn, Ross	
1806–1809	Conon, Ross and Cromarty	
1807–1811	Ballater, Aberdeenshire (replaced 1834/5#)	
1808–1811	Borlam, Inverness	
1808–1811	Kaun-a-Crock, Invernessshire (demolished)	
1808–1811	Torgoyle, Invernessshire	
1808–1819	Sark, Dumfries	
1809–1817	Cannich, Invernessshire	
1809–1817	Diak, Invernessshire	
1809–1817	Varrar, Invernessshire	
1809	Grantown-on-Spey, repaired	
1810–1811	Alford, Aberdeenshire	
1810–1811	Aultmore, Elgin	
1810–1811	Dualg, Inverness	
1810–1811	Nethy	

(* = iron, # = timber)

1810–1815	Allness–Ross	
1810–1815	Balnagown, Ross	
1810–1815	Roy, Invernessshire	
1810–1818	Kirklaggan, Invernessshire#	
1810–1812	Bonar, Sutherland*	T Rhodes
1810–1812	Helmsdale, Sutherland	
1811–1814	Lovat, Beauly, Inverness	
1811–1814	Potarch, Aberdeenshire	J Gibb
1812–1815	Craigellachie, Banff*	
1812–1817	Greystones, Caithness	
1812–1822	Ledwych, Salop	
1814	Tenbury, widening*	
1814–1817	Croe, Ross	
1814–1817	Shiel, Ross	
1815–1816	Ferness, Nairn	
1814–1815	Ken Bridge, New Galloway, rebuild	Rennie
1814–1820	Llanyblodwell, Salop	T Stanton
1815	Tyhill, Caernarvonshire	
1815	Waterloo Bridge, Betws-y-coed, Denbighshire*	J Sinclair
1815–1816	Pont Pen-y-Benlog, Caernafon	
1815–1817	Contin, Ross	
1815–1818	Aultnaharrar, Sutherland	
1819	Bannockburn	
1818–1826	Menai Suspension Bridge	W Provis and T Rhodes
1820–1822	Esk, Cumberland*	
1820	Glasgow Old Bridge, widening	
1821	Birkwood Burn, Lanark	
1821	Cander, Lanark	
1821	Fiddler's Burn, Lanark	
1821–1822	Cartland Crags, Lanark	H Welch
1821–1825	Elvanfoot, Lanarkshire (demolished)	
1821–1826	Conwy Suspension Bridge	W Provis
1822–1823	Stokesay, Salop	T Stanton
1823	Bath, widening	
1823–1826	Mythe Bridge, Tewkesbury*	W Mackenzie
1823–1826	Ver Bridge, Colne, Hertfordshire	
1825	Billingsley, Salop	T Stanton
1825	Hamilton, Lanark	

1825–1827	Soutra, Old Toll and Tunnel, Pathhead Road, Midlothian	
1825–1828	Over Bridge, Gloucester	T Fletcher
1825–1828	Kingston-upon-Thames	E Lapidge
1826–1829	Birmingham Canal • Rylands Aqueduct • Brass House Lane* • Galton* • Icknield Street • Lee Bridge • Oldbury • Pope's Lane* • Rabones Bridge* • Spon Lane • Spon Lane Aqueduct • Watery Lane	W Mackenzie, W Dalziell and F Jenkins
1826–1828	Stretford, Salop	T Stanton
1827–1830	Don, Aberdeenshire	J Smith
1827	Holt Fleet, Worcestershire	
1827–1831	Pathhead, Midlothian, Dean Burn, Fala Water, Cranston and Coty Burn	
1829–1831	Dean Bridge, Edinburgh	C Atherton
1831	Morpeth, Northumberland	J Cargill
1832	Onybury, Salop	T Stanton
1832–1833	Welsh Bridge, Shrewsbury, repairs	
1830–1833	Montrose Suspension Bridge, repairs	
1833	Clunsford, Salop	T Stanton
1833–1836	Broomielaw, Glasgow	C Atherton

Canals

DATE	PROJECT	RESIDENT ENGINEER
1793–1805	Ellesmere Canal (with W Jessop)	J Duncombe
1795–1800	Shrewsbury Canal	T Dunn
1803–1822	Caledonian Canal (with W Jessop)	T Rhodes, J Telford, A Easton and M Davidson
1804–1811	Glasgow–Paisley and Ardrossan Canal	D Henry
1807–1826	Weston Canal	S Fowls
1809–1833	Gotha Canal, Sweden	
1816–1823	Edinburgh–Glasgow Union Canal (consulted)	H Baird
1817–1823	Crinan Canal, reconstruction	J Gibb, W Thomson
1818–1825	Gloucester–Berkeley Ship Canal	T Fletcher, W Clegram
1822–1827	Harecastle Tunnel, Trent and Mersey Canal	J Potter
1824–1825	Bude Canal	A Easton
1824–1834	Birmingham Canal	W Mackenzie
1825–1834	Birmingham and Liverpool Junction Canal, aqueducts at Nantwich,* Weaver, Drayton and Albaston	A Easton

1825–1826	Macclesfield Canal	W Crosley
1826	Ulster Canal	T Casebourne
1828	Welland Canal, Canada	
1828	Ellesmere Canal, Midlothian branch	

Dock and harbour works

DATE	PROJECT	RESIDENT ENGINEER
1801–1813	Wick Harbour	
1801–1814	Tobermory, Island of Mull	
1801–1832	Aberdeen	J Gibb
1802	Kirkcudbright	J Gibb
1803–1821	Keils Ferry Pier	
1803–1821	Lochie Ferry Pier	
1803–1821	Corran of Ardgour Ferry Piers	
1803–1821	Small Isle Harbour, Jura	
1803–1821	Dornie Ferry	
1805–1810	Ardrossan Harbour	
1806–1833	Glasgow	
1808–1809	Burghhead, Elgin	
1808–1813	Kirkwall, Orkneys	
1813–1817	Fortrose, Moray Firth	
1813–1815	Avoch, Moray Firth	
1814–1821	Banff, Banffshire	J Gibb
1814–1832	Dundee Harbour, improvements including graving dock	D Logan and P Logan
1815–1817	Dingwall, Black Isle	
1815–1819	Gourdon, Kincardineshire	
1816	St Catherine's Ferry Pier	
1816–1825	Nairn, Moray Firth	
1817	Ballintraid Landing Pier	
1817	Folkestone Harbour	
1817–1819	Cullen, Banff	
1817–1820	Invergordon and Inverbeckie Ferry Pier	
1819	Channery Pier	
1819–1822	Tay Ferries, Dundee	
1821–1834	Holyhead Harbour	J Brown, J Provis
1821–1834	Howth Harbour	J Provis
1824–1828	St Katharine Docks, London	P Logan, J Hall, Abernethy, T Rhodes

1828–1834	Leith	
1829	Belfast	
1830–1832	Herne Bay Pier	
1830–1831	Greenock Harbour	
1831–1834	Dundee Dock	James Leslie
1833	Seaham	J Buddle
1834	Dover Harbour	

Drainage

DATE	PROJECT	RESIDENT ENGINEER
1808–1812	Loch Spyrie Canal	W Hughes
1818–1825	Eau Brink Navigation	T Townsend
1827–1830	Nene Outfall	W Swanborough
1830–1834	North Level of the Fens Drainage	W Swanborough, T Pear

Highland roads

DATE	PROJECT	RESIDENT ENGINEER
1801–1802	Surveys of the Highlands and coast	
1806–1811	Dunrobin, Sutherlandshire	
1809–1816	Dunbeath Road, Caithness	
1809–1813	Findhorn Road, Elgin	
1809–1812	Speyside, Elgin	
1813–1816	The Fleet Mound, Sutherlandshire	
1814–1817	Strathspey Road, Invernesshire	
1816–1817	Alford Road, Aberdeenshire	
1814–1825	Glasgow–Carlisle Road, 93 miles	W Provis, J Pollock

Holyhead Road

DATE	PROJECT	RESIDENT ENGINEER
1810	North Wales Roads, first survey	W Provis
1810	Chester to Corwen, survey	
1815–1829	Shrewsbury to Bangor	
1816–1829	Bangor to Chester	J Provis
1819–1829	Anglesea	W Dargan, J Provis
1820–1828	London to Shrewsbury	J Easton, J Provis, J Macneill
1821–1822	Stanley Sands Embankment	W Dargan

Other roads

DATE	PROJECT	RESIDENT ENGINEER
1802–1810	Carlisle–Portpatrick road surveys	
1819	Stirling Road	
1820–1823	Lanarkshire	J Pollock
1820–1830	Great North Road Surveys	
1820–1821	Loose Hill and Linton Hill, Kent	
1822–1830	London–Liverpool road surveys	T Casebourne
1823–1825	South Wales road surveys	H Welch
1824–1828	Carlisle–Edinburgh road surveys	H Welch
1825	Warsaw–Briesc	

Railways

DATE	PROJECT	RESIDENT ENGINEER
1821–1826	Stratford and Moreton Railway	J Rastrick
1828–1829	Clarence Railway	
1829–1833	Newcastle and Carlisle Railway	F Giles

River works

DATE	PROJECT	RESIDENT ENGINEER
1800	River Severn, consultancy	
1806–1832	River Clyde	
1809–1829	River Weaver, consultancy, including Weston Cut	S Fowls
1817–1829	River Dee Company, consultancy	T Wedge

Water supply

DATE	PROJECT	RESIDENT ENGINEER
1799–1802	Liverpool Waterworks	H Bell
1806	Glasgow	
1810–1822	Edinburgh	J Jardine
1827–1834	London Water Supply	J Mills etc

Data © ICE Publishing and reproduced by permission of Professor Roland A. Paxton.

NOTES

ABBREVIATIONS: ARCHIVE SOURCES
BL – British Library
ICE – Institution of Civil Engineers
IGM – Ironbridge Gorge Museum
NRS – National Records of Scotland
SCRO – Shropshire County Record Office

ABBREVIATIONS: CORRESPONDENTS
AL – Andrew Little
CvP – Count von Platen
JD – James Davidson
JL – James Little
JR – John Rickman
MD – Matthew Davidson
RS – Robert Southey
TD – Thomas Davidson
TR – Thomas Rhodes
TT – Thomas Telford
WJ – William Jessop
WL – William Little

INTRODUCTION

1   Joseph Pring, *Particulars of the Grand Suspension Bridge Erected over the Straits of Menai at Bangor*, Bangor, 1826, pp. 29–30.
2   Thomas Howell, *Shrewsbury to Holyhead Being a Historical and Descriptive Sketch of the Most Interesting Scenery on this Romantic and Beautiful Road*, Shrewsbury, T. Howell, 1827, p. 28.
3   Pring, *Particulars*, p. 2.
4   Ibid., p. 23.
5   William Provis to TT, 31 Jan 1826, ICE T/HO/113.

PREFACE

1   *Shropshire Journal*, 1 Nov 1968, p. 8.
2   *Shropshire Star*, 24 Oct 1968, p. 1.
3   Thomas Telford, *Life of Thomas Telford Civil Engineer Written by Himself*, London: Hansard and Sons, 1838, p. 1.

CHAPTER 1: THE EARLY SHEPHERD

1   Thomas Telford, 'Autobiography, Second version with annotations by John Rickman', IGM T/AUT.
2   Accessed via www.scotlandspeople.gov.uk.
3   Samuel Smiles, *The Life of Thomas Telford*, Teddington: The Echo Library, 2006, p. 109.
4   The story of the letters is told by Rickman, in Telford, *Life*, Appendix 1.
5   TT to AL, from Portsmouth Dock, 23 Jul 1784. All of Telford's letters to Andrew Little are found in ICE Acc 0006, Box 80, item 1.
6   TT to AL, from Shrewsbury, 16 Sep 1794.
7   TT to WL, 23 Dec 1832. All of Telford's lettes to other members of the Little family are in ICE Acc 0006, Box 80, item 2.
8   Lan D. Whyte, *Scotland Before the Industrial Revolution*, London, New York: Longman, 1995, p. 244.
9   R. D. Anderson, *Scottish Education Since the Reformation*, Dundee: Economic and Social History Society of Scotland, 1997, p. 5.
10   *St James's Chronicle or the British Evening Post*, 27 Feb 1776.
11   *The New Statistical Account of Scotland*, Edinburgh: Feb 1826, pp. ix, 432.
12   General Sir Charles W. Pasley, 'Narrative from Recollection', in David Irving, *The History of Scottish Poetry*, Edinburgh: Edmonston and Douglas, 1861, p. xxvi.
13   Suzanne Schwarz, *Slave Captain: The Career of James Irving in the Liverpool Slave Trade*, Liverpool: Liverpool University Press, 2008, p. 11.
14   TT to AL, from London, 12 Feb 1782.
15   TT to AL, from Portsmouth, 8 Jul 1786.
16   R. H. Vetch and John Sweetman, 'Charles Pasley', in *Oxford Dictionary of National Biography*, Oxford: Oxford University Press, 2004, pp. xlii, 969.
17   Pasley, 'Narrative', in Irving, *The History of Scottish Poetry*, p. xxvi.

18 Telford, *Life*, p. 14.

19 Robert Eric Frykenberg, 'John Malcolm', in *Oxford Dictionary of National Biography*, pp. xxxvi, 292.

20 William Kaye, *The Life and Correspondence of Major-General Sir John Malcolm G.C.B*, London: Smith, Elder and Co, 1856, p. 4.

21 Richard Graves, 'Columella; or, the distressed Anchoret', 2 vols, London, 1779, pp. ii, 44–5, in Paul Langford, *A Polite and Commercial People: England 1727–1783*, Oxford: Oxford University Press, 1998, p. 404.

22 E. Thompson, 'The Demi–Rep', London, 1766, p. 35, in Langford, *A Polite and Commercial People*, p. 407.

23 David Hume, *An Enquiry Concerning Human Understanding*, London, 1748, p. 256.

24 Telford, *Life*, p. 2.

25 TT to AL, from Salop, 8 Oct 1789.

26 Telford, *Life*, p. 2.

27 Ibid., p. 2.

28 Smiles, *Life*, p. 72.

29 Telford, *Life*, p. 15.

CHAPTER 2: ESKDALE TAM

1 Smiles, *Life*, p. 115.

2 TT to AL, from Portsmouth, 8 Jul 1786.

3 TT to AL, from Salop, 30 Nov 1799.

4 TT to AL, from Portsmouth, 8 Jul 1786.

5 TT to AL, from Salop, 11 Mar 1792.

6 TT to AL, from Salop, 7 Sep 1799.

7 TT to AL, from Salop, 13 Jul 1799.

8 TT to AL, from Salop, 11 Mar 1792.

9 TT to AL, from Portsmouth, 8 Jul 1786.

10 *The Weekly Magazine or Edinburgh Amusement*, 14 Apr 1779, pp. xliv, 68.

11 Ibid., 5 May 1779, pp. xliv, 151.

12 Ibid., 21 Jul 1779, pp. xlv, 96.

13 TT to AL, from Shrewsbury Castle, 21 Feb 1788.

14 Robert Burns, *Poems Chiefly in the Scottish Dialect*, Belfast: William Magee, 1793, pp. ii, 46–8. Earlier editions published in Scotland did not include Telford's work.

15 Thomas Telford, *Eskdale: A Descriptive Poem*, Shrewsbury: J. & W. Eddowes, 1795.

16 Robert Fitzhugh, 'The Paradox of Burns's Character', *Studies in Philology*, 1 Jan 1935, pp. xxxii–1, 110–19.

17 TT to AL, from Salop, 10 Oct 1792.

18 Robert Burns and James Currie, *The Works of Robert Burns, With an Account of His Life*, London: T. Cadell, Jun. and W. Davies, 1801, pp. i, 351.

19 TT to MD, from Salop, 9 Oct 1796.

20 TT to James Currie, 26 Jun 1800; available online: http://jamescurrie. gla.ac.uk/details.php?id=19.

21 George Turnbull, *Autobiography*, London: Printed by Cooke & Co., 1893, pp. 19–20.

CHAPTER 3: THE GREAT MART

1 Telford, *Life*, p. 15.

2 Available online: http://maps.nls.uk/towns/detail.cfm?id=413.

3 Telford, *Life*, p. 5.

4 Jerry White, *London in the Eighteenth Century*, London: The Bodley Head, 2012, p. 49.

5 Telford, *Life*, p. 19.

6 Daniel Defoe, *A Tour Through the Whole Island of Great Britain*, 3 vols, London: Strahan, Mears, Francklin, Chapman, Stagg, 1724–7, vol. 1, pp. i, 325.

7 White, *London*, p. 98. Kevan thrived in London eventually.

8 Peter Ackroyd, *London: The Biography*, London: Chatto and Windus, 2000, p. 187. Ackroyd suggests that today the words would be 'bloody' and 'fuckin'.

9 Smiles, *Life*, p. 74.

10 Telford, *Life*, p. 19.

11 TT to AL, from London, 12 Feb 1782.

12 White, *London*, p. 76.

13 Telford, *Life*, p. 19.

14 TT to AL, from London, Jul 1782.

15 Sir Archibald Alison, *Some Account of My Life and Writings*, Edinburgh and London: William Blackwood and Sons, 1883, p. 8.

16 TT to AL, from Portsmouth, 23 Jul 1784.

17 Ibid.

18 TT to AL, from London, 1 Feb 1786.

19 TT to AL, from London, 8 Jul 1782.

20 Telford, *Life*, p. 20.

CHAPTER 4: YOUNG PULTENEY

1 Maurice de Soissons, *Telford: The Making of Shropshire's New Town*, Swan Hill Press, 1991, foreword.

2 *Letters of Eminent Persons Addressed to David Hume*, Edinburgh and London: William Blackwood and Sons, 1848, p. 168.

3 William Pulteney, *Thoughts on the Present State of Affairs with America and The Means of Conciliation*, London: Printed for J. Dodsley and T. Cadell, 1778, p. 63.

4 TT to AL, 8 Dec 1798.

5 TT to AL, 3 Sep 1788.

6 TT to AL, from Salop, 8 Oct 1789.

7 TT to AL, 3 Sep 1788.

8 Douglas J. Hamilton, *Scotland, The Caribbean and the Atlantic World, 1750–1820*, Manchester: Manchester University Press, 2005, p. 76.

9 TT to AL, from Shrewsbury Castle, 21 Feb 1788.

10 TT to AL, from Salop, 8 Oct 1789.

11 Catherine Plymley's diary, SCRO 1066/20.

12 TT to AL, from Castle Salop, 27 Jan 1787.

13 TT to AL, from Salop, 8 Oct 1789.

14 TT to AL, from Salop, 16 Jul 1788.

15 Telford, *Life*, p. 26.

16 Barrie Trinder, 'Was there a Shropshire Enlightenment?', *West Midlands History*, i/1, 2013, pp. 26–9.

17 TT to AL, from Shrewsbury Castle, 21 Feb 1788.

18 Rick Turner, 'Thomas Telford the Archaeologist', *The Antiquaries Journal*, lxxxviii, 2008, pp. 365–75.

19 Alison, *Some Account*, p. 16.

20 TT to AL, 3 Sep 1788.

21 Jenny Uglow, 'Sexing the Plants', *Guardian*, 21 Sep 2002; available online: http://www.theguardian.com/books/2002/sep/21/features reviews.guardianreview30.

22 TT to AL, from Salop, 27 Jan 1787.

23 TT to AL, from Salop, 20 Oct 1792.

24 TT to AL, from Salop, 28 Jul 1791.

25 Thomas Telford, Rickman (ed.), *Life of Thomas Telford*, London: Hansard & sons, 1838, p. 282.

CHAPTER 5: SOMETHING LIKE BONAPARTE

1 Jenny Uglow, *In These Times*, London: Faber & Faber, 2014, p. 28.

2   Allison, *Some Account*, pp. 13–14.

3   Single sheet verse, *The Freeman*, Shrewsbury, 1796, BL Cu21.g.35/45.

4   TT to AL, from Shrewsbury, 29 Sep 1793.

5   TT to AL, from Derby, 6 Nov 1795.

6   TT to AL, from Salop, 6 Mar 1798.

7   TT to AL, from Salop, 13 Jul 1799.

8   TT to AL, from Salop, 30 Nov 1799.

9   TT to Charles Pasley, letter, 6 Sep 1815, in A. D. Harvey, *Collision of Empires: Britain in Three World Wars, 1793–1945*, London: Phoenix, 1994, p. 36.

10 TT to AL, from Derby, 6 Nov 1795.

11 TT to AL, from Salop, 8 Oct 1789.

12 TT to AL, from Derby, 6 Nov 1795.

13 TT to AL, from Salop, 7 Sep 1799.

14 Catherine Plymley, *Diary*, in J. B. Lawson, 'Thomas Telford in Shrewsbury', in Alastair Penfold (ed.), *Thomas Telford Engineer*, London: Telford, 1980, p. 2.

15 Douglas Grounds, *Son and Servant of Shropshire: The Life of Archdeacon Joseph Plymley*, Corbett: Longaston Press, 2009, p. 121.

16 TT to AL, 8 Dec 1798.

17 MD to TD, 24 Aug 1815, ICE Acc 0006, Box 81, item 1.

18 Quoted at http://www.shrewsburylocalhistory.org.uk/simpson.htm.

19 TT to AL, from Salop, 20 Aug 1797.

20 Robert Southey, *Journal of a Tour in Scotland in 1819*, London: John Murray, 1929, p. 110.

21 Smiles, *Life*, pp. 95–6.

22 Telford, *Life*, p. 28.

23 TT to MD, from Shrewsbury, 27 Oct 1796, ICE Acc 0006, Box 81, item 1.

24 Andrew Pattison, 'William Hazledine, Shropshire Ironmaster and Millwright', M.Phil. Thesis, University of Birmingham, 2011, p. 100; available online: http://etheses.bham.ac.uk/3358/1/Pattison12MPhil.pdf.

25 Uglow, *In These Times*, p. 206.

26 Barrie Trinder (ed.), *The Most Extraordinary District in the World*, Chichester: Phillimore, 1988, p. 40.

27 Jenny Uglow, 'Sexing the Plants'.

28 J. B. Lawson, 'Thomas Telford in Shrewsbury', in Penfold (ed.), *Thomas Telford Engineer*, p. 10.

29 John Newman and Nikolaus Pevsner, *The Buildings of England: Shropshire*, New Haven and London: Yale University Press, 2006, p. 162.

30 TT to AL, from Salop, 10 Mar 1793.

31 TT to AL, from Carlisle, 28 Nov 1794.

32 TT to AL, from Shrewsbury, 18 Mar 1795.

33 Plymley, *General View*, pp. 299–300.

34 Telford, *Life*, p. 42.

35 Barrie Trinder, *The Darbys of Coalbrookdale*, Chichester: Phillimore, 1974, p. 52.

CHAPTER 6: THE STREAM IN THE SKY

1 George Borrow, *Wild Wales*, London: John Murray, 1862, pp. i, 121. *Pendro* is Welsh for 'giddiness'.

2 Charles Dupin, 1816, trans. Barrie Trinder, in Wrexham County Borough and Royal Commission on the Ancient and Historical Monuments of Wales, *Pontcysyllte Aqueduct and Canal Nomination Document*, Wrexham County Borough Council: 2008, p. 35.

3 Robert Southey, 'Plaque at Caledonian Canal, Clachnaharry', 1822; available online at: http://www.ambaile.org.uk/en/item/item_photograph.jsp?item_id=43957.

4 William Wordsworth, 'Steamboats, Viaducts, and Railways', lines 10–14.

5 Charles Hadfield, *The Canals of the West Midlands*, Newton Abbott: David & Charles, 1966, p. 168.

6 Wrexham County Borough, *Nomination Document*, p. 78.

7 Calculated using http://www.measuringworth.com/ukcompare/relativevalue.php.

8 C. Hadfield and A. W. Skempton, *William Jessop, Engineer*, Newton Abbot: David and Charles, 1979, preface.

9 TT to AL, 3 Nov 1793.

10 Telford, *Life*, p. 34.

11 TT to AL, from Shrewsbury, 29 Sep 1793.

12 Charles Hadfield, *Thomas Telford's Temptation*, Cleobury Mortimer: M&M Baldwin, 1993, p. 16.

13 TT to AL, 3 Nov 1793.

14 Calculated using http://www.measuringworth.com/ukcompare/relativevalue.php.

15   TT to AL, 3 Nov 1793.

16   Hadfield, *Temptation*, p. 23.

17   Telford, *Life*, p. 34.

18   Hadfield, *Temptation*, p. 23.

19   *Edinburgh Encylopaedia*, Edinburgh: William Blackwood, 1817, pp. ii, 735–7. The unnamed author says he had known Jessop for more than twenty years. Telford would have been an obvious choice to write the piece.

20   Hadfield, *Canals*, p. 170.

21   L. T. C. Rolt, *Thomas Telford*, London: Longmans, Green & Co.,1957, p. 45.

22   Roland Paxton, 'Thomas Telford's Cast-Iron Bridges', *Proceeedings of ICE, Civil Engineering*, 160, May 2007, pp. 13–19.

23   This pioneering structure was demolished in 1970 to make way for Derby's inner ring road.

24   Hadfield, *Temptation*, p. 171.

25   WJ to TT, from London, 26 Jul 1795, IGM 1992.14918 DMD 9.

26   Telford, *Life*, pp. 41–2.

27   TT to MD, from London, 11 Feb 1796, 1992, 14919 IGM DMD 9.

28   Telford, *Life*, p. 41.

29   Hadfield, *Temptation*, p. 33.

30   TT to MD, from London, 11 Feb 1796, 1992.14919 IGM DMD 9.

31   TT to MD, from Salop, 14 Jan 1796, 1992.14919 IGM DMD 9.

32   TT to MD, from Salop, 9 Oct 1796, 1992.14919 IGM DMD 9.

33   TT to MD, from Shrewsbury, 7 Oct 1796, 1992.14919 IGM DMD 9.

34   TT to AL, from Salop, 20 Aug 1797.

35   TT to AL, from Salop, 6 Mar 1798.

36   Joseph Plymley, *A General View of the Agriculture of Shropshire*, London: Richard Phillips, 1803, p. 264.

37   Plymley, *General View*, p. 313.

38   TT to AL, 8 Dec 1798.

39   TT to MD, from London, 9 Mar 1796, 1992.14919 IGM DMD 9.

40   TT to AL, from Salop, 6 Mar 1798.

41   Thomas Telford, *For The Board of Agriculture on Mills*, unpublished manuscript, ICE 6 21.21 + 621.548.

42   WJ to TT, 26 Jul 1795.

43   Rowland Hunt, *Oration Delivered at Pontcysylte Aqueduct*, J. Eddows, Shrewsbury 1806, p. 26.

44   Hadfield, *Temptation*, p. 51.

45   Telford, *Life*, p. 45.

46   *The Gentleman's Magazine*, London, 1805, pp. lxxv, 2, 1228.

CHAPTER 7: THE APPEARANCE OF PLENTY

1   George Gilchrist, *Memorials of Westerkirk Parish*, Anan: G. Gilchrist, 1970.
2   TT to AL, from Salop, 7 Sep 1799.
3   TT to AL, from Liverpool, 9 Sep 1800.
4   Copy of letter from TT to AL, from Shrewsbury, 30 Nov 1801.
5   TT to AL, from London, 16 Jun 1802.
6   TT to AL, 18 Feb 1803.
7   TT to AL, from Castle Salop, 27 Jan 1787.
8   TT to AL, from Salop, 8 Oct 1789.
9   TT to AL, from Salop, 13 Jul 1799.
10 TT to AL, from Salop, 4 Dec 1797.
11 TT to AL, from Bridgnorth, 18 Oct 1799.
12 TT to AL, from Salop, 30 Nov 1799.
13 TT to AL, from London, 14 Apr 1802.
14 TT to AL, from London, 16 Jun 1802.
15 TT to AL, from Liverpool, 9 Sep 1800.
16 Rolt, *Telford*, p. 187.
17 *Oxford Dictionary of National Biography*, pp. ix, 864.
18 TT to AL on Campbell.
19 Alexander Gibb, *The Story of Telford*, London: Alexander Maclehose & Co., 1935, p. 46.
20 TT to AL, 13 May 1800.
21 Telford spelled his name with only one 's' as Douglas.
22 *Supplemental Report from the Select Committee on further Improvement of the Port of London*, 3 Jun 1801.
23 TT to MD, from London, 11 Jan 1801.
24 Draft letter, ICE T/LO.4.
25 Archibald Alison to TT, 6 May 1801, ICE T/LO. 35.
26 *The Monthly Magazine or British Register*, ii/74 ,1801, pp. 477–80.
27 TT to MD, from London, 11 Jan 1801, IGM 1992.14919 DMD 9.
28 TT to AL, from London, 16 Jun 1802.
29 TT to WL, from Ellesmere Canal Office, 24 Jan 1815, ICE box 80, ii.
30 TT to AL, from Castle Salop, 27 Jan 1787.
31 Telford, *Life*, pp. 663–94.

CHAPTER 8: MY SCOTCH SURVEYS

1   See http://www.futuremuseum.co.uk/collections/life–work/key–industries/mining–quarrying/gold–other–minerals/westerkirk–library–minute–book.aspx.

2 TT to AL, from Shrewsbury, 30 Nov 1801.

3 Christopher Wordsworth, *Memoirs of William Wordsworth*, London: Edward Moxon, 1851, pp. i, 216.

4 Samuel Johnson and James Boswell, *Journey to the Western Isles of Scotland* and *The Journal of a Tour of the Hebrides*, 1785; reprinted London: Penguin, 1993, p. 49.

5 Rolt, *Telford*, p. 69.

6 A. R. B. Haldane, *New Ways Through the Glens*, London: Thomas Nelson & Sons, 1962, p. 164.

7 Joseph Mitchell, *Reminiscences of My Life in The Highlands*, David & Charles Reprints, 1971, pp. i, 94.

8 Ibid., p. 189.

9 Jean Dunlop, *The British Fisheries Society 1786–1893*, John Donald Publishers, Edinburgh, 1978, p. 59.

10 John Knox, *A Tour Through the Highlands of Scotland and the Hebride Isles*, Edinburgh: The Mercal Press, 1975, preface.

11 Daniel Maudlin and Robert Mylne, 'Thomas Telford and the Architecture of Improvement: The Planned Villages of the British Fisheries Society, 1786–1817', *Urban History*, xxxiv/3, Dec 2007, pp. 453–80.

12 Telford, *A Survey and Report of the Coasts and Central Highlands of Scotland*, Lords Commissioners of His Majesty's Treasury: 5 Apr 1803, p. 4.

13 TT to AL, from Shrewsbury, 30 Nov 1801.

14 TT To James Currie, from Peterhead, 12 Oct 1801, ICE T/SC7.

15 Henry Cockburn, *Memorials of His Time*, Edinburgh, Adam and Charles Black, 1856, p. 56.

16 TT to AL, from Shrewsbury, 30 Nov 1801.

17 TT to Patrick Copland, from Shrewsbury, 3 Dec 1801, ICE.

18 TT to Patrick Copland, from Bath, 31 Dec 1801, ICE.

19 TT to AL, from London, 14 Apr 1802.

20 TT to Nicholas Vansittart, from Peterhead, 17 Oct 1801.

21 Nicholas Vansittart to TT, 1 Jul 1802.

22 Rolt, *Telford*, p. 67.

23 TT to AL, from London, 16 Jun 1802, and 14 Apr 1802.

24 TT to AL, 18 Feb 1803.

25 Telford, *Survey*.

26 *Commission for Highland Roads and Bridges*, First Report, Jun 1803, p. 6.

27 Haldane, *Glens*, p. 36.
28 Ibid., p. 112.
29 Telford, *Life*, p. 189.
30 Haldane, *Glens*, p. 103.
31 Orlando Williams, *Life and Letters of John Rickman*, Constable and Co.: 1912, p. 15.
32 Haldane, *Glens*, p. 39.
33 Ibid.
34 JR to RS, 18 Jan 1820, in Williams, *Life and Letters*.
35 Haldane, *Glens*, p. 102.

CHAPTER 9: TELFORD'S TATAR
1 Mitchell, *Reminiscences*, p. 73.
2 Ibid., p. 21.
3 TT to JR, from Edinburgh, 18 Oct 1803.
4 Haldane, *Glens*, p. 108.
5 *Commission for Highland Roads and Bridges*, First Report, Jun 1803, p. 37.
6 Haldane, *Glens*, p. 70.
7 JR to TT, 6 Mar 1806, in Haldane, *Glens*, p. 52.
8 Haldane, *Glens*, p. 53.
9 Telford, *Survey*, p. 9.
10 *Commission for Highland Roads and Bridges*, Third Report, 1805, p. 4.
11 Haldane, *Glens*, p. 95.
12 Ibid., p. 58.
13 Ibid., p. 76.
14 MD to Tom Davidson, 8 Mar 1809, ICE Acc 006, Box 81, item 1.
15 Haldane, *Glens*, p. 76.
16 Telford, *Life*, pp. 174–5.
17 Ibid., p. 177.
18 MD to TT, from Clachnaharry, 21 Aug 1809, NRS MT 1/3, 1809–10.
19 *Commission for Highland Roads and Bridges*, First Report, 1804, 10, Second Report, 1805, 11, Ninth Report, 1821, in Haldane, *Glens*, p. 119.
20 *Commission for Highland Roads and Bridges*, Sixth Report, 1813.
21 Maudlin and Mylne, 'Planned Villages', pp. 453–80.

CHAPTER 10: FROM SEA TO SEA
1 Hadfield, *Temptation*, p. 162.
2 Telford, *Second Survey*.

3 Haldane, *Glens*, p. 36.

4 Mitchell, *Reminiscences*, p. 26.

5 Haldane, *Glens*, p. 91.

6 TT to JR, from Salop, 31 Oct 1803, NRS MT 1/1, 1803–5.

7 Alexander Macdonell to TT, from Glengarry House, 7 Nov 1803, NRS MT 1/1, 1803–5.

8 Mitchell, *Reminiscences*, p. 50.

9 TT to Mr Hope, from Clachnacarry, 2 Nov 1820, NRS MT 1/6, 1819–25.

10 Hadfield, *Temptation*, pp. 134–5.

11 WJ to JR, from Newark 29 Sep 1803, NRS MT 1/1, 1803–5.

12 TT to JR, from Inverness, 27 Sep 1803, NRS MT 1/1, 1803–5.

13 TT to JR, from Shrewsbury 29 Nov 1803, NRS MT 1/1, 1803–5.

14 TT to JR, from the Salopian Coffee House, 24 Mar 1804, NRS MT 1/1, 1803–5.

15 TT to JR, from London, 6 Jun 1804, NRS MT 1/1, 1803–5.

16 Penfold (ed.), *Thomas Telford Engineer*, p. 139.

17 TT to WJ, 8 Jun 1804, NRS MT 1/1, 1803–5.

18 Haldane, *Glens*, p. 79.

19 Ibid., p. 80.

20 Mitchell, *Reminiscences*, p. 79.

21 Haldane, *Glens*, p. 83.

22 MD to TT, from Clachnaharry, 30 Apr 1810, NRS MT 1/4, 1811.

23 Mitchell, *Reminiscences*, p. 27.

24 Haldane, *Glens*, p. 89.

25 Ibid., p. 80.

26 Ibid., p. 83.

27 TT to JR, from Ellesmere, 17 Jul 1807, NRS 1/2, 1806–8.

28 TT to JR, from Inverness, 8 Oct 1811, NRS MT 1/4, 1811.

29 Telford, *Life*, p. 56.

30 Mitchell, *Reminiscences*, p. 81.

31 TT to JR, from Clachnaharry, 15 Sep 1805, NRS MT 1/1, 1803–5.

32 Telford, *Life*, p. 62.

33 Gibb, *Story*, p. 99.

34 TT to JR, from Shrewsbury, 29 Jun 1806, NRS 1/2 ,1806–8.

35 TT to JR, from Clachnaharry, 3 Jan 1808, NRS 1/2, 1806–8.

36 MD to JR, from Clachnaharry, 19 Jun 1813, NRS MT 1/5, 1812–19.

37 MD to JR, from Clachnaharry, 17 Aug 1813, NRS MT 1/5, 1812–19.

38 TT to JR, from Inverness, 26 Apr 1807, NRS 1/2, 1806–8.

39 TT to JR, from Inverness, 14 Nov 1808, NRS 1/2, 1806–8.
40 TT to JR, from Inverness, 10 Jun 1809.
41 TT to JR, from Inverness, 20 Oct 1809, NRS MT 1/3, 1809–10.
42 MD to TT, from Clachnaharry, 30 Apr 1809, NRS MT 1/3, 1809–10.
43 Haldane, *Glens*, p. 84.
44 TT to JR, from Ellesmere, 13 Jul 1810, NRS MT 1/3, 1809–10.
45 TT to JR, from Edinburgh, 21 Jul 1808, NRS MT 1/3, 1809–10.

CHAPTER 11: THE ENJOYMENT OF SPLENDID ORDERS
  1 CvP to TT, Apr 28 1808, ICE T/GC1.
  2 TT to CvP, from Inverness, 2 Jun 1803, ICE T/GC3.
  3 CvP to TT, 21 Jun 1808, ICE T/GC4.
  4 TT to CvP, from Inverness, 2 Jun 1803, ICE T/GC3.
  5 Samuel Bagge to TT, 22 Oct 1809, ICE T/GC.
  6 Rolt, *Telford*, p. 95.
  7 TT to JR, from Edinburgh, 27 July 1808, in Gibb, *Story*, pp. 120–21.
  8 Gibb, *Story*, p. 129.
  9 TT to JR, from Harwich, 10 Oct 1808, NRS MT 1/2, 1806–8.
 10 Calculated using http://www.measuringworth.com.
 11 CvP to TT, from Gothenburg, 3 Oct 1808, ICE T/GC9.
 12 Rolt, *Telford*, p. 99.
 13 See, for instance, http://www.gotakanal.se/sv/artiklar/Historia–ib/Thomas–Telford.
 14 CvP to TT, from Trugard, 23 Oct 1808, ICE T/GC10.
 15 CvP to TT, from Trugard, 4 Mar 1809, ICE T/GC16.
 16 See, for instance, https://www.regjeringen.no/en/the–government/previous–governments/ministries–and–offices/offices/governor–1814–1873/baltzar–von–platen/id479743.
 17 Earl of Kellie to TT, from Cambo House, Nov 6 1809, ICE T/GC24.
 18 TT to Sir John Sinclair, 12 Nov 1809, ICE T/GC26.
 19 TT to CvP, from Stonehaven, 31 Oct 1814, ICE T/GC.
 20 CvP to TT, Letter, 14 Nov 1809, ICE T/GC27.
 21 TT to WL, from Ellesmere Canal Offices, 24 Jul 1815.
 22 Telford, *Life*, p. 162.
 23 Ibid., p. 660.
 24 Ibid., p. 285.
 25 CvP to TT, Letter, 24 Nov 1808, ICE T/GC11.
 26 CvP to TT, 24 Apr 1809, ICE T/GC19.
 27 CvP to TT, from Stockholm, 15 Oct 1809, T/GC23.

28 CvP to TT, 9 Feb 1810, T/GC35
29 TT to CvP, 15 Mar 1810, T/GC40.
30 TT to CvP, 23 Dec 1811, ICE T/GC53.
31 TT to CvP, Draft reply from Bonar Bridge, Jul 1812, T/GC56.
32 CvP to TT, from Pomerania, 14 Aug 1814, ICE T/GC89.
33 TT to CvP, from Stonehaven, 31 Oct 1814, ICE T/GC.
34 TT to CvP, from London, 10 May 1801, ICE T/GC50.
35 Rolt, *Telford*, p. 102.
36 Gibb, *Story*, p. 132.
37 TT to CvP, from Stonehaven, 31 Oct 1814, ICE T/GC.
38 CvP to TT, 19 Sep 1815, ICE T/GC101.
39 CvP to TT, 24 Dec 1815, ICE T/GC113.
40 CvP to TT 11 May 1816, ICE T/GC134.
41 John Simpson to TT 24 Sep 1815, T/GC102.
42 Rolt, *Telford*, p. 103.
43 Ibid., p. 104.
44 TT to CvP, 1817, ICE T/GC.
45 Gibb, *Story*, p. 232.
46 CvP to TT, 6 Oct 1819, ICE T/GC199.
47 CvP to TT, 16 Dec 1819, ICE T/GC200.
48 TT to CvP, 4 Mar 1821, ICE T/GC.
49 Mitchell, *Reminiscences*, p. 88.
50 CvP to TT, from Harwich, 11 May 1822, T/GC237.
51 CvP to TT, from Gothenburg, 22 May 1822, T/GC240.
52 CvP to TT, from Mariestad, 29 Sep 1822, ICE T/GC244.
53 TT to CvP, from London 15 Dec 1824, ICE T/GC.
54 CvP to TT, from Christiana, 18 May 1829, ICE T/GC291.

CHAPTER 12: A HAPPY LIFE

1 Southey, *Journal*, p. 70.
2 JR to RS, 22 Jul 1816, in Williams, *Life and Letters*, p. 181.
3 TT to CvP, from Inverness, 10 Nov 1816, ICE T/GC.
4 TT to JR, Jul 1819, in Haldane, *Glens*, p. 103.
5 Southey, *Journal*, p. 2.
6 MD to TD, from Inverness, 1 Jan 1817.
7 'partial' MD to TD, from Inverness, 14 Nov 1812; 'my utter contempt' MD to TT, 17 Mar 1812.
8 MD to TD, 29 Apr 1809, ICE Acc 006, Box 81, item 1.
9 Archibald Alison to MD, 8 Sep 1809, ICE Acc 006, Box 81, item 1.

10 MD to TD, 8 Mar 1809, ICE Acc 006, Box 81, item 1.

11 John Davidson to TD, 19 Dec 1809, ICE Acc 006, Box 81, item 1.

12 MD to TD, 11 Mar 1811 ICE Acc 006, Box 81, item 1.

13 MD to TD, 11 Nov 1816 ICE Acc 006, Box 81, item 1.

14 MD to TD, from Inverness, 14 Apr 1812, ICE Acc 006, Box 81, item 1.

15 MD to TD, from Invernesshire, 11 Mar 1811, ICE Acc 006, Box 81, item 1.

16 MD to TD, 2 Jun 1818, ICE Acc 006, Box 81, item 1.

17 Alexander Easton to JR, from Corpach, 3 May 1819, NRS MT 1/5, 1812–19.

18 TT to JR, Letter from Clachnacarry, 6 Nov 1820, NRS MT 1/6, 1819–25.

19 JR to JD, 22 Sep 1822, NRS MT 1/6, 1819–25.

20 Note by TT, 25 Oct 1822, NRS MT 1/6, 1819–25.

21 Thomas Easton to TT, 28 Oct 1822, NRS MT 1/6, 1819–25.

22 JD to TT, 29 Oct 1822, NRS MT 1/6, 1819–25.

23 TT to JR, Letter from London, 2 Dec 1822, NRS MT 1/6, 1819–25.

24 TT to JR, Letter from London, 2 Aug 1826, NRS MT 1/6, 1819–25.

25 JD to JR, 19 Mar 1827, NRS MT 1/6, 1819–25.

26 JR to JD, 3 Apr 1827, NRS MT 1/6, 1819–25.

27 JD to JR, 10 Dec 1827, NRS MT 1/6, 1819–25.

28 JR to the Lords of the Treasury, 7 Aug 1828, NRS MT 1/6, 1819–25.

29 See: http://greatglencanoetrail.info/uploads/documents/caley_skippers_guide_2011_web_version_lo_res.pdf.

CHAPTER 13: THE COLOSSUS OF ROADS

1 See: http://www.cse.dmu.ac.uk/~mward/gkc/books/rolling.html.

2 Arthur Young, *A Six Months' Tour Through the North of England*, London: W. Strahan, 1770, p. 580.

3 Daniel Bourne, 'A Treatise upon Wheel-Carriages', London: 1763, in William Albert, *The Turnpike Road System in England 1663–1840*, Cambridge University Press: 1972, p. 8.

4 John Wright and George Agar-Ellis (eds), *The Letters of Horace Walpole, Earl of Orford, ii, 1744–1753*, London: Richard Bentley, New Burlington Street, 1840, p. 301.

5 H. C. B. Rogers, *Turnpike to Iron Road*, London: Seely, Service and Co., 1961, p. 55.

6 Thomas De Quincey, *Confessions of An Opium Eater and Other Writings*, Oxford: Oxford University Press, 2013, p. 193.

7 John Ogilby, *The Traveller's Guide or A Most Exact Description of the Roads of England Being Mr Ogilby's Actual Survey and Mensuration by the Wheel of the Great Roads from London to all the Considerable Towns in England and Wales*, London: Tim Child and Robert Knaplock, 1699, p. 15.

8 Charles G. Harper, *The Holyhead Road: The Mailcoach Road to Dublin*, London: Chapman and Hall, 1902, pp. i, 82.

9 See: http://thelionhotelbook.blogspot.co.uk/2011/07/150th-anniversary-of-end-of-era-for.html.

10 Harper, *The Holyhead Road*, p. 93.

11 Henry Brooke Parnell, *A Treatise on Roads*, London: Longman, Rees, Orme, Brown, Green, & Longman, 1833.

12 *Selection of Reports and Papers of the House of Commons*, London: Public Works, Volume 38, 1838, p. 24.

13 Holyhead Road Office, *Annual Report of the Commissioners*, First report, 1824, p. 7.

14 Parnell, *Roads*, p. 25.

15 Ibid., p. 175.

16 Albert, *Turnpike Road*, p. 146.

17 Parnell, *Roads*, p. 162.

18 Thomas Telford, *Third Annual Report on Holyhead Road*, quoted in Parnell, *Roads*, p. 265.

19 Telford, *Life*, p. 206.

20 Harper, *The Holyhead Road*, pp. i, 308.

21 Ibid., p. 18.

22 George Harrison to TT, 4 May 1810, ICE T/HO5.

23 Telford, *Life*, p. 209.

24 Edward Watson, *The Royal Mail to Ireland; or, an account of the origin and development of the post between London and Ireland through Holyhead, and use of the line of communication by travellers. . .With plates*, London: Edward Arnold, 1917, p. 155.

25 TT to 'William Little', from London, 4 Oct 1815.

26 Edward Mogg, *Paterson's Roads: Being an Entirely Original and Accurate Description of the Direct and Principal Cross Roads in England and Wales, with Part of the Roads of Scotland*, London: Longman, Orme, Brown, Green and Longman, 1829, 18th edition, p. 10.

27 Jamie Quartermaine, Barrie Trinder and Rick Turner, *Thomas Telford's Holyhead Road: The A5 in North Wales*, York: Council for British Archaeology, 2003, p. 1.

28 Draft report by Telford to commissioners, 12 Mar 1816, T/HO11.
29 Telford, *Life*, p. 210.
30 *Report of the Select Committee on the Holyhead Road*, 20 May 1830, British Parliamentary Publications, 1830, X.131.
31 Thomas Jones and Edward Morris to TT, 2 Jul 1817, T/HO23.
32 ICE TTMSS02.

CHAPTER 14: PONTIFEX MAXIMUS
 1 TT to WL, from Ellesmere Canal Offices, 2 Jan 1815.
 2 TT to Charles Pasley, from London, 6 Sep 1815, Pasley Papers, iii, BL Add MS 41963, ff 197.
 3 TT to Charles Pasley from Capel Curig, 15 Jul 1816, Pasley Papers, iii, BL Add MS 41963, ff 206.
 4 Telford, *Life*, p. 211.
 5 Report by William Jessop, 13 Jan 1784, ICE T/HO2.
 6 *The Times*, 14 Dec 1785, Issue 303, 3.
 7 William Provis, *An Historical and Descriptive Account of the Suspension Bridge Constructed over the Menai Strait in North Wales*, London, 1828, p. 5.
 8 TT to WL, from London, 24 Jun 1818.
 9 Report on Runcorn Bridge, 1814, IGM T/RB8.
10 TT to Charles Pasley, from London, 6 Sep 1815, Pasley Papers, iii, BL Add MS 41963, ff 197.
11 Provis, *Account*, preface.
12 Rolt, *Telford*, p. 130.
13 TT to Sir Robert Peel, from Edinburgh, 7 Dec 1817, BL Add MS 40272 ff 82.
14 Pring, *Particulars*, p. 2.
15 Provis, *Account*, p. 20.
16 Sir John Rennie, *Autobiography*, London, 1875, p. 201.
17 Quartermaine et al., *Holyhead Road*, p. 87.
18 See: http://rsta.royalsocietypublishing.org/content/373/2039/20140346.
19 *The Times*, 24 Jul 1824, Issue 12399, p. 2.
20 Smiles, *Life*, p. 161.
21 Telford, *Life*, p. 227.
22 TT to WL, from London, 24 Jun 1818.
23 TT to Thomas Wedge, from London, 8 Jul 1826, ICE T/HO.
24 William Provis to TT, Post Office Dublin, ICE T/HO.155.
25 TR to TT, 30 Dec 1825, ICE T/HO101.
26 TR to TT, 16 Feb 1826, ICE T/HO120.

27 Howell, *Sketch*.

28 ICE T/HO110.

29 TR to TT, 14 Jan 1826, ICE T/HO106.

30 *The Sunday Times*, 12 Feb 1826, Issue 174, p. 4.

31 John Wilson to TT, 20 Feb 1826, ICE T/HO117.

32 Pring, *Particulars*, p. 24.

33 John Aird to TT, from Dublin, 15 Feb 1826, ICE T/HO117.

34 Draft of Third Report on the Menai Bridge, 24 Apr 1819, IGM T/HR3.

35 TR to TT, ICE T/HO120.

36 TR to TT, from Menai Bridge, 8 Feb 1826, ICE T/HO114.

37 Extract from the Liverpool and Holyhead Time Bill of 1 Mar 1826, ICE T/HO127.

38 TR to TT, 8 Mar 1826, ICE T/HO129.

39 TR to TT, 14 Mar 1826, ICE T/HO131.

40 Henry Parnell to TT, 4 Apr 1826, ICE T/HO135.

41 TR to TT, 19 May 1826, ICE T/HO147.

42 Henry Parnell to TT, 24 Jul 1826, ICE T/HO165.

CHAPTER 15: 24 ABINGDON STREET

1 See www.british–history.ac.uk/survey–london/vol16/pt1/pp75–81.

2 *The Times*, 19 Nov 1834, Issue 15639, p. 4.

3 JL to his mother, from Langholm, 15 Dec 1825.

4 Mitchell, *Reminiscences*, pp. i, 83.

5 TT to WL from London, 4 Oct 1815, ICE Acc 0006, Box 81, item 1.

6 TT to Miss Malcolm, from London, 7 Oct 1830.

7 Ibid.

8 IGM T/M7.

9 BL Loan 96 RLF 4/3, 1826–9.

10 JL to Mr Tourle, ICE Acc 0006, Box 80, item 1.

11 JL to his mother, 15 Dec 1825; JL to Mr Tourle, ICE Acc 0006, Box 80, item 1.

12 Minutes of the Proceedings of the Institution of Civil Engineers, ICE.

13 Gibb, *Story*, p. 194.

14 Ibid., p. 196.

15 Ibid., p. 197.

16 Rolt, *Telford*, p. 191.

17 *The Morning Chronicle*, 11 Feb 1823, Issue 16790, p. 1.

18 TT to Charles Pasley, from London, 20 Apr 1830.

19 George May to JR, from Inverness, 21 Feb 1838, IGM T/POST 9.

20 R. A. Paxton, *Three Letters from Thomas Telford*, Rathalpin Press, Edinburgh, 1968, p. 6.

21 TT to WL, 25 Oct 1816.

22 Rolt, *Telford*, p. 149.

23 TT and Alexander Nimmo to W. H. Merritt, from London, 12 May 1868, BL Add MS 38756 ff 121.

24 Proposed construction of a causeway between Bombay and the island of Colaba, 172–3, BL IOR/F/4/643/17725.

25 Allan Maclean, *Telford's Highland Churches*, Isle of Coll: The Society of West Highland & Island Historical Research, 1989, p. 11.

26 Ibid., p. 15.

27 For the restoration of Berneray church see: http://www.channel4.com/programmes/the–restoration–man/on–demand/50459–009.

28 Hadfield, *Temptation*, p. 162.

29 Rolt, *Telford*, p. 149.

30 Telford, *Life*, p. 115.

31 James Bennett, *A Tewkesbury Guide*, Tewkesbury: James Bennett, c.1850.

CHAPTER 16: THESE RAILROAD PROJECTORS

1 Letter to Baron Charles Dupin from London, 22 Aug 1832, available online: http://www.npg.org.uk/collections/about/primary–collection/documents–relating–to–primary–collection–works/npg–25167a.php.

2 James Caldwell to TT, 2 Jun 1824, ICE T/TR18.

3 Thomas Telford, *Report By Mr Telford Relative to the Proposed Railway from Glasgow to Berwick-upon-Tweed*, Macneil: Edinburgh, 1810, p. 1.

4 Unnamed letter to JR, ICG T/POST33.

5 Rolt, *Telford*, p. 155.

6 JR to RS, 14 Apr 1838.

7 Prospectus of Liverpool and Manchester Rail Road Company, 29 Oct 1824, ICE T/LM1.

8 Mitchell, *Reminiscences*, p. 101.

9 David Ross, *George and Robert Stephenson: A Passion for Success*, Stroud: The History Press, 2010, p. 99.

10 William Holden to TT, 14 Nov 1828, ICE T/LM11.

11 Thomas Telford, *Report on the Liverpool to Manchester Railway*, ICE T/LM20.

12 Rolt, *Telford*, p. 158.

13 Thomas Telford, draft letter, 17 Feb 1832, ICE T/LM25.

14 *Selection of Reports and Papers of the House of Commons*, 1836, p. 101.

15 John Macneil to TT, from Stow, 20 Oct 1833, IGM T/SC1.

16 John Macneil to TT, from Coventry, 25 Oct 1833, IGM T/SC2.

17 Bryan Donkin to TT, from Dartford, 24 Oct 1833, IGM T/SC9.

18 *The Man*, 26 Oct 1833, Vol. 1, p. 16.

19 *Observer*, 3 Nov 1833, p. 3.

20 'Report of the Result of an Experimental Journey Upon the Mail Coach Line of the Holyhead Road in Lieut-Colonel Sir Charles Dance's Steam Carriage on 1 Nov 1833 1890', British Library Reference Collection *c*.6. (100).

21 Institution of Locomotion for Steam Transport and Agriculture, BL 1890, British Library Reference Collection *c*.6. (99).

22 John Macneil to TT, from Brickhill, 4 Feb 1834, IGM T/SC5.

23 William Vaughan Stone to TT, 22 Jan 1822, ICE T/TR4.

24 Undated reply, ICE T/TR5.

25 Draft report, 25 Mar 1822, ICE T/TR12.

26 Report, 10 Nov 1827, ICE T/TR70.

27 Graham Cousins, *The Building of the Macclesfield Canal*, Journal of the Railway & Canal Historical Society, Vol. 33, Pt. 2, No. 173, Jul 1999, pp. 63–76.

28 Preserved after Telford's death by John Rickman, the archive is now shared between the Ironbridge Gorge Museum Trust and the National Library of Scotland.

29 Joseph Priestley, *Navigable Rivers and Canals*, *A Reprint*, London: David and Charles, 1969, p. 251.

30 Palmer receipt to Telford, 16 May 1825, ICE T/LL8.

31 JR to W. C. Rickman, 13 Jan 1833, ICE Parcel 81 Telford Letters Misc File.

32 TT to Thomas Grahame, 24 Jan 1833.

33 Telford, *Life*, p. 86.

34 Draft letter, ICE T/TR41.

35 Rolt, *Telford,* p. 183.

36 TT to Mrs Little, 4 Feb 1832.

37 TT to JR, from London, 20 Sep 1827.

38 TT to JR, from London, 27 Sep 1827 and 3 Oct 1827.

39 TT to JL, 1829.

CHAPTER 17: A BLESSING ON HIM

1   Hadfield, *Temptation*, p. 176.
2   Marc Brunel, Diary, 26 Feb 1827, ICE TT.
3   Available online: http://www.cliftonbridge.org.uk/visit/history.
4   Geoffrey Body, *Clifton Suspension Bridge*, Bradford-on-Avon: Moonraker Press, 1976, p. 19.
5   Turnbull, *Autobiography*, p. 12.
6   Ibid., p. 25.
7   Smiles, *Life*, p. 179.
8   'Report judging designs for the Clifton Suspension Bridge – Mr Davis Gilbert, Prefatory remark', 16 Mar 1831, Brunel Collection, DM162/8/2/3/folio 4.
9   'Decision on Clifton Bridge, Mr Davis Gilbert Prefatory remark', 10 Mar 1831.
10 Marc Brunel, Diary, 13 Feb 1830.
11 *History and Construction of the Clifton Suspension Bridge*, Bristol: J. Wright & Co., 1864.
12 TT to Charles Pasley, ICE Add MS 41964.
13 TT to JL, from London, 5 Oct 1832.
14 JR to RS, 19 Feb 1838.
15 TT to Mrs Little, 28 Aug 1832.
16 TT letter to Mrs Little, post office, Langholm, 26 Dec 1832.
17 Turnbull, *Autobiography*, p. 12.
18 TT to Mrs Little, post office, Langholm, 28 Aug 1833.
19 JR to RS, 19 Feb 1838.
20 George May to JR, from Inverness, 20 Feb 1838, IGM T/POST8.
21 TT to JR, from Middlewich, 12 Jul 1828, SRO MTI/7, 1826–.
22 TT to JL, to Bombay, 1829.
23 TT to JL, from London, 5 Oct 1832.
24 Ibid.
25 TT to Mrs Little, 28 Aug 1833.
26 Turnbull, *Autobiography*, p. 16.
27 Ibid., p. 18.
28 Diaries and Memoranda: Sir C. W. Pasley, Vol. XXVII, ff. 68, Mar 1831–Mar 1835, BL Add MS 41987.
29 Turnbull, *Autobiography*, p. 17.
30 *Mechanics Magazine*, 1834, pp. xxi, 414.
31 TT to AL, Salop, 28 Aug 1797.
32 JR to RS, from Westminster, 21 Sep 1837.

33 JR to RS, 19 Mar 1838.
34 RS to Revd James White, from Keswick, 3 Dec 1834.
35 RS to JR, from Keswick, 4 Sep 1825.
36 JR to RS, from Westminster, 21 Sep 1837.
37 *London Quarterly Review*, Apr 1839, CXXVI, pp. 223–35.

SELECT BIBLIOGRAPHY

ARCHIVES

British Library, London (BL)
Documents from the India Office on Telford's proposals for Bombay; documents from the Pasley and Malcolm families and some Telford correspondence

Brunel Institute, Bristol
Marc Brunel and Isambard Kingdom Brunel's papers, in particular relating to the Clifton Suspension Bridge

The Institution of Civil Engineers, London (ICE)
Telford left most of his papers to the institution on his death and although some drawings and letters have been dispersed the archive remains the principal source. It now also holds the Little letters as well as the early minutes of the institution's meetings

Ironbridge Gorge Museum Library and Archives, Coalbrookdale (IGM)
Telford's business papers and multiple versions of the manuscript of his autobiography

The National Archives, Kew (TNA)
Documents relating principally to Telford's road and canal work, many inherited from the Ministry of Transport archive or canal company archives

National Records of Scotland, Edinburgh (NRS)
Extensive collection of Telford's business papers relating to Scotland and in particular his work on the Caledonian Canal, the British Fisheries Society and Highland Roads

Shropshire County Record Office, Shrewsbury (SCRO)
Documents on Telford's work in the county on bridges and churches as
   well as Catherine Plymley's diary

Miscellaneous papers related to Telford's work in county and city
   archives, including Birmingham, Bolton, Cheshire, the City of
   London, Cornwall, Cumbria, Derbyshire, East Sussex, Glamorgan,
   Gloucestershire, Hertfordshire, Kent, Lancashire, Northumberland,
   Somerset and Worcestershire

PUBLISHED SOURCES
Ackroyd, Peter, *London: The Biography*, London: Chatto and
   Windus, 2000
Albert, William, *The Turnpike Road System in England 1663–1840*,
   Cambridge: Cambridge University Press, 1973
Alison, Revd Archibald, *Essays on the Nature and Principles of Taste*,
   London: J. J. G. and G. Robinson; Edinburgh: Bell and Bradfute, 1790
Alison, Sir Archibald, *Some Account of My Life and Writings*, Edinburgh
   and London: William Blackwood and Sons, 1883
Anderson, R. D., *Scottish Education Since the Reformation*,
   Dundee: Economic and Social History Society of Scotland, 1997
Beaumont, H., *Shrewsbury Castle: A Brief History*, Shrewsbury: Wilding
   and Son, 1958
Body, Geoffrey, *Clifton Suspension Bridge*, Bradford-on-Avon: Moonraker
   Press, 1976
Borrow, George, *Wild Wales*, London: John Murray, 1862
Bracegirdle, Brian, and Miles, Patricia, *Great Engineers and their Works*,
   Newton Abbot: David & Charles, 1973
Brewster, David (ed.), *The Edinburgh Encyclopaedia*, Edinburgh: William
   Blackwood, 1830
Brooke, David, *The Diary of William Macken₃ie*, London: Thomas
   Telford, 2000
Burns, Robert, *Poems Chiefly in the Scottish Dialect*, Belfast: William
   Magee, 1800
— and Currie, James, *The Works of Robert Burns*, London: L. T. Cadell,
   and W. Davies, 1801
Burton, Anthony, *Thomas Telford*, London: Aurum Press, 1999
Butterworth, John, *Four Centuries at The Lion Hotel*,
   Shrewsbury: Stone, 2011

Calladine, C. R., 'An Amateur's Contribution to the Design of Telford's Menai Suspension Bridge: a commentary on Gilbert (1826), "On the mathematical theory of suspension bridges",' *Philosophical Transactions of the Royal Society*, Vol. 373, 2039 (2015), available online: http://rsta.royalsocietypublishing.org/content/373/2039/20140346

Carroll, Lewis, *Through the Looking Glass*, 1871; reprint, London: Penguin Books, 2007

Chambers Jones, Reg, *Crossing the Menai: An illustrated history of the ferries and bridges of the Menai Strait*, Wrexham: Bridge Books, 2011

Cochran, Peter, *Why Did Byron Hate Southey?*, http://www.newsteadabbeybyronsociety.org/works/downloads/byron_southy.pdf

Cockburn, Henry, *Memorials of His Time*, Edinburgh: Adam and Charles Black, 1856

Colley, Linda, *Britons: Forging the Nation 1707–1837*, London: Pimlico, 2003

Cossons, Neil, and Trinder, Barrie, *The Iron Bridge*, Bradford-on-Avon: Moonraker Press, 1979

Cousins, Graham, 'The Building of the Macclesfield Canal', *Journal of the Railway & Canal Historical Society*, Vol. 33, Pt. 2, No. 173, July 1999, pp. 63–76

Cruft, Kitty, Dunbar, John, and Fawcett, Richard, *Borders: Buildings of Scotland*, New Haven, Conn.; London: Yale University Press, 2006

Day, Thomas, 'Did Telford rely, in Northern Scotland, on vigilant inspectors or competent contractors?', *Construction History*, Vol. 13, 1997, pp. 3–15

Defoe, Daniel, *A Tour Through the Whole Island of Great Britain*, London: Strahan, Mears, Francklin, Chapman, Stagg, 3 vols, 172–7.

De Quincey, Thomas, *The English Mail Coach and Other Essays*, London: J. M. Dent & Sons, 1912

Dunlop, Jean, *The British Fisheries Society 1786–1893*, Edinburgh: John Donald Publishers, 1978

Fforde, Catriona, *The Great Glen*, Glasgow: Neil Wilson, 2011

Fielden, Kenneth, 'Samuel Smiles and Self-Help', *Victorian Studies*, 1 Dec 1968, Vol. 12 (2), pp. 155–76

Gibb, Alexander, *The Story of Telford*, London: Alexander Maclehose & Co., 1935

Gifford, John, *The Buildings of Scotland: Dumfries and Galloway*, London: Penguin, 1996

Gilchrist, George, *Memorials of Westerkirk Parish*, Anan: G. Gilchrist, 1970

Graham, Alexander, *A History of Freemasonry in the Province of Shropshire*, Shrewsbury: Adnitt & Naunton, 1892

Grounds, Douglas, *Son and Servant of Shropshire: The Life of Archdeacon Joseph Plymley*, Corbett: Longaston Press, 2009

Guedes, Pedro Paulo d'Alpoim, *Iron in Building, 1750–1855: Innovation and Cultural Resistance*, PhD thesis, University of Queensland, 2010

Hadfield, Charles, *The Canals of the West Midlands*, Newton Abbot: David & Charles, 1966

— *Thomas Telford's Temptation*, Cleobury Mortimer: M. & M. Baldwin, 1993

— and Skempton, A. W., *William Jessop, Engineer*, Newton Abbot: David & Charles, 1979

Hague, Douglas, *The Conwy Suspension Bridge*, London: National Trust, 1969

Haldane, A. R. B., *New Ways Through the Glens*, Colonsay: House of Lochar, 1995

Hamilton, Douglas J., *Scotland, The Caribbean and the Atlantic World, 1750–1820*, Manchester: Manchester University Press, 2005

Harper, Charles G., *The Holyhead Road: The Mailcoach Road to Dublin*, London: Chapman and Hall, 1902

Harvey, A. D., *Collision of Empires: Britain in Three World Wars, 1793–1945*, London: Phoenix, 1994

History and Construction of the Clifton Suspension Bridge, Bristol: J. Wright & Co., 1864

Holyhead Road Office, *Annual Report of the Commissioners*, HMSO, 1824–34

Howell, Thomas, *Shrewsbury to Holyhead Being a Historical and Descriptive Sketch of the Most Interesting Scenery on this Romantic and Beautiful Road*, Shrewsbury: T. Howell, 1827

Hume, David, *An Enquiry concerning Human Understanding*, London, 1748

Hyslop, John, *Langholm As It Was*, Sunderland: Hills & Co., 1912

Institution of Civil Engineers, *Thomas Telford: 250 Years of Inspiration*, London: 2007

Jenkinson, Andrew, *On the Trail of Thomas Telford in Shropshire*, Little Stretton: Scenesetters, 1993

Johnson, Samuel, and Boswell, James, *Journey to the Western Isles of Scotland* and *The Journal of a Tour of the Hebrides*, 1785; reprinted London: Penguin, 1993

Kaye, William, *The Life and Correspondence of Major-General Sir John Malcolm, GCB*, London: Smith, Elder and Co., 1856

Knox, John, *A Tour Through the Highlands of Scotland and the Hebridean Isles*, Edinburgh: The Mercal Press, 1975

Langford, Paul, *A Polite and Commercial People: England 1727–1783*, 1989; reissued, Oxford: Oxford University Press, 1998

McIntyre, Ian, *Robert Burns: A Life*, London: HarperCollins, 1995

Maclean, Allan, *Telford's Highland Churches*, Isle of Coll: The Society of West Highland & Island Historical Research, 1989

Malcolm, Major-General Sir John, *Life and Correspondence*, London: Smith, Elder, and Co., 1866

Matthews, Colin, and Harrison, Brian (eds), *Oxford Dictionary of National Biography*, Oxford: Oxford University Press, 2004

Maudlin, Daniel, and Robert Mylne, 'Thomas Telford and the Architecture of Improvement: The Planned Villages of the British Fisheries Society, 1786–1817', *Urban History*, Vol. 34, Issue 3 (Dec 2007), pp. 453–80

Mitchell, Joseph, *A Plan for Lessening The Taxation of The Country*, London: Edward Standfords, 1860

— *Reminiscences of My Life in The Highlands*, Vols I and II, Newton Abbot: David & Charles Reprints, 1971

Mogg, Edward, *Paterson's Roads: Being an Entirely Original and Accurate Description of the Direct and Principal Cross Roads in England and Wales, with Part of the Roads of Scotland*, London: Longman, Orme, Brown, Green and Longmans, 1829

Morris, Chris, *On Tour with Thomas Telford*, Longhope: Tanners Yard Press, 2004

— *Thomas Telford's Scotland*, Longhope: Tanner's Yard Press, 2009

Newman, John, and Pevsner, Nikolaus, *The Buildings of England: Shropshire*, New Haven and London: Yale University Press, 2006

Noble, Celia Brunel, *The Brunels, Father and Son*, London: Cobden-Sanderson, 1938

Ogilby, John, *The Traveller's Guide or A Most Exact Description of the Roads of England Being Mr Ogilby's Actual Survey and Mensuration by the Wheel of the Great Roads from London to All the Considerable Towns in England and Wales*, London: Tim Child and Robert Knaplock, 1699

Parnell, Henry Brooke, *A Treatise on Roads*, London: Longman, Rees, Orme, Brown, Green, & Longmans, 1833

Pasley, General Sir Charles W., 'Narrative from Recollection', in David Irving, *The History of Scottish Poetry*, Edinburgh: Edmonston and Douglas, 1861

Pattison, Andrew, 'William Hazledine, Shropshire Ironmaster and Millwright', MPhil. thesis, University of Birmingham, 2011

Pawson, Eric, *Transport and Economy: The Turnpike Roads of Eighteenth-Century Britain*, London: Academic Press, 1977

Paxton, Roland (ed.), *Three Letters from Thomas Telford*, Edinburgh: Rathalpin Press, 1968

Paxton, Roland, and Shipway, J., *Civil Engineering Heritage: Lowlands and Borders*, London: ICE Publishing, 2007

— *Civil Engineering Heritage: Highlands and Islands*, ICE Publishing, 2007

Penfold, Alastair (ed.), *Thomas Telford, Engineer*, London: Thomas Telford Ltd, 1980

Plymley, Joseph, *A General View of the Agriculture of Shropshire*, London: Richard Phillips, 1803

Pollard, Michael, *Walking the Scottish Highlands: General Wade's Military Roads*, Andre Deutsch, 1984

Priestley, Joseph, *Navigable Rivers and Canals, A Reprint*, London: David & Charles, 1969

Pring, Joseph, *Particulars of the Grand Suspension Bridge Erected over the Straits of Menai at Bangor*, Bangor, 1826

Provis, William, *An Historical and Descriptive Account of the Suspension Bridge Constructed over the Menai Strait in North Wales*, London, 1828

Pugsley, Sir Alfred, *The Works of Isambard Kingdom Brunel*, London: Institution of Civil Engineers, 1976

Pulteney, W., *Thoughts on the Present State of Affairs with America and the Means of Conciliation*, London: J. Dodsley and T. Cadell, 1778

The Quarterly Review, Vol. LXIII, London: John Murray, 1839

Quartermaine, Jamie, Trinder, Barrie, and Turner, Rick Thomas, *Telford's Holyhead Road: The A5 in North Wales*, York: Council for British Archaeology, 2003

Ransom, P. J. G., *Bell's Comet: How A Paddle Steamer Changed the Course of History*, Stroud: Amberley, 2012

Reader, William, *MacAdam: The MacAdam family and the Turnpike Roads 1798–1861*, London: Heinemann, 1980

Rennie, Sir John, *Autobiography*, London, 1875

A Reply to the Authorized Defence of the St Katharine's Dock Project, London: John Richardson, 1824

Rickman, John (ed.), *Life of Thomas Telford*, London: Hansard & Sons, 1838

Rogers, H. C. B., *Turnpike to Iron Road*, London: Seely, Service and Co., 1961

Rolt, L. T. C., *Isambard Kingdom Brunel*, London: Longmans, Green & Co., 1957

— *Thomas Telford*, London: Longmans, Green & Co., 1958

Ross, David, *George and Robert Stephenson*, Stroud: The History Press, 2010

Rothschild, Emma, *The Inner Life of Empires: An Eighteenth-Century History*, Woodstock: Princeton University Press, 2011

The Royal Society of Edinburgh, *The 250th Anniversary of the Birth of Thomas Telford*, Edinburgh: The Royal Society of Edinburgh, 2007

Schwarz, Suzanne, *Slave Captain: The Career of James Irving in the Liverpool Slave Trade*, Liverpool: Liverpool University Press, 2008

Scott, Sir Walter, *Waverley*, 1814; reprint, London: Penguin Classics, 2011

Skeat, W. O., *George Stephenson: The Engineer and His Letters*, London: Institute of Mechanical Engineers, 1973

Skempton, A. W. *Civil Engineers and Engineering in Britain, 1600–1830*, Aldershot: Variorum, 1996

— et al. (eds), *A Biographical Dictionary of Civil Engineers in Great Britain and Ireland*, Vol. I, *1500–1830*, London: Thomas Telford, 2002

Smeaton, John, Rennie, John, and Telford, Thomas, *The Reports by Smeaton, Rennie and Telford Upon the Harbour of Aberdeen*, Aberdeen: G. Cornwall, 1834

Smiles, Samuel, *The Life of Thomas Telford*, reprinted Teddington: The Echo Library, 2006

Smollett, Tobias, *The Expedition of Humphry Clinker*, London: Johnston & Collins, 1771

de Soissons, Maurice, *Telford: The Making of Shropshire's New Town*, Swan Hill Press, 1991

Southey, Robert, *Poetical Works of Robert Southey*, London: Longman, Brown, Green, and Longmans, 1847

Telford, Thomas, *Eskdale, A Descriptive Poem*, Shrewsbury: J. & W. Eddowes, 1795

— *Life of Thomas Telford Civil Engineer Written by Himself*, London: Hansard and Sons, 1838

— *To Sir John Malcolm on receiving his 'Miscellaneous poems'*, Edinburgh: Tragara Press, 1971

Tranter, Nigel, *Portrait of the Border Country*, London: Hale, 1972

Trinder, Barrie, *The Industrial Revolution in Shropshire*, Chichester: Philimore, 1973

— *A History of Shropshire*, Chichester: Phillimore, 1983

— *The Darbys of Coalbrookdale*, Chichester: Phillimore, 1987

— (ed.), *The Most Extraordinary District in the World*, Chichester: Phillimore, 1988

— *Britain's Industrial Revolution: The Making of a Manufacturing People, 1700–1870*, Lancaster: Carnegie Publishing, 2013

— 'Was there a Shropshire Enlightenment?', *West Midlands History*, 1, Vol. 1, 2013

Turnbull, George, *Autobiography*, London: Printed by Cooke & Co., 1893

Turner, Rick, 'Thomas Telford the Archaeologist', *The Antiquaries Journal*, Vol. 88, (Sep 2008), pp. 365–75

Uglow, Jenny, *The Lunar Men: The Friends Who Made the Future 1730–1810*, London: Faber, 2012

— *In These Times: Living in Britain through Napoleon's Wars, 1793–1815*, London: Faber, 2014

Wakelin, Peter, *Pontcysyllte Aqueduct and Canal World Heritage Site*, Milton Keynes: Canal & River Trust, 2015

Watson, Edward, *The Royal Mail to Ireland; or, an account of the origin and development of the post between London and Ireland through Holyhead, and use of the line of communication by travellers. . .With plates*, London: Edward Arnold, 1917

The Weekly Magazine, or Edinburgh Amusement, Edinburgh, 1768–79

West, Christopher, *The Story of St Katharine's*, London: The Cloister House Press, 2014

White, Jerry, *London in the Eighteenth Century*, London: The Bodley Head, 2012

Whyte, Ian D., *Scotland Before the Industrial Revolution*, London and New York: Longman, 1995

Williams, Orlando, *The Life and Letters of John Rickman*, Constable and Co., 1912

Wordsworth, Christopher, *Memoirs of William Wordsworth*, Edward Moxon, London, 1851

Wordsworth, William, *The Poetical Works of William Wordsworth, Vol. VII*, London: Edward Moxon, 1842

Wrexham County Borough and Royal Commissioners on the Ancient and Historical Monuments of Wales, *Pontcysyllte Aqueduct and Canal*

Nomination Document, Wrexham County Borough Council
2008

Wright, John, and Agar-Ellis, George (eds), *The Letters of Horace Walpole, Earl of Orford, Vol. II, 1744–1753*, London: Richard Bentley, New Burlington Street, 1840

Young, Arthur, *A Six Months' Tour Through the North of England*, London: W. Strahan, 1770

ACKNOWLEDGEMENTS

This book began with my parents, Emily and Ian Glover. They introduced me to history and archaeology, took me to Shropshire and, in the frozen winter of 1982, when we were snowed in at Coalbrookdale and had to drag coal by sled from the old ironmongers, Beddoes, as sheet ice flowed under the iron bridge, taught me to love the part of our history that produced this magical place. It was with my parents that I remember searching out the remains of lost iron furnaces around the Clee Hills; the wonder, then and now, being that somewhere so rural could breed such industry. Anything good in this book comes from the academic curiosity I so admire in them, and if I ever puzzled them by joining the worlds of journalism and politics rather than academia, then perhaps this book is a small apology.

My brother Adrian Glover and my sister-in-law Victoria Herridge, both scientists, have also encouraged me, by example, to respect the serious work of concentrated research over time and when, as a journalist with an evening deadline, I wondered why they took several years to write their doctorates, I now know the answer.

I owe a debt, too, to the history fellows who welcomed me as an undergraduate at University College, Oxford: Leslie Mitchell, Hartmut Pogge von Strandmann and Alexander Murray. There could not have been a warmer or more intellectually encouraging place to study. Leslie fixed in my mind a belief that the eighteenth was the best of all centuries and that the Whigs were the best of all rulers, and he has been tolerant of a student who strayed to write about an engineer who became something of a Tory. He asked a suitably Whiggish question when the text was done: 'Are there any jokes?' He made this book in more ways than he will ever know and, kindly, read it in draft.

I have been fortunate, too, in working with people who encouraged me to write: among them Alan Rusbridger and Martin Kettle at the *Guardian* and many others in politics. Sir John Major's persistence in asking when I was going to write a book, as he produced several of his own, fuelled a competitive spirit; David Cameron did not frown too much when his new member of staff announced he would be slipping out the back gates of Downing Street to visit the archives of the Institution of Civil Engineers; and Patrick McLoughlin has not only been the finest of transport secretaries and a friend, but unsparingly generous in allowing me time to investigate the infrastructure of the eighteenth century when I was supposed to be working on that of today. If speeches by Tory politicians in the years after 2010 contained excessive references to Telford, then I am to blame.

Others in politics and government and media have also helped with encouragement and comments, among them Andrew Adonis, Jonathan Dimbleby, Tristram Hunt, Ben Gummer, Andrew Marr and Rory Stewart, all of whom show that an interest in learning is not always a contradiction in politics. My fellow special advisers at the Department for Transport, Ben Mascall, Simon Burton, Tim Smith and Giles Bancroft never minded (or never mentioned) the many moments my mind strayed from the official grid. Tristan and Catali Garel-Jones were always generous hosts when I wrote at their home in Spain; Tristan's library in a tower above the orange trees is second only to the London Library as a place of contentment; the staff at the latter were unfailingly helpful.

Thanks should go, too, to the staff of the British Library, the Library and Archives at the Ironbridge Gorge Museum and the National Library of Scotland, as well as Robbie Smith, who helped with the notes and research in Bristol. Most of all, however, I should thank Carol Morgan, the magnificent keeper of the finest engineering archive of them all, at the Institution of Civil Engineers in London. Over several years she never once seemed to mind my unpredictable visits, or my habit of forgetting exactly which reference should apply to a particular quote.

In Langholm, Ron Addison, Brenda Morrison and Tom Hutton, among others, were helpful and keep alive an interest in the history of the place that bred Telford. In Westerkirk, Margaret Sanderson welcomed me to the wonderful library there. Steven Brindle provided good advice, at the start of my research, to a non-engineer, and I have been particularly grateful for the encouragement offered by some of the greatest champions of industrial history, including Professor Roland Paxton and Barrie Trinder, both of whose expertise I can only admire. Barrie Trinder read a draft and was kind about even my worst howlers. Peter Wakelin has been hugely supportive, reading and greatly improving the chapter on Pontcysyllte. He rightly persuaded me to be kinder to William Jessop. Chris Morris, the author of two wonderfully illustrated books on Telford, has been generous in allowing the use of some of his many outstanding photographs.

My publishers, Bloomsbury, have been the best of allies, never once complaining about deadlines. Michael Fishwick generously agreed to publish a book whose contents, at the start, I could not properly explain and both Michael and Anna Simpson brought expert judgement to improve my first draft. Anna's enthusiasm made the process of editing a pleasure and the success of the illustrations is all down to her. Also at Bloomsbury, thanks to

Marigold Atkey, Vicky Beddow and Rebecca Thorne. In Kate Johnson I was lucky enough to have not just a skilled and kind copyeditor, but someone who knew both north Wales and canals well. My thanks too to Catherine Best for reading the proofs and to David Atkinson for compiling the index.

Most of all, however, there are two people to whom I owe the greatest thanks and without whom this book would never have been written. Ed Victor has been the finest agent that could be imagined: he encouraged me to write, helped me shape the idea that became *Man of Iron* and was always quick with advice, always good, when I asked for it. Even more, my admirable, tolerant, clever and kind partner Matthew Parris encouraged me to stick at it, improved phrases and always appeared interested in Telford even when he must have been tired of the subject. If he was ever mystified by our many unexpected walks along towpaths, he never admitted it.

All mistakes are of course my own: corrections and comments welcome at julianglover@me.com

Gratton, Derbyshire, July 2016

INDEX

A NOTE ON THE TYPE

The text of this book is set in Fournier. Fournier is derived from the *romain du roi*, which was created towards the end of the seventeenth century from designs made by a committee of the Académie of Sciences for the exclusive use of the Imprimerie Royale. The original Fournier types were cut by the famous Paris founder Pierre Simon Fournier in about 1742. These types were some of the most influential designs of the eight and are counted among the earliest examples of the 'transitional' style of typeface. This Monotype version dates from 1924. Fournier is a light, clear face whose distinctive features are capital letters that are quite tall and bold in relation to the lower-case letters, and *decorative italics, which show the influence of the calligraphy of Fournier's time.*